Coherent Apertures in the Central Nervous System
A Model of the Internal Experience

Coherent Apertures
in the
Central Nervous System

A Model of the Internal Experience

Thomas D. Wason

• 2017 •

Author: Wason, Thomas D.

Title: Coherent Apertures in the Central Nervous System: A Model of the Internal Experience

Summary: This monograph proposes a coherent apertures model of the generation of the internal experience. This is not a computer model; it is a model of biological processes acting on biological structures. Cortical areas—apertures—synchronously bind together, instantiating aspects of information, thereby creating the internal experience. A test of the model is proposed. 843 references.

ISBN-13: 978-1548461577
ISBN-10: 1548461571

Printed by CreateSpace, an Amazon.com company.

Thomas D. Wason
www.tomwason.com
1421 Park Drive
Raleigh, North Carolina USA 27605
tdwason@ncsu.edu

About the Author

Thomas D. Wason received a BS in mechanical engineering from the Massachusetts Institute of Technology, and an MS and a PhD in experimental psychology from North Carolina State University. His thesis was *Auditory Autocorrelation: An experimental study of central nervous system processing of direct and reflected speech sounds*, and his dissertation was *Construction and evaluation of a three-dimensional display from a two-dimensional projection surface based on theoretic considerations of metrification and affine space*. He has coauthored papers in visual perception and auditory testing systems. The current work is an outgrowth of these studies.

He is a visiting scientist as a theoretical/computational neuroscientist in the Meitzen Laboratory in the Department of Biological Sciences at North Carolina State University. While an engineer he worked extensively in biotechnology, optoelectronics, sensor technologies, audio systems, digital and analog electronics, software development, and mechanics. He has 14 US and 4 foreign patents. He has worked in online learning systems and contributed extensively in the development of international online technology standards.

Dedication

To Marianne

This never would have happened without you.

Brief Contents

1. Introduction to the Coherent Apertures Model 1
2. A Framework 11
3. The Coherent Cortical Aperture 19
4. Information Instantiations in Apertures 49
5. The Aperture Operator 67
6. Integration of Coherent Cortical Apertures 81
7. Operations 119
Epilogue 145
Supplemental Materials 155
References 169

Contents

About the Author...v
Dedication..vii
Brief Contents..ix
Contents..xi
Preface...xv
Acknowledgments..xvii

1. Introduction to the Coherent Apertures Model 1
   1.1. Introduction 1
      Scope of the Model..*1*
      Précis..*2*
      Themes...*3*
      The Chapters...*4*
   1.2. Principles and Corollaries...6
      Principles...*6*
      Corollaries *6*
   1.3. History 7
      Development of the Model...*7*
      Background..*8*
   1.4. Research Models 8
      Computational Neuroscience..*8*
      Networks: ANNs..*9*
      Neurophysiological Models...*10*

2. A Framework 11
   2.1. Introduction 11
      Objective..*13*
   2.2. The Coherent Apertures Model Framework.....................................13
      Introduction..*13*
      Framework..*13*
   2.3. Summary 18

3. The Coherent Cortical Aperture 19
   3.1. Introduction 19
   3.2. Cortical Structure...19
      Parcellation...*19*
      Columnar Organization *20*
      Laminar Organization..*21*
      Circuitry...*21*
   3.3. Coherent Cortical Aperture 22
      Introduction..*22*
      Aperture Synchrony and Coherence...*22*
      Evidence of Cortical Aperture Coherence...*23*
   3.4. Aperture Coherence...24
      Introduction..*24*
      Network Synchronization Model..*26*
      Pumped AFP Model...*37*
      Laminar Coherence Model *44*
   3.5. Results 47
      Principles and Corollaries..*47*
      Summary..*47*
      Conclusion...*48*

4. Information Instantiations in Apertures...49
   4.1. Introduction..49
   4.2. Information Models...49
      Introduction...*49*
      Spike Encoding Models..*50*
      Field Models..*51*
   4.3. Information in Apertures: Dynamics.................................51
      Introduction...*51*
      Undecipherable Complexity..*52*
      Information Entropy...*52*
      Modulation..*53*
   4.4. Phase Structures..53
   4.5. Coherence Maps...54
      Introduction...*54*
      The Coherence Map..*56*
   4.6. Information in Coherence Maps..60
      Information Instantiation..*60*
      Local Phase: Information in Coherence Maps........................*62*
      Aperture Entropy..*64*
   4.7. Coherent Information Structure (CIS)..............................65
   4.8. Results...65
      Principle and Corollary..*65*
      Summary..*65*

5. The Aperture Operator 67
   5.1. Introduction..67
   5.2. Functions and Functionals 67
      Properties..*67*
      Functions...*68*
   5.3. Coherent Functions, Functionals, and Coherence Maps.........72
      Coherent Functions and Functionals.....................................*73*
      Synchronous Functions, CMs, and CISs..................................*75*
      Functions over Multiple Apertures..*75*
      Fuzziness and Quantization..*76*
   5.4. Modulation 76
      Evidence..*76*
      What Is Modulation?..*77*
      Mechanisms of Modulation...*77*
   5.5. Results...79
      Principle and Corollaries...*79*
      Summary..*79*

6. Integration of Coherent Cortical Apertures 81
   6.1. Introduction..81
   6.2. Oscillations, Synchrony, Micro- and Macrocoherence 82
      Introduction...*82*
      Oscillations..*82*
      Synchronous Associations...*84*
   6.3. Apertures in Cortical Integration 86
      Introduction...*86*
      Development of the Internal Experience...............................*86*
      i)   Coherent Apertures Produce CISs with Keys................*87*
      ii) Coherent Apertures are Nodes.....................................*88*
      iii) CIS Exchanges are Enabled by Aperture Coherence States.........*90*
      iv) Ensembles Form through Bidirectional Exchanges of CISs........*93*
      v) Bidirectional Exchanges Regulate the Composition of Ensembles.........*106*

    vi) An Ensemble Seeks a Low Entropy State. *108*

    vii) The Internal Experience Emerges. *115*

  6.4. Results 117

    Principles and Corollaries *117*

    Summary *118*

7. Operations 119

  7.1. Introduction 119

  7.2. Operations Model 120

    Introduction *120*

    General Network Structure and Aperture Classes *120*

    Process *121*

  7.3. Perception 124

  7.4. Memory 128

    Introduction *128*

    Memory in Coherent Apertures *130*

    Memory in the CM *132*

    The Memory System *136*

  7.5. Results 143

    Principle and Corollaries *143*

    Summary *143*

Epilogue 145

  Introduction 145

  Principles and Corollaries 145

  Issues 147

    Biophysics of Neurons and Local Circuits *147*

    Synchronization in the Aperture *147*

    Information Instantiation *148*

    Multi-Aperture Processes *148*

  A Test of the Aperture Coherence Model 149

    Introduction *149*

    The Experiment *149*

  Speculations 152

    Decisions *152*

    Seizures *152*

    Uncoherent Instantiations *152*

    Phase Conjugate Mirrors *152*

    Evolution and the Composite Node *153*

  Future Explorations 153

    Introduction *153*

    Domains *153*

    Research Topics *154*

Supplemental Materials 155

  Introduction 155

  Ch. 3. The Coherent Cortical Aperture 155

    Array Synchronization *155*

    Monopole Model *159*

  Ch. 6. Integration of Coherent Cortical Apertures 161

    Laser Models *161*

    Intermediate Synchronizer *161*

    Common Synchronizer *162*

    Peer Synchronization *163*

  Glossary 165

References 169

Preface

Alas, dear reader, this is a rather lengthy piece. If you wish, you can read Ch. 2, *A Framework*, to get an overview of the model (*what*) and Ch. 7, *Operations*, for its application to perception, memory, recognition, and recall (*so what?*). The intervening chapters describe the internal workings of the coherent apertures model (*how*) that is derived from the literature. These chapters are in a logical sequence, each building on the previous ones. Such a house of cards can be fragile. I have attempted to buttress this one with verifiable evidence and logic. Ch. 6, *Integration of Coherent Cortical Apertures*, describes the mechanisms of cortical integration, the main result. In the *Epilogue* I consider what this model has taught me. The Epilogue also explores issues raised, a test of the model, and possible future research. *Supplemental Materials*, which includes a Glossary, provides technical details not essential to the main flow.

This work in theoretical neurobiology draws from many domains beyond neuro-biology. Concepts in radiant energy systems proved useful in many ways, the aperture being a fundamental concept. The model of Yoshiki Kuramoto on synchronization in complex networks, brought to my attention by Aleksandr Davydov, was invaluable in understanding the process of aperture cohering. Information entropy was useful for describing the nature of information instantiated in the CNS without actually specifying the detailed form of the instantiation, particularly important in dealing with what I came to recognize as undecipher-able complexity. Since the coherent apertures model addresses biological structures and processes, it calls into question the idea of "encoding." Information processing models, outside of this model, are abstractions, with no physical reality. The CNS performs biological processes within physical structures that are, therefore, processes that are physical. Information—whatever that is—is physically instantiated in the CNS, subject to biological processes.

This work proceeded from the bottom up. I started by working to understand how a cortical area could become coherent, leading to a model of cortical operation within a limited scope. I did not delve into neuronal circuit models since they do not capture concepts embodied in this model. For example, it appears that aperture cohering involves mechanisms beyond neuronal circuits, notably the local field potential and its source. The role of the glia, particularly the astrocyte, is now more fully appreciated. Phase also appears to be an impor-tant factor in instantiating information. I have approached the problem of the operation of the CNS by limiting the scope of this exploration to early stage sensory processing, specific-ally to declarative objects in perception and memory, and their access in recognition and recall. My intent was to reduce complexity such that, without loss of validity, I could understand meaningful aspects of the CNS processes. I have proposed principles that embody some of those aspects, perhaps contributing to future work. The potential for technological implementation, an attractive test of an abstraction of the model, can be addressed in the future.

Thomas D. Wason
12 July 2017
Raleigh, North Carolina, USA

Acknowledgments

This project has been an undertaking of many years, and I am pleased to express my sincere gratitude to those who have given support and guidance throughout the process. I wish to thank those who agreed to read drafts of the monograph and offer comments:

- James Kalat, Professor Emeritus, Dept. of Psychology, North Carolina State University
- Gerald Katzin, Professor Emeritus, Dept. of Physics, North Carolina State University
- Joseph Lappin, Professor Emeritus, Dept. of Psychology, Vanderbilt University
- Donald Mershon, Professor Emeritus, Dept. of Psychology, North Carolina State University
- William Troxell, retired, CIA, assigned to the U.S. National Reconnaissance Office as Director, Research and Technology, Imagery Systems Acquisition and Operations Directorate.

Jim Kalat and Don Mershon, my PhD and MS advisors, encouraged me to undertake this project, which is an extension of my graduate work. Jim provided significant guidance in expressing the concepts clearly. Will Troxell made invaluable technological comments and gave considerable encouragement, validating the soundness of some key technical ideas. I had many prolonged discussions with Jerry Katzin on the concepts proposed and terminology; in the process he taught me a great deal.

Marianne Wason, my wife, contributed much to this work with her substantial editorial and proofing skills. Most of all, I am deeply appreciative of her unflagging support and encouragement.

1. Introduction to the Coherent Apertures Model

1.1. Introduction

The coherent apertures model proposes an explanation of how biological structures and processes in the central nervous system (CNS) generate the internal experience. The internal experience includes awareness, consciousness, working memory, the global neural workspace, attention, and perception; I have not differentiated among them. How does the internal experience occur? Where does it occur? The internal experience occurs primarily in the cortex. I offer a model primarily focused on the operation of cortical areas and their dynamic binding into ensembles [1, 2, 3, 4, 5, 6] that contribute to the internal experience. The cortical areas and limited subcortical structures work together to instantiate, store, and subsequently access information, moving back and forth from working memory to the various forms of memory. The cortex is composed of many distinct areas, defined here as *apertures*. Apertures instantiate information; apertures communicate and store. Of central importance, apertures can <u>synchronize</u>, enabling information instantiations through <u>coherence</u>, hence becoming coherent apertures. Groups of coherent apertures can communicate, modulating each other, in the process generating the internal experience. The group's collective information instantiations can be stored as memories, and later retrieved as an internal experience by resurrecting the group. In this document I have considered processes related to structured dynamic sensory information in perception, memory, recognition, and recall as the sources of the internal experience.

Steven Rose [7] remarked that neuroscience is "data rich, but theory poor." Churchland and Abbott [8] and Sejnowski et al. [9] similarly noted a need for theoretical work in the neurosciences to provide explanations derived from, and tested by, data. Models of brain operation have long reflected the technology of the time.[1] Descartes proposed a hydraulic model in the 1660s; in the 1920s Weiss proposed a radio model. Although the computer is the dominant technology today, I do not consider the CNS to operate like a computer. The CNS does not process information, which is an abstraction, but performs biological processes on biological structures: it's physics. The coherent apertures model incorporates biological and technological points of view, incorporating synchrony and coherence in both.

Scope of the Model

The coherent apertures model addresses three fundamental issues in the neurosciences:

- What is the internal experience?
- How does cortex work?
- How is information instantiated in the cortex?

The coherent apertures model of the CNS consolidates a broad range of topics into a theoretical framework, attempting to span from cellular to global scales. This model is a behavior of materials model—dynamic granular materials. The elements of the cortical material are minicolumns characterized by the behavior of their pyramidals. The cortex and subcortical structures operate on

---

[1] This historical perspective provided by James Kalat.

different principles.[2] The (neo)cortex of mammals fills multiple needs, effective over species and modalities, having developmental flexibility and malleability. This leads one to think that it has common principles with common underlying mechanisms.[3]

Considerable experimental and theoretical work has shown that the cortex is parceled into discrete areas—apertures—that can have synchronous local field potentials (LFPs). This raises the question, why might this be useful for the operations of the CNS? Of necessity the problem of the instantiation of information in the aperture is addressed. The framework hypothesizes a relationship between how and why a cortical area becomes synchronous. Synchronization provides a substrate for the instantiation of information through coherence. An argument is made here for the role of coherent cortical apertures in the CNS through discussions of the synchronization and cohering of apertures, the information contained in and exchanged among those apertures, aperture functionals, cortical integration, and in some operations performed within the CNS. I provide specific operations to illustrate the processes of the framework.

Although some cortical circuits are discussed, this model is not a circuits model, but rather a framework of biological structures and processes. The coherent apertures model reveals a number of plausible common principles while still maintaining "irreconcilable differences" among individuals. The more difficult aspects of CNS function such as behavior and motivation must be addressed at a later time or, more appropriately, by others. My intent is to explore a system through which behaviors may operate. I have not given consideration to the limbic system, arousal, genetic expression, or signaling pathways—major components in the functioning of the CNS. They also affect the mechanisms of the CNS including profound impacts on the behavior of animals, all the way down to insects' mating rituals. It is not my intent to provide a model that defines the specific activities of specific neurons when the cortex responds to specific inputs; such a model would be able to perform this action on any plausible input, producing a predicable result, a test proposed by Yamins and DiCarlo [10]. Given the variations among individuals and the continuous changes in the CNS, the success of such a model is unlikely.

Following a précis, I discuss the major themes within the coherent apertures model. This is followed by notes on the chapters, an enumeration of proposed principles, and a brief history of the model development. I have sketched some of the current research models. A Glossary is provided in Supplemental Materials. In the Epilogue I propose a test of the coherent apertures model and discuss some issues, speculations, and possible future research.

Précis

The coherent apertures model is a theoretical framework for the operation of cortical areas for the generation of the internal experience, principally from early stage sensory inputs. Synchrony provides binding within an aperture and among apertures. Coherence instantiates information within those synchronized contexts. In this model, synchrony refers to the gross behavior of a population, an aperture. Coherence describes the participation of a subpopulation of elements in that synchronization, and timing correlations among their behaviors, their phase differences. Thus coherence in an aperture occurs in the <u>context</u> of its synchronization, forming a coherent aperture. Instantiations are resolved to low energy, and hence entropy, forms in both the synchronous and coherent contexts, producing the internal experience, which can be stored as a memory, and recovered. I have adopted the term "instantiation" of information rather than "encoding" as the

---

[2] The subcortical structures provide much of the drive for cortical operations. Essentially the cortex is a consistent structure that can perform many functions; the different subcortical structures each have specific functions. In computational analogies, the subcortical functions could be modeled with a modest (1,000–2,000) set of rules; the cortex is a "programmable" structure able to perform many complex functions. The two systems interact. The cortex is evolution's crown jewel.

[3] "Mechanism" implies both a structure and associated processes.

latter implies that, with the appropriate tools, one might be able to decode something, extracting information, which I feel is unlikely, given the apparent undecipherable complexity of the processes and their results.

Modeling a set of elements—mini- and macro-columns—as comprising a single aperture supports the concept of projections among apertures. Information is instantiated in an aperture as a phase field, called a coherence map (CM) here to avoid confusion, and in coherent information structures (CISs), the active outputs of CMs, that are projected among apertures. Beyond the thalamus and thalamic reticular nucleus I have not included the subcortical structures in this paper, other than a reference to those structures as influencing the underlying natural frequencies of cortical apertures. The relationship of an aperture to its corresponding thalamic area and the intervening sector in the thalamic reticular nucleus (TRN) is reduced to a composite node (CN), an abstraction useful in describing how various cortical apertures could form a temporary synchronous network, an ensemble, bidirectionally exchanging these CISs. The bidirectional causation within the ensemble enables all its constituent apertures to contribute to its evolving resolution through information entropy minimization, differing from a flow-through, or waterfall, model.

The ensemble constitutes a working memory. An ensemble may form through perception or through access to a memory. Memory (I have reserved the term for the latent states) resides in associated latent CMs in multiple apertures. CMs are formed through various modulatory mechanisms such as synaptic remodeling and glial interactions. CMs may have both strength-of-coupling and phase-modulation mechanisms, the latter becoming more predominant during memory consolidation. An aperture may contain multiple CMs, each of which will be responsive to different input projections. When properly stimulated, a subset of CMs in multiple apertures can cause the re-establishment of the ensemble—resurrection—in recall or recognition. The associations for stabilized declarative memory are maintained in the medial temporal lobe (MTL, e.g., the hippocampus), but are maintained within the apertures themselves for consolidated memory. The MTL provides a stabilized memory bridge between the ensemble and the formation of a consolidated memory.

Themes

The coherent apertures model incorporates several themes that reveal underlying consistencies, which I presume are actually found in the CNS. These themes emerged as my understanding of the CNS evolved:

- Radiant energy concepts
- Networks
- Information and entropy.

A radiant energy system manipulates energetic waves such as light, microwaves, radio waves, and sounds. In the coherent apertures model, radiant energy concepts apply to many circumstances, with the property of coherence particularly significant. In this model, radiant energy is analogous to neural activity. Radiant energy will emanate from a source, potentially projecting through an aperture, a "hole," within which the waves may be transformed or modulated in some manner, depending upon the characteristics of the aperture. The aperture may be a single "hole" or it may be formed from multiple elements that may or may not be adjacent or have some fixed pattern. Reflective and refractive concepts were particularly useful (e.g., thick mirror, phase conjugate mirror, dual phase conjugate mirror, hologram, Fourier optics, point spread function, and superposition), some of which were suited to the instantiation of information in the aperture. The consolidated instantiation of information was modeled as analogous to a phase hologram, a phase-modulating latent coherence map. Phase quantization, applied in digital holography, enhanced the development and recovery of CMs. The pumped AFP model is analogous to a laser. An aperture-wide field potential (AFP) reflects the degree to which the LFPs are synchronous across the aperture. A coherence path length is the distance and/or time over which photons in a projection are coherent, here a measure of the distance across an aperture over which the LFP is coherent, and in the case of the LFP, synchronous. Coherence path length is useful in understanding the coupled array behavior of an aperture. The coherence of radiation projected from multiple points in one aperture to a point on another aperture can be described through application the van Cittert-Zernike theorem (VCZ), which allows a coherent aperture to be treated as a node in an ensemble.

A cortical aperture, in addition to being modeled as a radiant energy system, is often described as a flat lattice of coupled elements—minicolumns—to which the network dynamics concepts initiated by Kuramoto [11] have been applied, particularly with respect to the critical point for network synchronization. Applying the work of others, in Supplemental Materials I have extended Kuramoto's basic concepts of phase-modulated synchronization. An ensemble, although initiated from a limited source, recruits multiple areas, reflecting the network connections. Perception can be modeled as establishing an ensemble, the internal experience, incorporating both relatively stable and malleable apertures. Memory is instantiated among a network of distributed apertures with latent connections that can establish, reinforce, or re-establish an ensemble.

Information entropy provides a useful way of describing information in an instantiation, while still supporting a physical model. Information entropy and energy are associated, since higher degrees of freedom (more complex states) have higher entropies, requiring more energy. Information entropy, as determined by complexity, can be used as a relative measure of instantiation efficiency. The converse of complexity is orderliness, resulting from underlying biophysical processes. Increased efficiency is a reduction in the complexity of the instantiation that maintains the most important information. A decipherable neural code at all but the ends of the input and output subgraph apertures seems unlikely. Undecipherability does not mean that information is lost, only that it can't be directly recovered. I did not find undecipherable complexity a comforting idea. I could not find a reasonable alternative.

The Chapters

The coherent apertures model is theoretical, built on experimental evidence in the literature. I have attempted to provide plausible concepts as rationally constrained components of a useful operational framework. Each chapter has its role in the framework, introducing concepts that relate to coherent apertures and their collective and individual operations. Beyond the framework, the coherent aperture is the obvious starting point.

2. A Framework

The coherent apertures model has two frameworks: anatomical and conceptual. The system composed of the cortex, thalamus, and thalamic reticular nucleus provides a framework for integration of cortical areas—apertures—into synchronous subpopulations—ensembles—that are involved in a common task, creating the internal experience. Chapter 2 describes this system and its general operation.

3. The Coherent Cortical Aperture

Chapter 3 explores the evidence that a cortical area—an aperture—may become essentially synchronous over its extent. Why is the mechanism of aperture synchronization important? The mechanism(s) of synchronization can reflect information instantiation (Ch. 4). A synchronized aperture may contain a coherent phase structure, which is composed of local phase differences relative to the overall synchronization. What good is coherence, anyway? The answers may help differentiate synchronization from coherence, as the relationship may be (and I would maintain, is) significant. The concept of an element in the aperture is less straightforward than expected. The pyramidals provide the outputs from minicolumns, and the outputs from apertures; I have merged the terms for efficiency, as this is a model, not a strictly neuroanatomical discussion.

Gross aperture synchronization is a prerequisite of coherence. Measures of synchronization may differ: neural spikes or the LFP. They are not mutually exclusive, and may reflect a particular model under consideration, although an aperture-wide synchronized LFP will be considered the indicator in this model unless stated otherwise. I refer to an aperture-wide synchronized LFP as the Aperture Field Potential (AFP).[4] Ising-like network arrays of minicolumn or pyramidal nodes underlie many aperture synchronization models, usually derived from the work originating with Yoshiki Kuramoto [11] describing phase modulation models, with a critical point in coupling strength leading to synchronization. In addition to modification of the Kuramoto model, I propose two additional synchronizing models. The pumped AFP model considers an aperture as a node connected to all of the elements. The laminar coherence model explores the possible effects of laminar differences in frequencies. I have attempted to unify this discussion of aperture cohering through expansion of the Kuramoto model, including the pumped AFP, in Supplemental Materials.

4. Information Instantiations in Apertures

The network, pumped AFP, and laminar aperture coherence models all contribute to information instantiation in the aperture through synchronization and subsequent coherence, described in Chapter 4. Cortical apertures synchronize. What is important about an aperture being synchronous, able to support coherence? How might this relate to information instantiation? The instantiation models are speculative, needing deeper modeling supported by experimental work and the analysis of large data sets arising from the simultaneous recording of many neurons within an aperture. I found no useful way to define particular information in an aperture. First, information was disseminated among the apertures of the ensemble. Secondly, as there were intermediate nodes that maintained relationships among CMs, such associations would be undecipherably complex, best characterized by information entropy. The relationships between memory types, their instantiations, and access are addressed in Ch. 7. Suffice it to say that a uniquely identified CM proves to be a useful, if abstract, concept.

5. The Aperture Operator

The model of an aperture operator is discussed in Ch. 5. The aperture in the coherent apertures model is a cortical area with a de facto operator that responds to inputs, producing outputs that are projected to other apertures. The operator of an aperture is composed of one or more functionals. Functionals are composed of functions, which respond to a particular facet or feature of input projections, e.g., edges. Functions are the local responses in the aperture, embodied in minicolumns and macrocolumns.

6. Integration of Coherent Cortical Apertures

It is well known that different sets of cortical areas synchronize depending on the task. In Ch. 6, I undertake modeling the binding of a set of specific synchronized cortical areas into an ensemble, with the exclusion of unsynchronized apertures. I elected to call such an active synchronous subgraph an "ensemble" for ease of reference. I favor the metaphor of the pieces playing together

---

[4] I shall use the acronym LFP when used as such in the literature cited, although it may in fact be the AFP. It is presumed to be an aperture-wide synchronization unless noted otherwise.

harmoniously, as in an ensemble. The set of all apertures and their connections is a graph; an ensemble is a subgraph of apertures with synchronized LFPs, limited by its connections within the graph. The ensemble can be formed from perception, internal processes (e.g., recall), or a combination of them both. Three synchronizing models support the bidirectional causation between apertures, hence within the entire ensemble, resolving to a solution or consensus. I propose that an ensemble will resolve by minimizing its entropy, hence its energy expenditure. The ensemble, and its process of resolution, constitutes the internal experience.

7. Operations

Chapter 7 was fun, an exploration of the operations of perception, memory, and memory access of simple declarative sensory input. The coherent apertures model plausibly fits the situations and requirements. The formation of an ensemble and its resurrection reflected the coherence of apertures and their interchanges. A surprise was the emergence of potential outcomes as shaping the processes by which an ensemble operated: a face is recognized as being a face. We tend to see or do the expected. The resolution of an ensemble was biased by the most likely outputs and the early feature responses. The existence of three memory states—working, stabilized and consolidated—is well documented, their operation initially unclear. They all are, or were derived from, the ensemble, which is working memory.

1.2. Principles and Corollaries

Principles

Some principles for time-structured activity in the CNS, e.g., object perception, emerged while developing the coherent apertures model:

- The cortex is composed of functionally distinct areas—apertures (parcellation).
- A cortical area may synchronize.
- A synchronous aperture can produce a coherence map (CM), a phase field of a pattern of synchronized and disordered elements.
- The responses of the cohered functions within the CM produce one or more coherent information structures (CISs) that are embedded in the projections of activity from an aperture.
- Cortical apertures form a network graph.
- Communications in the cortical graph are bidirectional (with a few exceptions).
- A cohered aperture is a node in a subgraph.
- An aperture that is synchronized or about to synchronize will filter to accept CISs that are in the appropriate phase.
- Synchronized apertures seek a low net information entropy.
- An aperture may contain a pattern that, when active, is appropriate for activating one or more CMs, including those in other apertures.

Corollaries

Corollaries follow from the principles:

- Not all elements need to synchronize for the cortical area, or aperture, to synchronize.

- Synchronized elements may have local phase differences, i.e., they cohere.
- Aperture coherence may occur in response to coherent inputs.
- Aperture coherence may emerge.
- Concepts from radiant energy systems may be applicable.
- A CM may be characterized by its information entropy.
- A functional of an aperture is composed of elements performing some essentially common functions.
- The CIS instantiates some aspect of information in the synchronized portion of the aperture's projection.
- There is no coherent communication if both apertures in a pair are not cohered.
- A disordered aperture does not synchronize with other apertures.
- A subgraph of synchronized apertures will synchronize into an ensemble.
- Cohered apertures that are synchronized in an ensemble will comodulate through bidirectional causation.
- The internal experience is defined as the current ensemble.
- An active or latent CM may provide associations between or among other active or latent CMs.
- A CM may be latent, able to contribute to the self-assembly of an ensemble when receiving a critical level of appropriate excitation.
- Outcomes with higher probabilities of occurrence influence the resolution of the ensemble.
- Not all apertures in an ensemble may have instantiating CMs when the memory is latent.

1.3. History

Development of the Model

I did not begin by attempting to figure out how a resonant cortical network forms and works. I began with the cortical area. It was clear from the visual phenomena of (1) common fate [12, 13], in which a group of moving dots will perceptually separate from a stationary field of random dots, and (2) structure from motion [14, 15, 16], that the primary visual cortex would, and probably must, cohere. That led to the question of "how?" The cortical area does indeed cohere, and I found mechanisms supporting "how." Aleksandr Davydov pointed me to work initiated by Yoshiki Kuramoto [11, 17, 18] on the synchronization of coupled arrays and networks that was important. This expanded to include a more field-oriented model of a pumped local field potential, and a laminar differentiation of synchronization. This led to explorations of how information could be instantiated in such areas, how instantiations could be formed and modulated through local functions, how the instantiations could be communicated among cortical areas, how a group of areas could bind together around a common task, and how such a group could perform various operations, resolving to some conclusion. The overall model of synchronous groups of synchronous cortical areas emerged, almost of its own volition. It is a result of an exploration of coherent cortical apertures, not a starting point.

Background

This work has grown from many sources. Having received an undergraduate engineering degree (BSME, Massachusetts Institute of Technology) I worked in optoelectronics research and development for a large US corporation, including coherent optics and electronics. Following premedical studies (Duke University) I worked again in optoelectronics for industrial automation, leading to R&D in high accuracy electronic measuring systems involving electronic hardware, including custom digital and analog integrated circuit design, software, and statistical methods. During this period I undertook studies in perception, eventually leading to an MS and a PhD in experimental psychology (North Carolina State University), with theoretical and experimental work in hearing (thesis) and three-dimensional perception (dissertation). I have patents in auditory displays and a 3D display technology. I worked under grants and contracts from federal entities developing FAA audio displays, and 3D visual displays of molecular models and airspaces through software. Some of the work, and good collaboration on spatial perception, was done in the laboratory of Joseph Lappin at Vanderbilt University. I developed auditory displays in my company's laboratory (Allotech, Inc.), patenting one display method.

While working at Intelligent Automation, Inc. (Rockville, MD), Aleksandr Davydov introduced me to the work of Yoshiki Kuramoto [11] on the synchronization of large networks for consideration as a model of the behavior of the cortex. Kuramoto mentions this possibility himself. I had come to the conclusion that cortical areas could be considered as apertures and that they should self-cohere. I was not aware of Kuramoto's model, so his work was a joy to discover. I have incorporated it, and the considerable subsequent work, into my own in multiple scales and embodiments.

In addition to network modeling, I have been introduced to many ideas in radiant energy, from large array telescopes to lasers, that have had a significant input to this model. The concept of the coherent cortical aperture is a logical outgrowth, predating my dissertation (1993). Phase conjugate mirrors [19] and the thick mirror model together model both the cortical transform and memory. The pumped aperture field potential (AFP) model has some conceptual roots in the laser. Using the van Cittert-Zernike theorem to describe the projection of radiation from one coherent aperture to another [20] came to me while I was exercising (I guess it _is_ good for you). Research by others on the coherence and synchronization of multiple lasers in various configurations led to the realization that a transient (ad hoc) network of such nodes could form a synchronous network during working memory that could decompose into a latent memory, and re-form from it. The use of Shannon and Weaver's [21, 22] work in information entropy reflects my exposure to statistics in work and academia, allowing consideration of information instantiated in various forms to be described by its entropy.

A rather remarkable result from modeling an active bidirectionally coupled subgraph was the realization that all nodes in such an ensemble would influence the behavior of the ensemble; thus, past the initiation stages, the potential outcomes would influence the trajectory of the entire ensemble: you tend to see faces in clouds. My future work is undetermined. There are many paths to follow, many questions left unanswered, concepts in need of testing. An abstract model and potential technological implementations are topics for a separate document.

1.4. Research Models

Computational Neuroscience

Computational neuroscience is theoretical research, employing models as representations of a system, intended to further the understanding of brain function. The CNS is considered to be composed of structures that collectively provide information processing capabilities. Computational neuroscience reaches beyond models of symbolic processing [23] into modeling of neural activity, attempting to describe the operating principles of the central nervous system and its components. Models differ in their relationships to neurophysiology, from computerized large scale neuron-based architectures (e.g., Eliasmith et al. [24], Merolla et al. [25]) to more abstract systems (e.g., Grossberg [26]). Attempts to model the brain with tools such as NEST [27] computationally stress the limits of computers [28]; in 2013 a supercomputer required 3 minutes to calculate the operations of a model of about 1 mm^2 of cortex over one second [29]. Computational power is increasing but the finite limits of computers probably limit the success of such modeling. Different paradigms have attempted to model CNS function from operational standpoints rather than neurological modeling. De Garis et al. [30] have reviewed such work.

Technology [30, 31] has supported the growth of computational neuroscience, creating both a great opportunity for more complete knowledge of the activities in the CNS and frustration over what to do with this wealth of data. Neural circuit models [32, 33, 34] have been overwhelmed despite of massive computational capabilities, as more data compounds the problem [35]. Computational neuroscience is in a state of flux—nothing new in the neurosciences—as new hypotheses emerge. The very concept of computing is changing. Computation beyond Turing machines is being explored [36]. Goldin et al. [37] propose that beyond the Church-Turing model of computation, communication and computation occur together as one might expect in the mammalian central nervous system (CNS). In a review Hepp [38] describes approaches to a global workspace (GW) [39] of coherently linked neurons. Liquid state [40], analog computer [41], and quantum mechanical [42] models of the CNS have been proposed. Ideas abound.

Networks: ANNs

Network models, a class of computational neuroscience, are currently of considerable interest, although complex [43]. Synaptic timing, strength, and plasticity provide the bases for many artificial neural network (ANN) models [33, 34, 44, 45, 46, 47, 48, 49]. Such neural network models are incorporated into higher level models [26, 50, 51] either directly or by implication, acknowledging the division of the cortex into many interacting areas [52]. The network models of the CNS are long standing. Hebbian models [53] of changes in synaptic strengths through long–term potentiation (LTP) and long-term depression (LTP), particularly under conditions of critical timing, have been applied to ANNs operating in fixed and flexible architectures. Spike timing dependent synaptic plasticity (STDP) is often incorporated in ANNs [49]. Such ANNs have had widespread applications, but have little direct relationship with current models of the CNS. Cortical ANNs (CANNs) [54] attempt more cortically oriented models to explore hypotheses of cortical function. Consistent with empirical evidence, such models presume a similarity across all areas within the cortex, against which differences are interpreted. Neural network models may employ the columnar structure of the cortex to provide the nodes. The generation of such columns may itself represent self-organization in the cortex [55, 56]. The outputs of these cortical models are the activity patterns of the elements as a corpus [57]. Hierarchical convolutional neural networks (HCNNs) [10] attempt to maintain significant neurophysiological validity by modeling cortical areas as layers performing increasingly abstract functions. Feedback loops create goal-driven aspects of such modeling.

In most models the architecture (connectivity) is predetermined (e.g., Edelman 1982 [45]), although Risi and Stanley [58] propose an evolved connectivity. The CNS is composed of a well organized set of neural elements with connectivity described at multiple scales. That is not to say that the connections between the elements are obvious—great effort has gone into elucidating these patterns [59, 60, 61]; more is yet to be done [62]. In principle the CNS can be mapped, although there are always individual differences. Potentially the CNS is a deterministic system [63, 64] with sensitive locally definable functions, making it difficult to predict results from inputs, as such systems are often chaotic [65, 66]. Models that attempt to use operational regularity face the challenges of deterministic systems that have simple behaviors that are exercised in a large number of instances—in the CNS in the form of a great many neurons or minicolumns. Given the increases in computing power and multiprocessor architectures, very large scale models incorporating synaptic plasticity within hierarchical network architectures are being developed [24, 25, 30, 67, 68, 69]. Consistently, these models are based on synaptic actions, although Steck et al. [70] propose an ANN model with an optical field in a bulk non-linear continuum.

Neurophysiological Models

The coherent apertures model is basically a neurophysiological model with a network emphasis. The Hodgkin–Huxley model [71, 72, 73, 74, 75] of neural action potential generation underlies higher level models and is reflected in the pumped LFP model of synchronization in the coherent apertures model. Neurophysiological models of the CNS describe functions and operations in terms of biological structures and actions. They are based on anatomy, physiology, and activity flows. Peripheral inputs to thalamic nuclei are modulated and subsequently relayed to primary perceptual cortical areas. Higher order cortical areas have corresponding thalamic nuclei that have no peripheral inputs. The majority of inputs to the thalamus are not of peripheral sensory origin. Destexhe [76] proposes that the majority of activity is the modulation of internal dynamics of the cerebral cortex, consistent with the models of vision of Jehee et al. [77] and Lamme et al. [78] in which layers or levels of visual areas reciprocally interact to refine function. Synchrony in dynamically networked cortical areas is considered a hallmark of consciousness [79, 80, 81, 82]. Boustani and Destexhe [3] and Muller and Destexhe [83] propose that such synchrony is large-scale, while local cortical activity incorporating complex information has more overtly chaotic behavior [84]. Dehaene et al. [85] and Dehaene and Changeux [86] support the concept that consciousness arises from the long-range synchronization of cortical areas into a Global Neuronal Workspace (GNW) with a subsequent taxonomy of consciousness. Along that line, Henke [87] proposes that CNS modes for memory formation and consciousness differ.

The thalamic reticular nucleus (TRN) is a component of some models. The TRN is a sheet-like structure of inhibitory neurons receiving collaterals from both thalamocortical and corticothalamic projections. Relevant to the framework suggested in the coherent apertures model, Min [80] and Drover et al. [88] proposed an operation of CNS function focused on the TRN as providing the underlying synchronization among cortical regions and their corresponding nuclei in the thalamus through loopbacks. Destexhe et al. [89] propose the TRN as capable of generating oscillations within its population. The TRN is an important structure, but perhaps not of such singular importance as Min proposes. It has an inherent natural frequency, being an active component of a more complex system.

There are many model types employed in theoretical descriptions of the operation of the CNS. The large number, type variety, and complexity of structures make any model a simplification. What is gained by the simplification may be lost through excessive reductions. The end result of any modeling must be amenable to some tests that can be performed with real, living creatures. Otherwise it is self-referential; perhaps a useful exercise in thinking about how the CNS works, but still a speculation. Some tests are post hoc—how well does the model fit actual data? Are there data that refute it? What tests can be proposed? Can they tease apart to true and false aspects of the model? I have addressed some of these issues through extensive references and a proposed test of the model.

2. A Framework

2.1. Introduction

The coherent apertures model addresses three fundamental issues in the neurosciences:

- What is the internal experience?
- How does cortex work?
- How is information instantiated in the cortex?

Thus, how does the process of instantiating information in the cortex generate the internal experience?

The model integrates a number of interacting concepts:

- The cortex is divided into many areas, *apertures*.
- A cortical area may become synchronized.
- Coherence occurs in the context of synchrony.
- Information is instantiated in a spatiotemporal structure within a synchronized aperture as a *coherence map* (CM) in a *coherent aperture*.
- An *ensemble* is a synchronous subgraph of coherent apertures, constituting the internal experience.
- CMs have multiple cross-linkings among apertures within an ensemble.
- Information instantiated within and among apertures in an ensemble may be characterized by its information entropy.
- Resolution of an aperture to greater orderliness creates a lower information entropy, hence greater information instantiation efficiency.
- Through bidirectional causation among its apertures, an ensemble will move toward a minimum entropy, and thus a consensus.
- An ensemble, with CMs with their multiple cross-linkings, can be stored in a latent state as a memory.
- A memory can be retrieved by resurrecting its ensemble through exercise of its CM links.

Three complementary models of aperture synchronization and cohering will be presented: Kuramoto networks, local field potential pumped synchronization, and laminar synchronization-/coherence propensity.

The human cortex has several hundred functional areas. Different constellations of areas form for different tasks. The coherent apertures model addresses the formation and operation of these constellations. The coherent aperture is at the core of the model. An aperture is a bounded region through which radiant energy (e.g., light, microwaves, sound) flows. In the coherent apertures model, the "radiant energy" is neural activity flowing among apertures, projected over bundles of axons (tracts). Coherence, in the sense of radiant energy, is a spatio-temporal correlation: order, as opposed to disorder. A coherent aperture is able to maintain some spatio-temporal aspects of its inputs in its outputs, such as a telescope lens that is able to project an image from the light falling across its area, its aperture. Cortical aperture coherence occurs in the <u>context</u> of the aperture's synchrony, instantiating some aspect of information. A transform may occur through an aperture, as a lens with light passing through its aperture may focus an image. The functions of the element responses within a cortical aperture form its transform, which may be complex and modulated.

The importance of the coherent aperture is supported by the known behavior of the cortex. Although we may consider the human cerebral cortex to be a continuous sheet crumpled up to fit in the cranium, in reality it is composed of between 200 and 400 functionally distinct areas. Cortical areas collaboratively produce the internal experience, the putative output of the CNS. It is well known that under the appropriate conditions, e.g., stimulus or task, a cortical area will exhibit grossly synchronous periodic activity as reflected in the EEG, which in turn reflects the net LFP over that area, when multiple cortical areas can become synchronous as a group. We assume that meaningful information is being exchanged during this synchronous behavior. Information is distributed, instantiated among apertures. Apertures that are synchronous (hence coherent) form the synchronously bound subset, an ensemble, through the participation of some subcortical structures (Figure 1). The ensemble performs biological processes on the biological instantiations of information through mutual interactions among, and the response characteristics of, the apertures. The ensemble forms the internal experience in short-term memory. Perception, memory, recognition, and recall are the results.

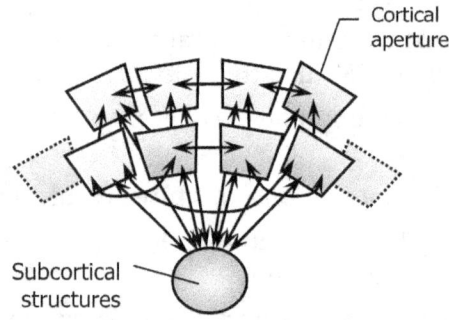

Figure 1. Multiple cortical apertures with an active ensemble (white).

Several interrelated concepts underlie the model: cortical apertures, aperture coherence, information instantiations, and synchronous ensembles. Instantiation is not encoding; there is no evidence that higher level information can be extracted, a priori, from patterns of activity. There is only the detection of known information-activity correlations. Existing and new models have been combined here to create a unified model of the operation of the cortex, thalamus, and thalamic reticular nucleus. Many behaviors of components and their collections are emergent. I shall consider the internal experience as the output of the CNS. Ultimately it appears that what the brain does best is make decisions. All day long we make decisions fluidly, many of them so small we are unaware of them: What shall I wear? Have for breakfast? How many eggs? How much pepper? Which egg will I eat first? Where do I fork up the first mouthful? We are constantly resolving to solutions. When possible, we use pre-established patterns, only deciding to "launch" them.

Objective

One must strive for a balance between simplicity and adequacy in explanations. The model is described through a progression of topics, following this chapter that provides a framework, starting with the coherent aperture and concluding with the operation of the CNS during perception, memory, and the access to memory. You may choose to first read this chapter (2. *A Framework*) summarizing the model (*what?*), and then Ch. 7 (*Operations*) that describes some representative operations (*so what?*), turning to intervening chapters as needed for more complete explanations (*how?*). Each chapter provides a basis for subsequent chapters. Each chapter raises questions that I have tried to address in succeeding chapters. Ch. 3, *The Coherent Cortical Aperture*, describes coherent apertures and how they arise through synchronization. Ch. 4, *Information Instantiations in Apertures*, describes how information could be instantiated in apertures. Ch. 5, *The Aperture Operator*, summarizes how functions determine the responses of apertures to inputs. Ch. 6, *Integration of Coherent Cortical Apertures*, provides models of how multiple coherent apertures can be bound into a synchronous subgraph focused on a task. The Epilogue summarizes the issues, principles and results; provides concluding remarks and speculations; and mentions the possibilities for future research. In addition to technical notes, I have provided a Glossary in *Supplemental Materials*. In this chapter I describe the framework of the model, the principles and results that ensued from this effort, and a hypothesis. Technological implementations are considered elsewhere.

2.2. The Coherent Apertures Model Framework

Introduction

Edelman and Gally [*90*] invoke the principle of *degeneracy* in biological systems: "the ability of elements that are structurally different to perform the same function or yield the same output." *Elements* are <u>logical</u> constituents. This is consistent with the proposal by Marder and Taylor [*91*]: "Studying a population of models with different underlying structure and similar behaviors provides opportunities to discover unsuspected compensatory mechanisms that contribute to neuron and network function." Such a "population of models" potentially reduces the complete array of possible structures to a hopefully adequate subset of functional models in which each model conveys a particular class with a similar behavior[5]. They argue that given the current abundance of data, a more complete set of models can be developed, while testing for commonality of behaviors, particularly of individual neurons. A model of the CNS incorporating these concepts of degeneracy and adequate populations will simplify complexity in an attempt to expose some principles of operation. Such reductions entail some loss of accuracy. I have attempted a reduction such that losses are not serious.

Framework

Architecture

A soccer ball (Figure 2) provides a rough illustration of the CNS cerebral cortex. If expanded to a sphere, it will have a size roughly equivalent to a soccer ball [*92, 93*]. The cerebral cortex is 2–4 millimeters thick, characterized by an apparently consistent, although not completely uniform, general structure, having six layers. It has been crumpled up to fit in the skull. At the bottom of the ball is a big lump, the subcortical region. The subcortical region is an altogether different, and more

---

[5] Conversely, I suggest that essentially the same structure may perform different functions, e.g., cerebral minicolumns in visual, auditory, and somatosensory areas.

complex, matter. It has many characteristics of ganglia, although with (somewhat) greater organization. It is a snarly mess. It has no consistently repeated pattern, although the same architecture appears consistently across species.

Figure 2. Soccer ball model of cortex.

The cortex has distinct regional variations that are the result of genes, development, and experience, particularly early experience [94, 95]. The cortex has 200–400 distinct cortical areas [96, 97], the cortical apertures, corresponding to the soccer ball's patches. As 70% of the cortical surface is buried in the folds (sulci), the crumpling provides rich opportunities for connections among cortical areas and with the subcortical structures. The apertures are richly, although not completely, bidirectionally interconnected within the ball. The apertures comprise a graph of vertices (the apertures and other structures) and edges (the connections among the vertices), currently being elucidated for humans in the connectome project [98, 99, 100, 101]. The coherent apertures model addresses primarily the cortex and, within the subcortical region, the thalamus

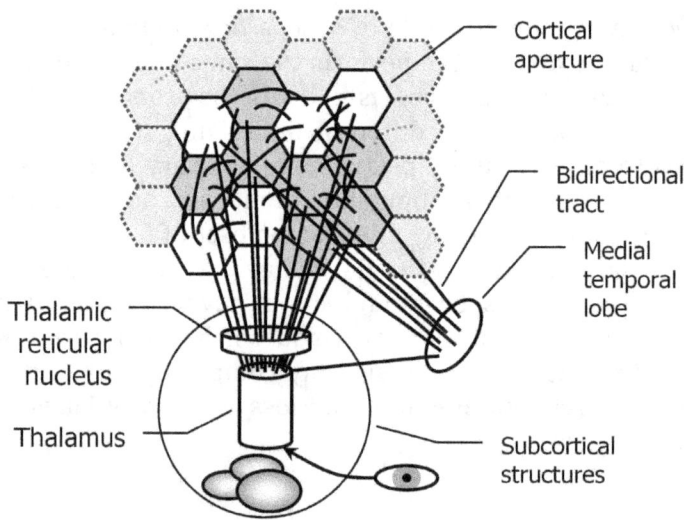

Figure 3. Basic CNS structure of the model. All cortical apertures are bidirectionally connected to central structures of the thalamus and thalamic reticular nucleus. The medial temporal lobe is a specialized section of cortex connected to other cortical apertures. Apertures in white are a synchronous group (ensemble).

and its attendant thalamic reticular nucleus (TRN) lying between the thalamus and the apertures (Figure 3). Synchronous networks of cortical areas are not unfamiliar, frequently being observed in

EEG and fMRI studies. The coherent apertures model attempts to understand how and why. The model is only concerned with adult mammals.

The Coherent Cortical Aperture

What is an aperture?

A cortical aperture, for example the primary visual cortex (Brodmann 17, V1 [*102*]), is functionally similar across its extent, with functionally distinct borders that may not be evident morphologically, the borders being revealed during single-unit measurements under various stimuli. A cortical aperture is composed of macrocolumns containing 40–80 minicolumns. There may be up to 750 million neurons organized in up to 9 million minicolumns in up to 125,000 macrocolumns in a cortical aperture. Elements are modeled as potentially active components, with outputs projecting from a minicolumn over axons from a few pyramidal cells in an infragranular layer; consequently, for modeling purposes, a minicolumn and a pyramidal neuron are usually considered synonymous, a simplification with restrictions when laminar differences are considered.

Synchrony and coherence

What is the structure of an aperture's activity? If all the elements in an aperture are behaving identically, then the opportunity for communicating different information is extremely limited, as only the gross activity would matter. Surely nature evolved such that the lines of communication are not so massively redundant. Thus there must be differences among the aperture's neurons' projecting axons that are meaningful. This implies—nay, requires—that locales within a cortical area differ such that among them they instantiate information. Two aspects of activity present themselves for immediate consideration: which elements have appropriate activity, and what is that meaningful activity? An aperture is capable of performing both of these functions in concert. The meaningful differences in activity can occur in the context of a gross synchrony. Coherence is an attractive concept supporting meaningful relationships among separated cortical elements, e.g., minicolumns.

The synchrony of an aperture is evidenced by a LFP that changes nearly uniformly over its area. Differing subpopulations of elements may participate in the synchrony, with variable timing (phase) relationships amongst them. Such a subpopulation is coherent. Thus the aperture that contains it is a *coherent aperture*. A coherent aperture can dynamically instantiate information; thus the relationships between synchrony and coherence are significant. Aperture synchronization and cohering are discussed in Ch. 3. Three mechanisms are proposed for synchronization, their being potentially complementary: network, pumped local field potential, and laminar coherence propensity. The three models are complementary.

Information Instantiations in Apertures

The instantiation of information in apertures is discussed in Ch. 4. Information instantiations are physical; however, I do not expect to deduce a "neural code" [*103*], if such a thing exists. *Undecipherable complexity* describes the state of an instantiation from which information cannot be extracted through external means. An undecipherable physical instantiation can be manipulated by biological processes without regard to its content. Information in a cortical aperture is modeled as instantiated in a *coherence map* (CM), a field with two states:[6] a synchronous, or potentially synchronous (i.e., latent), distributed subpopulation of elements, and a background of disordered elements. Instantiations may be characterized by their *information entropy*, an expression of quantity, complexity, and orderliness. A CM is a state of increased orderliness. Entropy reduction

---

[6] Technically a phase field.

represents greater efficiency and usually lower energy expenditures. This will prove significant during cortical integration, which is concomitant with the operations of the CNS.

The Aperture Operator

The major neural projections enter and exit the aperture from the inside of the cortex, conceptually a thick mirror with internal functions. A function is a local relationship between an input field and the resulting output, an aperture's functions collectively creating its operator, discussed in Ch. 5. A minicolumn's response behavior, its function, is expressed in the output of its pyramidals (layers V, VI). The active responses of the local functions to the facets of the inputs, e.g., oriented edges, across an aperture create its active CM, which produces outgoing *coherent information structures* (CISs). The functions may be modulated, evidenced by accessible memory. Short term modulation will contribute to the immediate operations of the CNS. Generally modulation progresses from a strength of coupling (weighting) to a timing (phase) mode, although the modulation mechanisms may overlap.

Integration of Coherent Cortical Apertures

Experientially we <u>are</u> what our brains are doing in certain conditions. In the coherent apertures model the CNS produces an internal experience by integrating (Ch. 6) coherent cortical areas (Ch. 3) that contain CMs (Ch. 4). During the internal experience the instantiation flow is from a disordered network of apertures to a synchronously bound subset of synchronous apertures, an *ensemble*, with subsequent entropy minimization to create a unified percept or construct that is emergent as a whole. I have considered the internal experience to encompass awareness, consciousness, working memory, the global neural workspace, attention, and perception, without differentiating among them.

CNS integration includes: i) binding of separated areas, ii) communications, and iii) communications encoding. All the cortical apertures comprise a graph with bidirectional connections. Apertures are recruited into, and excluded from an ensemble based on their ability to exchange CISs. Communications among the apertures, exchanging CISs, occur in the context of ensemble synchrony. The coherence of a CM's projected CIS and the coherence, or propensity to cohere, of the receiving aperture form a filter set enabling the CIS acceptance. The functions of each aperture relate to a common task. The outputs of a specific subset[7] of the elements comprising a CM can activate or coincide with CMs in target apertures, linking the appropriate CMs among apertures. A disordered or functionally mismatched input will not create a CM, excluding the aperture from the ensemble.

Through application of the van Cittert-Zernike theorem a coherent aperture is considered a node (*VCZ node*) in an ensemble. During working memory, VCZ nodes become synchronously bound into an ensemble with an underlying frequency, as evident in the EEG. I have suggested three modes of synchronization of apertures into an ensemble, the modes flowing one into another in stages: mediated synchronization, common drive synchronization, and bidirectional synchronization (peer). The continuous bidirectional activity results in *bidirectional causation* with the comodulation of the apertures' functions (Ch. 5), sharpening the CMs, lowering the entropies in both apertures and of the ensemble, as the correlation between apertures increases organization. The system continually resolves toward entropy minimization through the synchronies within and among apertures. The non-participating apertures have decreased activity, further lowering system entropy. The internal experience, defined here as the "output" of the brain, is the result of the continually emerging formation of these synchronous ensembles, with the accompanying entropy minimization.

---

[7] A surrogate for the CM, potentially useful in computer modeling.

Operations

The coherent apertures model explores early stage sensory processing of declarative stimuli, e.g., visual objects, discussed more fully in Ch. 7. The operations of perception, memory, recognition, and recall coincide in common structures, modeled as intimately related through subgraphs of apertures containing coherence maps (CMs) that instantiate information. Sensory information is distributed among apertures in features instantiated in CMs within apertures, and in the relationships among those instantiations in multiple apertures. Apertures fall into several classes, according to their roles in ensembles. The separation is somewhat artificial, as apertures may serve multiple changing roles:

- **Primary sensory cortex** such as V1 or the retina
- **Feature responsive (sub)apertures**, which may lie within another aperture
- **Domain apertures**, intervening apertures that serve one or more domains such as vision and hearing
- **Association apertures** that support interaperture CM relationships
- **Key registry apertures**, association apertures in the medial temporal lobe (MTL), including the hippocampus, parahippocampal cortex, perirhinal cortex and entorhinal cortex, that incorporates natural keys (Ks) linking CMs
- **Outcome apertures** dedicated to an output domain such as the premotor cortex.

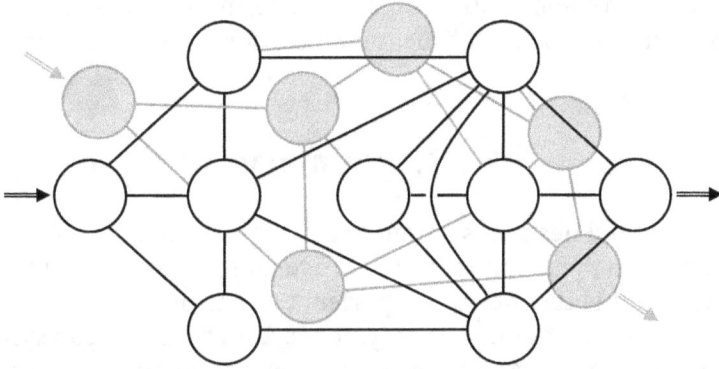

Figure 4. A network of cortical apertures. Dark outlines designate the ensemble. Grey elements have been excluded from the ensemble.

A CM performs two functions: it instantiates some facet(s) of information in an aperture, and it instantiates associations with CMs in other apertures. An active CM is a response pattern. A latent CM instantiates facets of memory, including CM relationships. Aperture functionals may extract features from inputs, transforming them into more invariant forms, the relationships among which are themselves information that is instantiated within CMs. Fundamental CMs within some apertures (e.g., phonemes) typically have high persistence and may comprise a library of components from which a complex CM is produced. Similarly, a set of associated apertures may connect preferentially as a motif. A *natural key*, K, is an abstraction representing the specific relationships of the CMs in domain apertures; thus it puts the right mouth in the right place in the face, so to speak. Ks may reside in a specific MTL aperture or be included in CMs in multiple apertures, including association apertures, depending upon the memory state involved.

Perception occurs in an ensemble. Seeing an object is a process starting with image stimuli on the retinas that results in activity over many apertures, integrated in an ensemble that resolves to a solution—a percept in working memory, the internal experience. The ensemble solution may be biased by existing potential outcomes, as their persistent CMs have low information entropies. Invariants, as persistent latent CMs, constitute de facto intermediate outcomes that influence the process of perception. You quickly recognize faces you know; you may see a new face as resembling someone you know. Perception requires a dynamic input, as the entropy is continuously minimized. Memory is best understood relative to the ensemble, the internal experience. I suggest three sources of the ensemble—working memory, stabilized memory, and consolidated memory, which correspond to when it is accessible—now, today, or tomorrow:

- Working (now): the internal experience, short term, including perception, which can become;
- Stabilized (today): no longer in working memory, but accessible until sleep, during which it either decays or becomes;
- Consolidated (tomorrow): long–term memory that becomes fused during sleep.

A memory does not reside in a single site, being distributed among CMs in multiple apertures. Memories are accessed by resurrecting the CMs and their relationships. A memory, in addition to its component CMs, requires the maintenance of the associations among those CMs. During a working memory, an active ensemble manifests the associations. Within a stabilized memory, most of the associations of the previously active ensemble are maintained in some structure in the MTL (e.g., hippocampus), although some relationships may exist in consolidated motifs of apertures with associated CMs. A distributed consolidated memory incorporates the associations within the CMs in apertures.

2.3. Summary

The coherent apertures model describes a framework for the biological processing of declarative information instantiations in the mammalian CNS, focusing on the cortex, thalamus, and thalamic reticular nucleus. The result is the internal experience. Cortical areas are modeled as apertures, bringing to bear concepts from radiant energy systems. A cortical area can synchronize when only subset of its elements synchronize, and cohere. Such a subpopulation, although somewhat noisy, will become approximately synchronous through several mechanisms, creating coherence maps (CMs) of coherent elements in a sea of disordered elements. The associations among the CMs in the ensemble constitute the glue binding a particular internal experience produced by specific CMs in specific apertures. The creation of a memory entails capturing those associations and those CMs. Not all CMs form dynamically. They may exist in a persistent latent (inactive) form. The new or malleable CMs may become latent through several mechanisms that include coupling modifications and phase modulations with various persistences. The associations may be incorporated in a dedicated structure such as those found in the medial temporal lobe (MTL) such as the hippocampus, or may be incorporated within some of the ensemble's CMs. Matching enough of the association CMs, either directly or indirectly, can result in the resurrection of the memory, either completely in a recall of the ensemble or sufficient to recognize an input. Perception, and memory and its access, are bound together through information instantiated in coherent apertures and ensembles.

3. The Coherent Cortical Aperture

3.1. Introduction

A coherent cortical aperture is the underlying concept in this model. It instantiates [104] information, enables the formation of task-specific working groups of cortical areas, provides the internal experience, and creates and accesses memory. In the CNS, information is instantiated in cortical areas. It is these instantiations that are operated on. A cortical area with one or more essentially uniform functionals, e.g., visual edge, orientation, or motion response, is an *aperture* [105]. In radiant energy systems, such as electromagnetics [106, 107, 108], optics, and acoustics [109], an aperture is a bounded region through which energy flows, potentially being transformed. A coherent aperture maintains some spatio-temporal features of that radiant energy. The flow of radiant energy is analogous to the flow of neural activity, spatio-temporal activity providing a medium for information instantiation. The aperture model consolidates frequently used concepts including distributed information, field, projection, transform, and point spread function (divergence and convergence).

This is a complex chapter. Some repetition is inevitable, as different structures play multiple roles. For example, the macrocolumn, a collection of minicolumns, provides structural organization via local circuits, enhances local synchronization, and organizes a complementary set of local response functions. Such intertwining is a well-established characteristic of the brain. The three synchronization models are not independent. The network and field models have a converging commonality, a critical point, which is described primarily in Supplemental Materials, since it includes potentially distracting mathematics. The network and field models are both components of the laminar cohering model. Further, the relative weights of the processes which are described in each model change over time (Figure 8).

A cortical aperture may become synchronous, an aperture-wide state. Coherence is an information instantiation in the context of synchrony. Therefore, the term "coherent aperture" is significant, for it denotes a structure that <u>can</u> instantiate information. The aperture must be capable of synchronization. This chapter primarily addresses the mechanisms of synchronization, specifically in the context of the ability to instantiate information through a coherent subpopulation of elements.

3.2. Cortical Structure

The structure of the cortex underlies the synchronization and cohering models. The human cortex is a crumpled sheet with an area of ~2600 cm^2 and a thickness of 3–4 mm containing up to 28×10^9 neurons [110]. The organization of these neurons is far from random. A brief review of cortical structures is in order: parcellation, columnar organization, laminar organization, and circuitry.

Parcellation

Rakic [111] specifies cortical areas as functional units, i.e., the parcellation of the cortex [43] into what are defined here as apertures. Gross anatomical features are not reliable indicators of area boundaries. Fox et al. [112] found functional evidence of distinct boundaries. Sporns cites Cohen et al. [*113*], who were able to produce maps of bounded functionally defined areas in the cortex using resting state functional connectivity (rs-fcMRI). Kaas [*114*] had proposed that more than 100 such areas evolved in the human cortex. Power et al. [*97*] propose 264 such areas, Glasser et al. [*96*] report 180 per hemisphere. Vision uses approximately 32 areas, 25 solely for vision [*52, 115, 116*]. The work of Frostig et al. [*117*] and Kajikawa and Schroeder [*118*] calls into question the sharp borders of cortical areas, although the frequency-independent spread reported by Kajikawa and Schroeder is inconsistent with a passive model of conduction. Frostig et al. propose such "border crossings" as due to active horizontal connections. In all cases, one must consider the state of the experimental animal, as thalamic actions are important and subject to changes due to awakeness and anesthesia. The larger body of evidence supports a relatively discrete parcellation of the cortex.

Columnar Organization

The cortical plate is an array of columns oriented normal to the cortical plane (radially). The meaning of "column" varies in the literature. I shall use the terms "macrocolumn" and "minicolumn" to avoid confusion. Each *macrocolumn* contains 40–80 *minicolumns*—the cortical array *elements*. The macrocolumns [*110*] are defined as *hypercolumns* in visual cortex [*119*], *segregate columns* in somatosensory cortex [*120*], and *barrels* in barrel somatosensory (for the vibrissae) cortex of the rodent [*121*]. Cortical apertures may be as large as 40 cm^2 (primary visual cortex, V1) with 750 million neurons [*122, 123*] organized into about nine million minicolumns that form about 125,000 macrocolumns of functionally closely related minicolumns. Such a large array of coupled elements in a network, be they neurons, minicolumns, or macrocolumns, approaches being a continuum, i.e., an aperture.

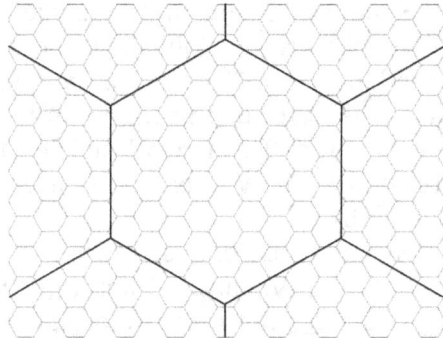

Figure 5. Macrocolumn comprised of minicolumns (hexagonal approximation).

Minicolumns and macrocolumns are arranged in a roughly hexagonal array [*110, 124*] (Figure 5) spanning layers II–VI, providing an estimated 0.5–9 million minicolumns per cortical area, varying by area (estimated from Braendgaard et al. [*125*], Henery and Mayhew [*93*], Rakic [*111*]), and Blinkov and Glezer [*123*]). Rakic [*111*] describes the minicolumn in a "radial unit model." Mountcastle [*110*] and Buxhoeveden and Casanova [*120*] make the case that the minicolumn [*110*] is the fundamental element in the cortex (see Buxhoeveden and Casanova for a review of the generally accepted "minicolumn element hypothesis" [*120*], although Horton and Adams [*126*] disagree with this interpretation). Buxhoeveden and Casanova generally define the number of minicolumns in a macrocolumn in terms of the number of currently active minicolumns, so the number varies with time. Jones [*127*] notes considerable variation in the interpretation of anatomical and physiological evidence of the minicolumn, since it is difficult to separate the structure of pyramidals and double bouquet cells from a minicolumnar or microcolumnar interpretation. Some cells span multiple minicolumns, which further confuses the issue. The minicolumn hypothesis is useful, however, in modeling some neural interactions, as will be described below.

Laminar Organization

The cortex has a laminar structure of nominally 6 layers (I–VI), with supragranular (SG) layers I–III and infragranular (IG) V–VI layers, respectively above and below an input layer, IV. Layer I contains very few neuron cell bodies, being composed of a felt of axons providing a rather diffuse pattern of innervation. The mini- and macrocolumns participate in the cortical laminar organization in addition to the columnar structure [128, 129], bound into what Buxhoeveden and Casanova [120] refer to as a "temporal column," although the frequency profiles and coherence of the layers differ [130, 131, 132].

Circuitry

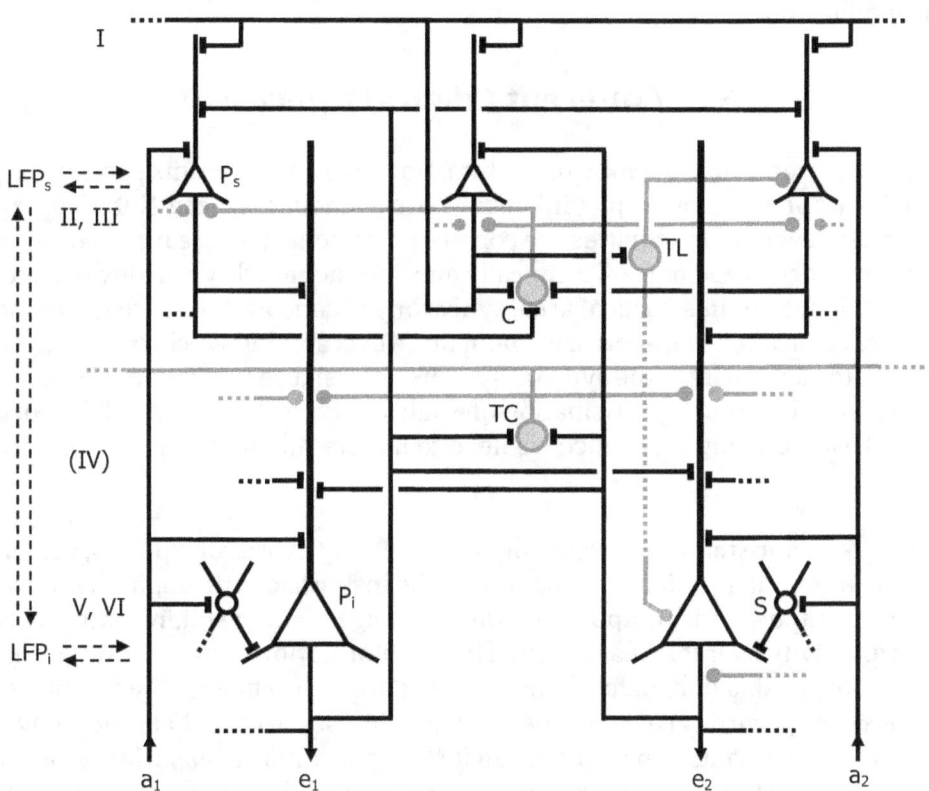

Figure 6. A suggestive canonical cortical circuit. Black = excitatory, grey = inhibitory, a_i = afferent, e_i = efferent; numerals indicate cortical layers. Some general neuron types (from DeFelipe et al. (2013) [286]: P_i = infragranular pyramidal, P_s = supragranular pyramidal, C = chandelier (a local type), S = stellate, TL = translaminar, TC = transcolumnar. Glia are not shown. LFP_i = infragranular local field potential, LFP_s = supragranular LFP (adapted from Douglas, R., et al. (2004) [140].

The cortex has both laminar and planar network organizations. Figure 6 is a sketch of a representative canonical cortical circuit illustrating representative neural types that form mini- and macrocolumns with a laminar structure. Glia have been omitted. Pyramidal cells are aligned with the axes of the minicolumns. Pyramidal somas are in broad SG and IG laminar categories. Those with their somas in SG are smaller, with dendrites and axons more local; those in IG are larger, span more lamina, and are the source of output projections. The behavior of the minicolumn elements is important, as the behavior of a cortical area is the result. Minicolumns are composed of 80–120 neurons, glia, and an extracellular matrix. Cortical synaptic coupling—excitatory and inhibitory [133, 134]—occurs in the context of local synaptic circuits [135] that include excitatory and inhibitory interneurons, as suggested in Figure 6. Excitatory pyramidal neurons comprise 70–80% of the neurons within a column, their apical dendrites aligned with the axes of the columns [136, 137]. There are a few primary inputs in the middle layer, IV [120], which may spread laterally [138] over a few minicolumns. Some IV inputs directly excite inhibitory neurons in V [139], others rising to II–III. The output from a minicolumn is typically one or a few layer V–VI pyramidal axons. Thus the entire minicolumn's input and output are carried on only a few axons, condensing the external manifestation of 80–120 neurons into a few axons expressing the behavior of the cortical minicolumn element. This behavior is part of a local function of a macrocolumn. A minicolumn is consistent with the "canonical cortical circuit" of Douglas et al. [140] and Shepherd [141, 142]. Harris and Shepherd [143] similarly propose a modular cortical circuit repeated across the cortex, with area-specific modifications.

3.3. Coherent Cortical Aperture

An orchestra is synchronized to a tempo or rhythm. Not all the instruments participate all the time, the particular subset of instruments playing changes throughout the piece. Not all instruments play the same notes; however, their activities are correlated, or coherent, creating harmonious music; they do not make disorganized noise. Technically one does not need synchrony to have coherence (as correlations); however, in a cortical area, synchrony is concomitant with coherence. As in an orchestra, coherence in a cortical aperture is built on an overarching synchrony. Not all elements in an aperture need to participate in the synchrony; thus an aperture that is coherent may instantiate information, our music, through participation of a subset of the elements, and the correlations of their activities. To understand coherence we need to understand how cortical synchrony occurs.

Introduction

A cortical aperture is more than a collection of independently expressing points; rather it is both a network and a continuum in which information can be instantiated through relationships among points. Activity in an aperture is composed of time-varying LFPs and spiking patterns. An aperture that has uniform activity over its area is limited in the information it may instantiate, lacking fine-grained modulation. In this document "synchrony" is gross, intended as "essentially isochronal synchrony" (zero lag synchrony) over an aperture. We shall see in Ch. 6 that synchrony is key for an aperture to participate in a multiaperture group forming the internal experience. Synchrony is an LFP synchronous <u>across</u> the area, whereas coherence implies some correlated pattern of activity <u>within</u> the aperture. In the coherent apertures model, synchrony is required for coherence. A *coherent aperture* will be grossly synchronous. Such a pattern may be changing in local temporal structure and element membership across the aperture. Coherence is expressed as an aperture-wide coherence map that can instantiate information [144], which is considered in Ch. 4.

Aperture Synchrony and Coherence

Functionally defined cortical areas have synchronized local field potentials (LFPs) across their areas when participating in the performance of a particular task, e.g., visual perception. The aperture field potential (AFP) is the mean LFP over an aperture. When the aperture is synchronized the LFP is the same as the AFP at all locations. What causes this synchronization? How does it relate to the performance of the task? Synchrony and coherence are related: in the coherent aperture model, coherence occurs in the context of synchrony. I include the coherence capabilities that support task performance in the synchronization models without describing the actual processes which are discussed in subsequent chapters. The expression "coherent apertures" denotes the ability to sustain coherence.

Within an aperture, the LFP can be synchronous over a range of 5–10 mm, extending coherence over the aperture, perhaps mediated by more central structures. Coherence describes the relationships among activities at different points [145], quantified by the cross-correlation of activity between two points separated in time and/or space squared. Together these define the *mutual coherence function*. The longest interval, e.g., distance, over which points are coherent is the *coherence path length* [146] (Figure 9). A coherent aperture has a coherence path length that is equal to the largest dimension of the aperture [147]. A synchronized subgroup of elements within an aperture that have small relative timing differences is coherent. The participants in the subgroup, and their relative timing (phases), may change while the aperture maintains its synchrony. A coherent aperture can dynamically instantiate information. As the coherence of synchronized elements occurs within a small AFP phase window, I shall consider coherence and the coherent aperture to include synchronization. Synchrony [131] does not necessarily mean oscillation [148], as temporal patterns may be complex. A coherent aperture's information potential is greater that the simple sum of the information at each of its points, as it includes the relationships among the points.

Evidence of Cortical Aperture Coherence

In 1979, James Blackburn, the prosecutor in the Jeffrey MacDonald murder trial in North Carolina, said that he did not have to prove that MacDonald was the kind of person who could have committed the murders if he could prove that he did commit the murders [149]: facts trump hypotheses. I don't need to show that an aperture could cohere if others have demonstrated that a cortical aperture (area) does cohere. Neurophysiological evidence illustrates coherence (and gross synchrony) over cortical areas. Models attempt to explain its significance. Frequently, measures of coherence and synchrony are conflated. Through recording within the cortex itself, Parameshwaran et al. [150], Thiagarajan et al. [151], Petermann et al. [152], Maier et al. [132], and Gray and McCormick [153] collectively illustrated the coherence of neural activity across areas in the cortex [4] related to the local field potential (LFP). Developmentally, cortical networks have a proclivity to synchronize [154]. Neurons spike under conditions of probability more than of certainty [155], although Tiesinga et al. [156] consider precise spike timing to be significant, an apparent difference discussed below in the context of local inhibitory circuits. Cortex is frequently poised to synchronize [157, 158, 159], potentially initiated by noise [160], being reflected in neuronal avalanches of activity, hence coherent activity [152]. Synchronization may occur rapidly. Nobili [161] found that "firing synchronization over extended brain areas often appears to be established in about 1 ms, which is a small fraction of any EEG frequency component period." Coherent input to sensory cortices can induce coherent activity [162], presumably in cortex that is near cohering or synchronization [163] (cohesive). Temporally structured sensory input to the cortex [164] comes almost exclusively through the thalamus, olfaction being the exception. Coherent inputs into apertures may arise from other apertures.

3.4. Aperture Coherence

Introduction

Why is aperture coherence significant? In Ch. 4 I describe how aspects of information may be instantiated in apertures. If synchronization is to be maintained under conditions of changing content, rapid communication across the aperture is important; thus synchronization should occur rapidly, incorporating coherence as it does so. In this chapter I address how such coherence occurs within the context of synchrony. The cortex contains a vast number of components with complex changing relationships that operate in the context of noise. I suggest that an attempt to provide a model that specifies how any specific neuron will respond to any arbitrary input is unlikely to be fruitful. It is my intent to elucidate some properties of cortical areas and to provide models of how these arise. An aperture model has particular relevance for information instantiated in distributed phase structures. The introduction of phase concepts is particularly useful. A set of synchronized elements may operate as a unit [165, 166], synthesizing a single aperture, such as with a multi-dish radio telescope. The many aspects of apertures contribute to a framework for the integration of multiple interacting cortical areas to perform early-stage processing of sensory input, generating the internal experience.

An aperture may progress from a disordered to a synchronized state. A cortical aperture is both an electrical continuum and a network of locally interconnected active elements. I propose three complementary models of aperture synchronization:

1. network synchronization
2. AFP pumping
3. laminar cohering.

All three models support a transition to a synchronized aperture with a long coherence path length, incorporating the LFP. 1) I expand the Kuramoto network model of synchronization to include multiple coupling mechanisms among elements with differing time delays. This expanded model also includes synchronization of a subset of a network. 2) In my pumped AFP model, the aperture-wide synchronous LFP is a common node among the elements (see Supplemental Materials). 3) In the laminar cohering model, I consider the layered structure of the cortex as a substrate for coherence, progressing through synchronization in the supragranular layers (I–III), a coherence propensity state, to a cohered aperture incorporating the infragranular layers (V–VI). The synchronization models are not mutually exclusive.

One can speculate on a progression from a disordered aperture to a coherent aperture (Figure 7) over several paths according to the state of the aperture and the strength of the input(s):

i) Noise brings a disordered aperture to synchronization through increased activation of pyramidals and synaptic activity, potentially leading to coherence in response to a coherent input;

ii) A moderately strong coherent input both brings an aperture to synchronization and then to coherence; or

iii) Strong, cohered inputs drive aperture-wide synchronization and coherence at the same time.

The relationship among complementary processes of synchronization and coherence in the three models is illustrated in Figure 8. An aperture may also become coherent directly when driven by sufficiently strong inputs.

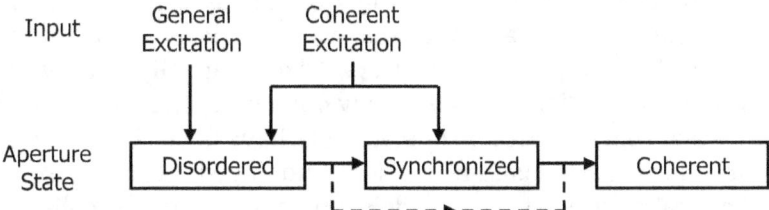

Figure 7. Aperture state progression. If the coherent input is strong enough, the intermediate synchronized (coherent propensity) state is bypassed.

Synchrony is bistable: it is or it isn't. There is no intermediate state. During synchronization the LFP is essentially synchronous across the electrical continuum with some dominant frequency spectra. In considering cortical activity, we must consider two general ranges of frequencies: those inherent in the stimulus, and those in the aperture. Additionally, we must consider how the higher synchronizing frequency appears, in phase, almost simultaneously at separated cortical points that have been stimulated appropriately. A coherent aperture will synchronize in the gamma band (30–80Hz), frequencies typically higher than those encountered directly from a primary visual stimulus; thus synchronization is the result of processes largely within the CNS.

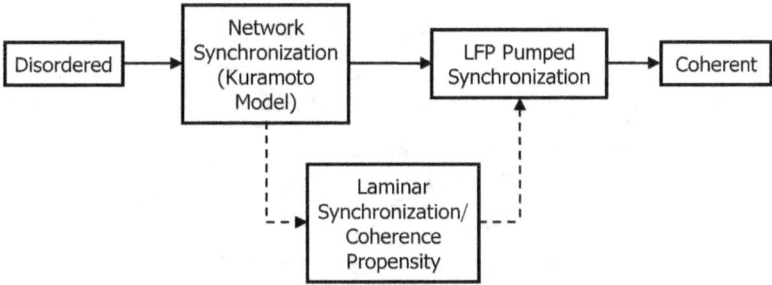

Figure 8. Progression from a disordered to a coherent aperture state through complementary synchronization models with a laminar model as an adjunct.

An aperture is a planar network with complementary direct and diffuse couplings that can create synchrony over its extent with a virtually infinite coherence path length despite finite coupling delays, through some combination of reciprocal coupling mechanisms modeled in i) network dynamics and ii) the pumped LFP. These bring a subpopulation of the minicolumnar elements' spiking activity into near synchrony, with a synchronous LFP across the cortical aperture, an (AFP). This remarkable phenomenon is the result of the phase sensitivity of neurons to the LFP, and of a moderate level of local connections among elements. The iii) laminar coherence propensity establishes a state of readiness-to-synchronize in the I–III layers with temporal gating windows that establish lower frequency profiles in the V–VI pyramidals.

Network Synchronization Model

The cortical network

The transient rhythmic (synchronous) activity over cortical areas makes it appealing to apply network models that sometimes express area-wide (domain) synchrony within various frequency bands [156, 167]. Network models have a strong presence in cortical modeling. In the broadest sense, artificial neural networks (ANNs) [33, 34, 44, 46, 47, 49, 58] are partially or fully scale-free; any node may connect to any other node [17]. They may or may not include synchrony. ANN models have not been successful in characterizing cortical properties, but have been useful as computational tools.

In graph theory, networks consist of vertices and edges. Vertices are also referred to as elements or nodes. Cortical neural network models typically employ the columnar structure of the cortex to provide the elements [120] in a flat lattice (Figure 9), with lateral coupling as the edges. The generation of such columns may itself represent self-organization in the cortex [55, 56]. Here we run into the macro- mini-column problem, the distinction between them often being unclear in the literature. Are minicolumns truly separable elements from a network standpoint? Their independence on a small-world scale [168] is not complete. They exist within the context of the macrocolumn, which could also be considered an element. However, I shall consider the mini-column as an element because it has an identifiable pyramidal output (at least one) from layer V or VI. As discussed below, the macrocolumn may develop its own synchrony; hence the independence of these minicolumnar elements collapses under macrocolumnar synchronization. The same can be said of the entire aperture in an ensemble.

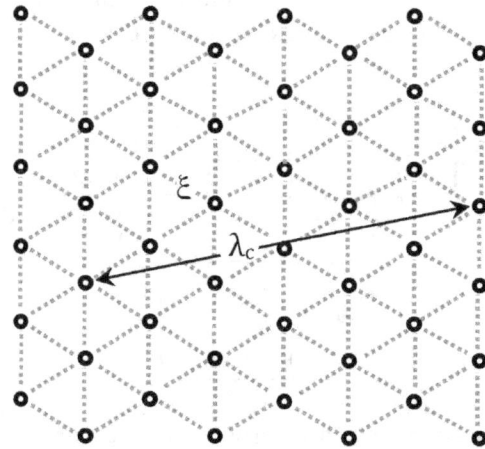

Figure 9. A regular triaxial flat lattice of locally coupled (ξ) nodes with a coherence path length λ_c.

The edges of the cortical network are composed of various coupling mechanisms; some coupling is point-to-point, e.g., synapses and gap junctions. Other coupling is more diffuse, such as the LFP and the extracellular space. These do not have a direct correspondence with edges but are important, as will be seen in the pumped LFP model below. For a uniform non-scale-free array, the number of nodes that each node is coupled to is the *degree* of the network. In most cortical models the nodes are connected to nearest neighbors, a low degree of connections. Cortical models are architecturally constrained [169], typically using Ising-like models [158]—flat lattices with direct coupling—that have neurons or (mini)columns as oscillator nodes that provide the outputs of the networks [57].

Synchronized networks

The aperture synchronization network model presented here is based on the model of Kuramoto and subsequent others. A synchronized network can characterize some of the behavior of an aperture. Abrams et al. [170] have provided a useful guide to synchronizing networks in their introduction to a focus issue in *Chaos*. A central driving source may not be needed to achieve synchrony [171] although synchronization may reflect an external influence [172]. Low degree networks with weak coupling can synchronize [48, 17, 173] with all elements participating, a model potentially applicable to cortical models. Kuramoto [11, 174] and colleagues [55], and others who subsequently built on his work [159, 175, 176, 177, 178], provide self-synchronizing neuronal network models.

Kuramoto's general model of a scale-free (any-to-any) network of oscillator nodes achieves synchrony quickly with relatively weak coupling among nodes. Kuramoto models [18] (KMs) self-synchronize through reciprocal coupling of nodes providing phase modulation of similar oscillators, supporting coherence across a network with little distance delays [160, 179, 180]. In 1975 Kuramoto [11] proposed large networks of coupled oscillator nodes as models for cortices that achieve synchronization when a threshold level of coupling exists, i.e., a *critical point*. A population of nodes, with some specific degree of coupling and a specific coupling strength (ξ, Figure 9), will have a critical point, analogous to the Curie critical temperature in ferromagnetic materials [181], at which the nodes' oscillators become synchronized with a coherence path length (λ_c, again, a term from radiant energy systems) that approaches infinity (Figure 9), spanning the network no matter what the size. The critical point reflects a transition between disorder and synchronization. Network synchronization is bimodal—it occurs or it does not occur. It snaps from one mode to another. There is no stable intermediate state. It is a robust phenomenon. Synchronization may occur with pulse [159, 174, 182, 183], inhibitory [183], and delayed [184, 185] couplings. These coupling characteristics are obviously significant for cortical lattices or networks.

The KM cortical model is a 2-D flat lattice model (Figure 9), such a (Ising-like) lattice having been reduced to an analytic description by Onsager [186] in 1944. As initially demonstrated by Kuramoto and colleagues [11, 187, 188, 189], a low-degree array, or lattice, of coupled non-linear oscillators may self-entrain, or self-cohere [190], even with weak coupling, if the nodes have similar natural frequencies [64]. Oscillator models include pulses that have natural frequencies reflecting a normal repetition rate [191] or inactive interval [192] (refractory period). The nodes within neural arrays can be considered oscillators with stated or implied natural frequencies that are similar, but need not be identical [193]. Nodal sinusoidal activity is not required [187] to achieve synchronization, nor is actual oscillation [148]. Presuming that nodes spontaneously oscillate, inhibitory coupling is more stable than excitatory coupling, which could quickly run to saturation [192]. Thus local, weak, inhibitory coupling among randomly oscillating nodes with similar natural frequencies can result in network-wide synchrony. Galan [160, 169] also proposed coherent network models incorporating synchronization and phase locking of neural activity. By different methods, Faugeras et al. [50] came to a similar coupled network model. In short-range locally coupled systems, coherent phase locking without synchronization is the more common case [187], although nearly synchronous excitatory inputs with a positive non-linear summation (the norm in a neural context) argues for both phase locking and coincidence across a network. In this case the excitation or coupling of most of the network has been raised to near a critical point. Tkačik et al. [194] found that Ising models of networks of neurons (retinal) would become highly correlated as network size increased due to only first-order coupling effects.

It is difficult—if not impossible—to make a testable mathematical model of a cortical network that has the potential to cohere. Referring to the work of Kuramoto, Crawford, and others, Strogatz [195] discusses the great difficulty in proving stability in synchronized networks, even with uniform coupling and almost identical nodal oscillators with only slight variations in their natural frequencies. Numerical and statistical methods have been used to demonstrate the behavior of large networks composed of many nodes [*17, 195*]. Arenas et al. [*17*] and Strogatz [*195*] have reviewed synchronizing networks extensively, noting that a population of oscillators may split into two subpopulations: synchronized and disordered [*196, 197*]. Mirollo and Strogatz [*159*] used a continuum model to analyze the stability of steady (cohered) solutions. The nearest-neighbor coupling of an Ising array can be expanded to include some superadjacency, increasing the mathematical complexity.

Inhibitory KM phase modulation may be the mechanism that supports pyramidal synchrony [*160*], resulting in the LFP. This is consistent with the model of King et al. [*198*] that predicts that only a small number of inhibitory cells relative to excitatory cells is required for such a web (Figure 12). Yoshimua and Callaway [*199*] found that fine-scale inhibitory connections, including pulsatile coupling [*159, 174, 184*], can create near synchrony of nearby cells through cross connections even with coupling delays [*185*], as all pyramidal–chandelier circuits have about the same delay time. Lovett-Barron et al. [*200*] describe a similar CA1 hippocampal circuit of interneurons and pyramidal cells providing a synaptic substrate for tuning pyramidal cell output through interactions in the local inhibitory network. Interestingly, Couey et al. [*201*] demonstrate a similar web circuit in the hippocampus composed of stellate cells and fast-spiking inhibitory (GABAergic) neurons. The involvement of inhibitory neurons in synchronization would prevent runaway oscillations that would probably be engendered with solely excitatory coupling.

The coherent aperture network model

A coherent aperture may be formed as a lattice of coupled elements with both spatial and temporal correlations [*202, 203, 204, 205*], the correlations occurring in the context of network synchronization, which may be restricted to subpopulations [*195*]. For a flat cortex of locally coupled elements in a two-dimensional array (Figure 9), an Ising-like model of a flat triaxial array[8] of locally coupled oscillators is an appropriate initial model either by direct reference [*157, 158, 206, 194, 207, 208*] or inference [*187*], decreasing the number and span of couplings from a more general scale-free network. Synaptic coupling, pulsatile with delays, can drive synchronization but lacks speed and resolution of modulation, particularly important if information is instantiated in small local phase modulations, i.e., coherence. Weak electrical coupling (ephaptic) produces small rapid phase modulations through threshold shifts. As discussed in Supplemental Materials, different coupling strengths, polarities, and delays may be incorporated into a model. The underlying concepts of self-synchronization, correlations, and a critical point are still appropriate. An aperture can have a subpopulation of synchronized elements that have small phase differences. This is a coherent aperture.

Coupling

Coupling mechanisms and modulations both distinguish among and unify the three aperture synchronization models within the coherent aperture model. In addition to architectural features [*209*], rapid coupling forms the basis for synchronization as discussed above. For an array to synchronize there must be some activity coupling the elements, the communication reflecting the strength of the coupling. Calcium signaling, although significant [*210, 211, 212*], will not be considered here as a dynamic component of coupling, although it has short-term modulatory effects [*211, 213, 214*]. Coupling is important both in synchronization and in modulating the network to instantiate

---

[8] An Ising model is usually a flat two-dimensional square lattice. In the coherent aperture model, it is a flat triaxial array, a computationally more difficult structure as there are more neighbors that are not arranged along orthogonal axes. This is a hexagonal array.

information. The complex nature of coupling makes modeling network behavior problematic. Mathematical sketches of the concepts are suggested in Supplemental Materials.

The potential for synchrony controls an aperture's readiness to accept coherent inputs, its *coherence propensity*. An aperture must be near a critical point for a coherent input, e.g., a visual moving edge that coherently stimulates a subpopulation of primary visual cortical neurons, to cause the aperture to synchronize—and cohere. In the absence of active coupling, a network will not be able to cohere or synchronize. Although the transition to synchronization is sharp, the difference in coupling strength for the two states is small [17, 161, 215]. Synchrony may occur quickly, according to Nobili [161], with times approaching 1 ms. He implicates astrocytes, potentially through electrotonic coupling. Such coupling could also travel through the ECS, just as any responsive communication system constantly exchanges some form of activity among its nodes even when it is not exchanging information.

Coupling modes

The primary couplings among neurons are wiring transmission and volume transmission [216]. Coupling has three factors of interest here: type, span, and delay. Synaptic coupling can be relatively strong, as connections may be polysynaptic, but has delays of both synaptic action and of axonal propagation time. Excitatory and inhibitory synapses have been extensively researched [133, 134, 217, 218, 219, 220, 221, 222]. A neuron may provide and receive a thousand or more synapses. This raises caution with respect to circuit models. The elements are coupled with multiple processes over various distances, not necessarily symmetrical. Brunel [223] found bidirectionally coupled pairs of neurons have stronger synaptic coupling. Coggan et al. [224] found that neurotransmitters may act at a distance from the synaptic cleft, providing modulation. Relative to synaptic coupling, electrical (e.g., electrotonic, ephaptic) coupling is weak but fast. Cortical networks also include electrical gap junctions, a weak intermediate coupling system.

The most notable activity of neurons is electrical, related to the generation, propagation, and synaptic action of the action potential [71, 72, 73, 74, 75, 225]; thus the electrical interactions of neurons are significant [226, 227]. In 2009 McCraig et al. [228] remarked that "the potential impact of extracellular electricity in the brain is vast and remains underexplored." As described above, the electrical activity of the neuron produces, and is responsive to, the LFP through pumping (see below). The LFP pumping produces an electrical coupling. Electrically, cells in the cortex communicate directly through gap junctions and electrotonically through the surrounding medium of extracellular space (ECS), with its extracellular matrix (ECM) and cerebrospinal fluid (CSF), and glia. Glial cells, particularly astrocytes [229, 230], form both an electrical network through direct gap junctions with neurons, and gap junctions among astrocytes. Gap junctions provide electrical conduction ("electrotonic synapses" [231]) [232, 233] among cells. Gap junctions are direct connections between neurons, between glial cells, and between glial cells and neurons that provide the direct interchange of molecules and ions [210].

Astrocytes are well suited to implement and retain phase relationships among neurons, important in coherence and information instantiation retention. The gap junctions among astrocytes, and astrocytes and oligodendrocytes [234], extend the local network over larger distances [235, 236]. The astrocytes form a species-specific syncytium in which the neurons are embedded [211, 237, 238]. Each astrocyte has an exclusive domain, enfolding the soma of 4–8 neurons [239], therefore enhancing electrical connections, and ion and molecular distribution [210, 211, 240, 241], among these neurons with synaptic involvement through extensive branching. Astrocytes can change their configurations on long- and mid-time scales, changing the extracellular volume [242, 243] and ion redistribution speeds. At longer time scales they release, take up, and redistribute various molecules: at mid term, neurotransmitters and ions; at a short term, electrical potentials and current flows both within the field of the individual astrocyte and through gap junction connections among astrocytes. Thus glia differ from general electrical conduction through the extracellular space, having more specific electrical connections.

Electrical conduction through gap junctions has a resistive/capacitive exponential decay; the initial response to an input is essentially instantaneous [244]. Gap junctions among neurons are typically dendro-dendritic [235, 232, 245] although axo-axonal junctions are present [246], together serving to modulate relative spike timing, and as a result supporting network synchrony [175, 247, 248, 249] and local phase structures. Although some propose that gap junctions only participate in low-frequency synchronization [177], the gap junctions appear critical in controlling the relative timing of spike initiations in populations of neurons [250], a potential participation in aperture cohering. Oligodendrocytes (glia) provide the myelin coating (analogous to the insulation on a wire) over CNS axons in short segments, wrapping multiple layers around an axon, wrapping up to 50 segments among multiple axons [251]. Conceptually the oligodendrocytes could create a net neutral ion flow by bringing the traffic along the axons into balance through conduction modulation (phase locking [252]) with subsequent myelin thickness modulation.

The ECS that surrounds the neurons and glia comprises 20% of the cortical plate volume [253]. Electrical potential gradients and currents in the ECS can have a significant impact on the action of neurons singly and in concert [254], although knowledge in this area is far from complete. The ECS is composed of an extracellular matrix (ECM) [255, 256] and CSF [228], in part locally modified by the glia [257, 258]. The ECM and CSF are sometimes conflated in studies; it is difficult to separate the electronic contributions of each. Ephaptic electrical coupling of neurons through the ECS can affect spike thresholds and timing [252, 259, 260]. The ECM is not homogeneous, having differing characteristics associated with different parts of the neuron, most particularly at the axon initial segment (AIS) of the pyramidal cell [261, 262] where action potentials initiate [263, 264]. The ECM modulates synaptic plasticity [265, 266, 267]. The ECS is not constant in tortuosity or volume [253] as the astrocytes dynamically change shape [239, 242, 268, 269, 270], hence the electrical characteristics of the ECS. The ECS is more than passively resistive, its impedance also reflecting the capacitance effects of the cell membranes [271].

Noise

Noise increases coupling. For example, noise causes adjacent retinal ganglion cells (specialized cortex) to have correlated spiking [272], a coupling. Counterintuitively, noise alone can cause a network to synchronize [160, 188]. Noise may effectively increase the coupling strength and subsequently <u>lower</u> the threshold to synchronization [160] up to some level that overwhelms synchrony. In the absence of cohering inputs, noise modulates the network's proximity to a critical point [157, 158]. Diffuse noise inputs may arise from subcortical structures. Noise injected into a cortical area can increase the natural frequencies of pyramidals (f_n), increasing the pumping of the LFP (see below). See Supplemental Materials for a discussion of the relationships among network behavior, noise, and the LFP. The sources of noise are also involved in the integration of individual cortical apertures into dynamic assembled intercortical networks (Ch. 6).

The coherent macrocolumn

Introduction

Macrocolumns are significant in the realization of response functions within an aperture. They also contribute to intra- and inter-aperture synchronization. Macrocolumns are structures intermediate between the minicolumn and the aperture. The emergence of the cohering macrocolumn as a small world of minicolumns was unexpected during this model development. It is a natural outgrowth of the exploration of the pyramidal-chandelier circuit and LFP coupling. The coherent macrocolumn may be considered analogous to a subaperture in an array aperture (e.g., dishes in a multi-element radio telescope). Each cortical macrocolumn is composed of 40-80 minicolumns (Figure 5) that respond to similar features. Bressloff [56] and Chawanya et al. [55] describe the minicolumns within the macrocolumn as collections of neurons that may self-organize through mutual synchronization, although the reports mention phase differences; so in the context of this paper, "coherence through mutual synchronization" may be a more apt expression. The minicolumns within a macrocolumn are synaptically cross-connected; thus the macrocolumn comprises a small-world network as described by Watts and Strogatz [168], summarized by Arenas et al. [17]. Newman and Watts [273] discuss the relationship between the small world correlation length, infinite correlation length, and critical point with the conclusion that a small world may achieve its local critical point for its correlation (coherence) path length, which can then percolate into large world coherence as synchronization. Differentiation among emergence, percolation, and cascading [274] is beyond the scope of this document.

Pyramidal-chandelier circuit

The pyramidal-chandelier circuit provides important behaviors in the macrocolumn. Within the context of a larger circuit (Figure 6), the excitatory pyramidal and inhibitory chandelier neurons form circuits that perform two interacting operations: phase quantization of minicolumns, and local phase modulation among minicolumns. These processes support coherence within a macrocolumn and may subsequently percolate coherence across the aperture as gross synchronization. Gray and Singer [275] found that minicolumns that fired in response to a stimulus were tightly correlated with the LFP. Is there an underlying circuit that creates the macrocolumnar coherence, or is it an interaction with the LFP? Both probably apply. A local cortical circuit may contribute to macrocolumnar coherence. I shall describe an aspect of an archetypal macrocolumn, *phase quantization*, which also encompasses the small local phase differences in coherence, the result being a window of synchronicity.[9] A window of synchronicity is a period during which elements essentially spike together, excluding elements outside of this window; thus the phases of spikes are quantized as synchronous or disordered. Although modulation results in timing changes over a phase continuum, phase quantization reduces activity to essentially two states: in- and out-of-phase. In-phase activity is coherent within a narrow window; out-of-phase is disordered. Following a model of a pyramidal-chandelier circuit (PCC), I shall address phase quantization, as it clearly illustrates the relationships among pyramidal and chandelier cells. Phase modulation will be described relative to these relationships.

Inhibitory interneurons play important roles within and among macrocolumns [276]. Chandelier cells (Figure 10) are a particularly important kind of inhibitory interneuron. Each chandelier is local in its macrocolumn [277], synapsing onto 200–300 pyramidal axon initial segments (AISs) [278] with cartridges of multiple boutons over a 200–400 μm field [279], which is significantly larger than a minicolumn's 30–60 μm diameter [120], but within a macrocolumn, which has a diameter of 300–600 μm [110, 280]. Thus a chandelier spans a macrocolumn, but is essentially local to it. Pyramidal cells (Figure 11) are sparsely interconnected; chandeliers form a dense network receiving and making many synapses with the pyramidals. Each pyramidal receives inputs from about four chandeliers (3–5); thus there is a ratio of about 75 pyramidals to one chandelier cell, forming a local web [199, 281, 282], sketched in Figure 12. Marin-Padilla [283] reports 60–80 pyramidal contacts per chandelier.

In the supragranular layer [284], the pyramidal cells and the chandelier neurons form a gating circuit through cross- and self-inhibition[142, 285, 199] (Figure 13, neurons P_1, P_2, C). Given the high ratio of pyramidals to chandeliers, there is a high probability that a chandelier has reciprocal synapses with any given pyramidal within a macrocolumn, therefore is self-inhibiting. A pyramidal,

---

[9] A two-category quantization would be dichotomization. In the phase quantization here, intermediate phases can exist in one phase.

P_1, may be cross-inhibited with another pyramidal, P_2, through intermediating chandeliers, C. P_1 synaptically excites C, which in turn inhibits P_2. Conversely, P_2 inhibits P_1 through an inter-mediating chandelier. These may be the same or different chandeliers. If the excitatory spikes input to P_1 and P_2 are nearly synchronous, both will produce spikes in their distal AISs.

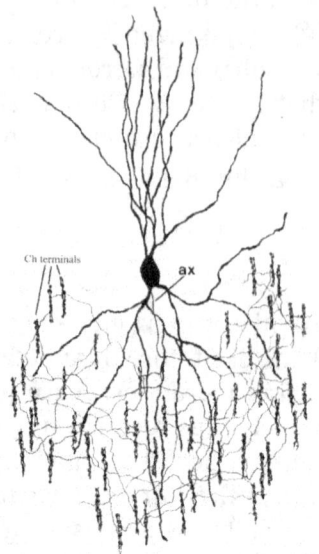

Figure 10. Chandelier cell, an inhibitory neuron; from DeFelipe (1999) [*281*].

Reproduced by permission from Macmillan Publishers Ltd: *Brain Research 122*(10), 1807-1822, copyright 1999.

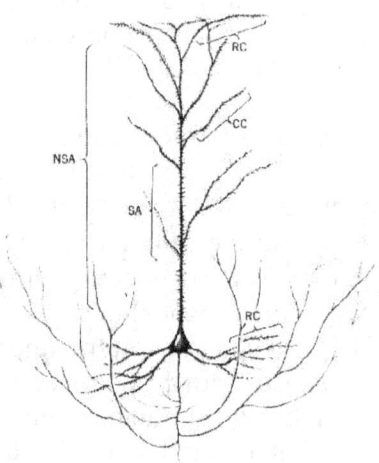

Figure 11. Pyramidal cell, an excitatory neuron. The fine process at the bottom is the axon with its recurrent collaterals ascending. The apical dendrite, NSA, rises from the pyramidally shaped soma; the basal dendrites spread from the soma. The section of the axon between the soma and the first collateral is the axon initial segment (AIS).

Original source undetermined.

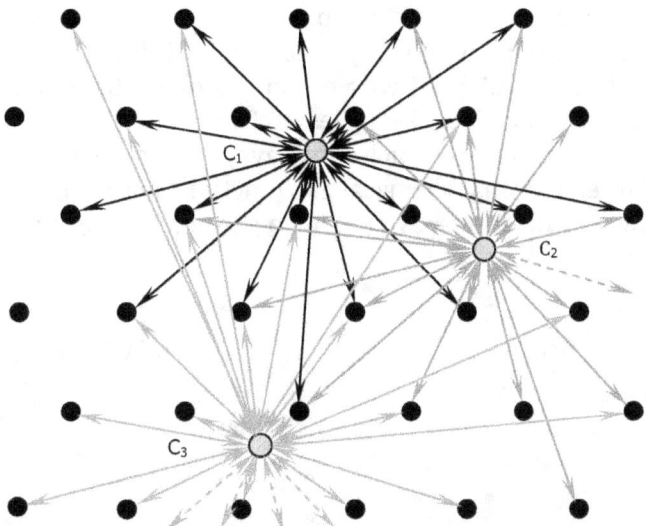

Figure 12. Pyramidal–chandelier web of a pyramidal array (black) within a macrocolumn reciprocally coupled with a small number of chandelier cells, C1, C2, & C3 (gray).

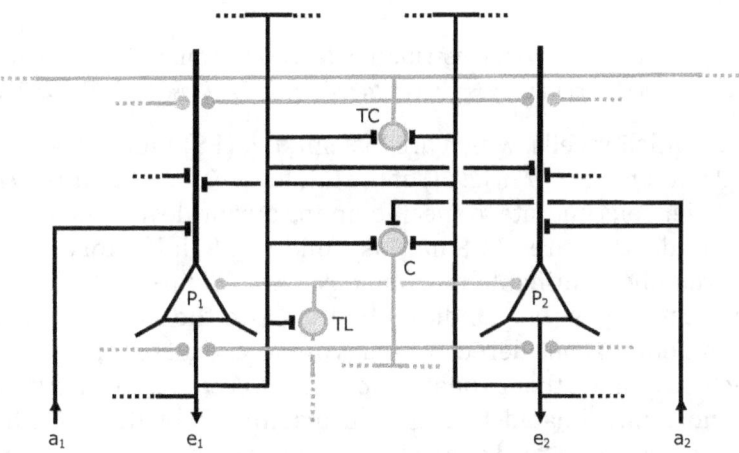

Figure 13. Pyramidal–chandelier circuit. Black = excitatory, grey = inhibitory, P_1, P_2 = pyramidals, C = chandelier, TC = transcolumnar inhibitory neuron, TL = translaminar inhibitory neuron.

Although there are many types of inhibitory cells defined, they have been reduced to a smaller, morphologically described set. A classification scheme for inhibitory interneurons was developed by DeFelipe et al. [286] based on axonal field features (Figure 14). Axon fields were defined as having inter- vs. intracolumnar and inter- vs. intralaminar spans. Local cells have low intralaminar and low intracolumnar spans. There is a generally inverse relationship between columnar and laminar axonal spans. Inhibitory cells, and their axonal fields, may be laterally displaced. Almost all inhibitory neurons except the chandeliers (local), Martinotti (translaminar, intracolumnar), horse-tail, and neurogliaform, are generally classified as basket cells, comprising 50% of the cells, spanning multiple columns [287] with mainly perisomatic targets. According to Stepanyants et al. [288], each minicolumn receives 92% of its excitatory synapses from sources external to the minicolumn, and 76% for a macrocolumn. Approximately 80% of synapses are inhibitory. This is roughly consistent with the report by Buxhoeveden and Casanova [*120*] that 95% of the synapses from outside a minicolumn are inhibitory. This is also consistent with the modeling by Hansel et al. [*192*] and King et al. [*198*] who found it difficult to stabilize a network with only excitatory coupling. In their modeling, Wang et al. [*289*] found synchronization is favored by neurons' inhibition when there are conduction delays (e.g., the pyramidal-chandelier circuits). Traub et al. [*249*] found interneuronal electrical gap junctions among interneurons, with evidence that this fast coupling with a threshold can enhance gamma synchrony through an interneuron network.

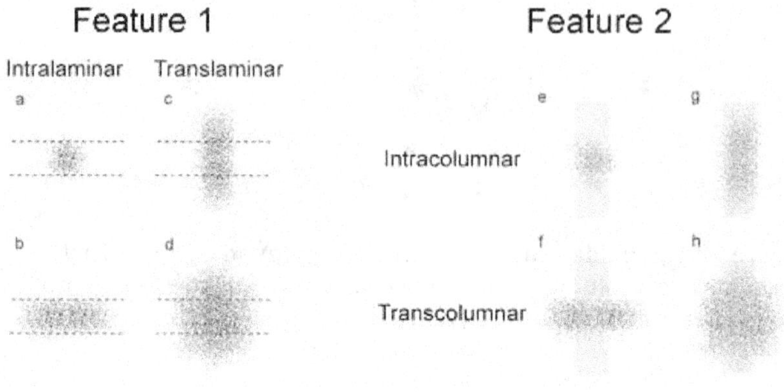

Figure 14. Inhibitory neuron axonal distribution features, from DeFelipe et al. (2013)[*286*].
Reproduced by permission from Macmillan Publishers Ltd: *Nature Reviews Neuroscience 14*(3), 202-216, copyright 2013.

Although the chandelier cells, which are fast spiking (FS) inhibitory (GABA) interneurons, make up a very small percentage (<7% [*290*]) of the inhibitory population, they have an inordinately large effect. Chandelier cells migrate to specific lamina during development [*285*], integrating into circuits with the pyramidals (Figure 11). Small fast spiking (FS) inhibitory (GABAergic) chandelier cells have a unique, readily identifiable morphology [*286*] (Figure 10). The terminal boutons of chandelier cells form cartridges (or "candles") that enclose the AIS of the pyramidal cell in the pineau (Figure 15), sometimes considered examples of axo-axonal synapses. The AIS of a pyramidal is a specialized structure unlike other axonal regions [*291*], creating a remarkable spike generation process. It is thin and unmyelinated [*262, 292*], projecting from the axon hillock (AH) on the pyramidal soma, with an asymmetrical distribution of voltage-activated Na+ channel types (Na1.6, Na1.2) [*293, 294, 295, 296, 297, 298, 299*], the distal Na1.6 having about a 7mv lower threshold than the proximal Na1.2 [*300*]. This geometry and channel distribution cause the action potential [*225*] of the pyramidal cell to be generated in the distal portion of the AIS [*263, 301*] (as opposed to a non-specific Na+ channel density model of AIS spike generation [*302*]). The distally initiated spike travels antidromically back up the AIS to the AH. The approximately exponential horn shape [*303*] of the AH and its high concentration of Na+1.2 channels [*304*] subsequently couple the spike into the soma, causing the cell body to undergo a spike [*264*], propagating into the dendrites, which can affect the efficacy of synaptic inputs. The spike also propagates normally down the axon [*291*].

A collateral of the pyramidal axon (typically at the first node of Ranvier) makes excitatory synapses onto nearby chandelier cells [*199*] (Figure 13). These synapse back onto the pyramidal's AIS [*305*] (and others) in a negative feedback loop, inhibiting subsequent spikes at the distal end of the AIS. A single chandelier input could significantly inhibit antidromic propagation. Dugladze et al. [*306*] found that the distal AIS could produce action potentials at a high frequency, while inhibition in the pineau due to chandelier cell action could prevent backpropagation of spikes into the soma.[10] Thus a pyramidal may produce spikes that are synchronous with the local network activity, producing one or more spikes correlated with the LFP. Because chandelier cells synapse on the axon rather than the soma or dendrite, it is reasonable to expect the chandelier to modulate the probability of <u>when</u> a neuron will spike, but to have little effect on the probability of <u>if</u> it will spike. It modulates phase not activity.

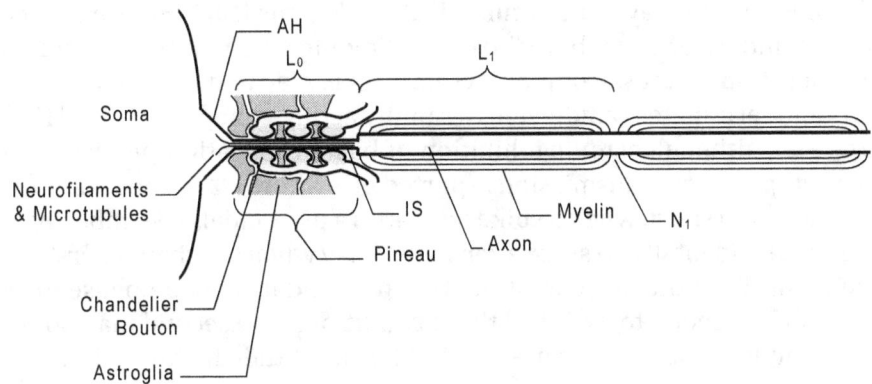

Figure 15. The axon initial segment: AIS. Simplified diagram of the pyramidal neuron soma, axon hillock (AH), initial segment (IS), axon proper, 1st myelinated segment (L_1), 1st node of Ranvier (N_1), neurofilaments and microtubules, and chandelier cell boutons.

The interactions of the pyramidals and chandeliers can produce gating windows during which spike initiation is favored versus periods outside this window when spike initiation is inhibited or disordered, effectively *quantizing* the phase responses of pyramidals within a macrocolumn into in-phase and out-of-phase components. In reality, the out-of-phase activity is simply disordered, as the in-phase gating window associated with the LFP is relatively narrow (Figure 17), while the out-of-phase activity is broadly distributed. This phase quantization will prove significant in my model, as it supports separation of a particular information instantiation from among other instantiations in the same aperture.

Normal neural action potential refractory periods limit the minimum period between spikes of a neuron, a de facto gating function that may range from 1 to 10 ms depending on the cell type [153, 307]. The burst spiking chattering pyramidals [153] in supragranular layers have short refractory periods (~1 ms), while the large pyramidals in infragranular layers V–VI, which provide the outputs from the minicolumns, have longer refractory times (~10 ms). Given the difference in layer LFP frequencies, it appears that phase quantization occurs in the supragranularly moderated gamma band while the underlying coherent input is reflected in the lower frequencies (theta, alpha), presumably of thalamocortical origin. A coherent aperture has natural frequencies in the beta (12–30 Hz) and gamma (30–80 Hz) bands, a superposition potentially supported by the laminar coherence model, below.

The pyramidal-chandelier circuits provide a gating window for the relative spiking activity among nearby pyramidals through cross-inhibition (Figure 13) that extends down into layers V–VI [139]. The infragranular layers project spikes that may profile lower frequencies (theta, beta), gated

---

[10] The relationships among the AH, AIS, and chandeliers needs further exploration with respect to this finding of continued distal spiking without somatic spiking. The AH might retain a potential sufficient for additional distal spiking but low enough not to backpropagate spikes into the soma, which may have a low degree of ionic motility, hence have been "discharged" to a low effective potential by the first spike.

at the higher SG gamma frequencies. The delay from the spike of a pyramidal (P_1) to the initiation of a chandelier postsynaptic inhibitory neurotransmitter (GABA) release at the AIS is 1–1½ ms, a delay uniform in the cortex. The gating period (Figure 16) is the interval between the spike occurring on one pyramidal, e.g., P_1, due to an excitatory synaptic input, e.g., a_1 (Figure 13), and the period during which a chandelier, C, cross-connected pyramidal, e.g., P_2, can produce a spike and the converse: a spike on P_2 will have the same delayed effect on P_1. Thus an inverse of a phase relationship produces the inverse gating result. Consequently the gated period of receptivity during which spikes from P_1 and P_2 may both occur is I, about 2-3 ms [307] (Figure 16), consistent with Swadlow's ±1ms "window of excitability" [308]. Following this period the threshold for soma-sourced spike generation by cross-coupled pyramidals is elevated considerably. The chandelier inhibition occurs and terminates rapidly due to rapidly acting voltage-activated K^+ channels in the AIS [309, 310, 311, 312], although ectopic inhibition of backpropagation persists [306], there being evidence of such ectopic neurotransmission reported elsewhere [224]. The gating window does not have sharp boundaries. In short, when a spike occurs on a pyramidal, a neighboring pyramidal may i) spike essentially coincidentally, ii) spike after some delay when the chandeliers' inhibition ceases, or iii) not spike at all. Thus the outputs of the two pyramidals may be phase locked if both are active. Research will be needed to verify all three conditions; the second may not be a reliable, or meaningful, behavior, or it may contribute to stimulus amplitude instantiation.

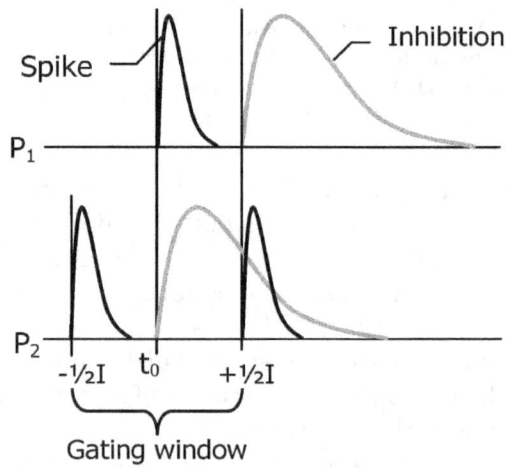

Figure 16. Cross inhibition of pyramidals. Spike for P_2 can occur over the span -½I to +½I relative to a spike in P_1.

Other systems such as feedforward inhibition [308, 313] may produce windows of excitability. The concentration of chandeliers is greater in layers II–III than V–VI (Figure 6), although this varies significantly with cortical area [290, 314]. The greatest concentration is in layer IV, the putative input layer. Coherent thalamocortical activity may serve to gate the spikes of multiple pyramidals, as some thalamocortical inputs synapse directly or indirectly onto the chandeliers [139], with particularly strong innervation of inhibitory neurons in layer IV [315, 316]. Presumably multiple coherent projecting inputs will synapse on multiple chandeliers. As it requires multiple chandelier AIS bouton activations to completely inhibit a pyramidal, coherent thalamocortical—or other—inputs will create nearly simultaneous chandelier activations, gating the outputs into discrete windows [317], quantizing phase responses [318], resulting in a coherent output. Coherent inputs will raise the natural frequencies of multiple pyramidals coherently, hence causing them to synchronize (i.e., the Kuramoto model [11, 17]) over some period, reflected in the LFP. The result will be a low frequency envelope of high-frequency nearly synchronously gated (coherent) spikes [*319*]. Further study is needed.

Phase quantization is particularly significant for phase-instantiated information. Small phase modulations in the pyramidal–chandelier system may be reduced to dichotomous phase states through phase quantization, moving the spike of a particular pyramidal within, into, or out of the gating window. Hofer et al. (*220*) have shown that the activity of inhibitory neurons (e.g., chandeliers) is less correlated than that of pyramidals, which are correlated with the input stimulus. As might be expected for local phase modulation, only subpopulations of the chandeliers are locally creating gating, and those participating more in the aperture synchrony than the input coherence. This illustrates that the aperture synchrony is separate from the stimulus input coherence. This interpretation needs further study. The AIS pineau (Figure 15) is a complex dynamic system able to affect spike initiation probability and timing [*320*]. The chandelier boutons and AIS have high densities of fast-acting voltage-activated K$^+$ channels resulting in brief action potentials [*310, 311*] well suited to fine-grained phase modulation, well within the gating window, potentially supporting population-referenced local phase structuring of action potentials within a population of neurons. Given the small differential spiking threshold at the distal end of the AIS, minor changes in the external potential or inhibition of the AIS can have significant effects on the spike timing [*264, 311*] and thus on its occurrence. See Debanne et al. [*291*] for an in-depth discussion of this topic. The chandelier's GABA may, under some circumstances, actually cause reverse polarization, having an excitatory effect [*321*]. The ECS, with its ECM [*261, 262*], provides malleable transient electrical coupling within and among minicolumns. Pyramidal spike initiation timing can be modulated by changes in the threshold due to ECS levels of neurotransmitters [*320*] and ions (K$^+$, Na$^+$, Ca^{++}, Cl$^-$), and by the LFP [*254, 322*], that are reflected in the state of the distal portion of the AIS. These changes reflect recent activity [*291*], glial (astrocytic [*237, 257*]) actions, astrocytic electrotonic and gap junction coupling [*230, 323*], and neural-neural gap junctions [*231*]. In addition to the cross-inhibition, synaptic input coincidence and spike backpropagation timing are known to affect both current and future neural responses.

Macrocolumnar cohering

Chandeliers have been shown to provide inhibitory coupling supporting synchronization, particularly through phase modulation resulting in phase quantization. As chandeliers lie within a macrocolumn and form a dense network within the macrocolumn, macrocolumns will cohere in response to inputs specific to any subgroup of minicolumns. Macrocolumnar small worlds differ from the hubs described by Bullmore and Sporns [*324*], as they are not connected in a hierarchy. The macrocolumnar small world coheres through both the LFP and local network dynamics, the coherence of macrocolumns percolating [*273*] to synchronize the aperture. The role of electrical gap junctions with astrocytes is unknown, although the exclusive domains of astrocytes suggest involvement. A macrocolumn performs a function through the behaviors of its minicolumns; thus their coherence is significant. Evidence of the relationship among synchronization, coherence, and functions is expanded in Ch. 5. In short, care must be used in interpreting experimental evidence, but there appears to be adequate evidence for the consideration of minicolumns as elements in coherent populations comprising macrocolumns, and macrocolumns as coherent structures relative to the aperture-wide synchrony. A coherent macrocolumn increases LFP pumping (see below).

Pumped AFP Model

Introduction

Synchronization onset and changes can occur over the aperture more rapidly than synaptic coupling and spike propagation speeds would suggest [*161*], suggesting an electrical component, the

LFP. The AFP, as a field, has virtually instantaneous communication, electrically influencing the thresholds for spiking, hence the statistics of element activity across the aperture. As described in Supplemental Materials, as an aperture coheres via the LFP, the aperture itself may be considered a node that is coupled to all elements as part of a process during which an increase in element activity can lead to synchronization. In addition to general (noise) input, reciprocal coherent interactions among apertures will raise the activity levels. Via the pumped LFP model, a subpopulation may be noncontiguous, coherent with fine scale phase differences within the subset. The pumped LFP model I have developed is essentially a field model. The LFP occurs in a network context. The LFP is the voltage in the extracellular space in the cortex, typically measured with an electrode using a 300 Hz low pass filter: the low pass filter frequency has repercussions on the conclusions reached about the LFP and its source. There is a relationship between the LFP and aperture synchrony, although hypotheses about this relationship differ. The EEG is a gross measure of an underlying AFP, often reflecting synchronous (hence coherent, the coherence window being small) activity in a large number of neurons. This activity has been considered to be the result of neural avalanches [325] of interacting neurons creating correlated activity over distance [326].

A cortical aperture may be considered as composed of subapertures of macrocolumns or elements of minicolumns [116, 327] or as a continuous medium as a limiting case of a very large network or array. This relationship between large discrete networks and a continuum is useful in modeling cortical activity, as it provides a link between the LFP and element activity. What descriptive model can explain the synchronizing behavior of cortex? Hoppensteadt and Izhikevich [328] consider cortical columns (presumably minicolumns) to be weakly coupled "autonomous oscillators." Given the extremely large number of cortical elements and their complex multifaceted couplings, the behavior of large networks is applicable, although from the standpoint of the LFP a cortical area approaches a continuous aperture, which can be modeled as a field. The cortical aperture, as an array or network of active elements, has the potential to develop spatially synchronous or coherent activity, thus creating a spatially synchronous LFP. Astrocytes, forming essentially an electrical syncytium through gap junctions, contribute to the synchronization of the LFP, a hypothesis supported by recent findings of Lee et al. [237]. Berens et al. [329], beim Graben and Rodrigues [330], Parameshwaran et al. [150], Peterman et al. [152], and Thiagarajan et al. [151] suggest that the LFP itself may carry or reflect information, perhaps interacting with spiking probabilities. Caution must be exercised in interpreting the LFP as containing information: the amount of information across an area is most likely too complex to be represented in a single waveform of such low bandwidth.

Source of the LFP

The mechanism(s) for generating the time-varying electrical field external to the cells is not well understood. The LFP has been widely investigated experimentally and theoretically. It must arise from the electrical activity of the cells in the cortex, generally considered to be the neurons. According to Berens et al. [331], the LFP is indicative of underlying synaptic, dendritic and (by extension) spike processes, carrying no inherent information, but reflecting its presence. Three models of the sources of the LFP have been considered: current dipoles, spike monopoles, and spike trains. These are not mutually exclusive [332]. The LFP has frequency components of less than 140 Hz, far less than the Fourier components of an action potential, although the low-pass filter normally used for such recordings (300 Hz) could remove higher frequency components. The amplitude of the LFP is maximally 500 μv peak-to-trough [118]. It is small but reliably present. There is a large number of synapses, not all of them leading to immediate spiking, so synaptic currents provide an attractive model. Waldert et al. [333] presume that spikes can be disconnected from LFPs for analytical purposes, but they do not resolve possible action potential contributions to the LFP.

Destexhe and Bedard [*334*] have reviewed models of sources of the LFP, raising the question that the "standard" model of spikes as dipole sources may not be sufficient, the data being consistent with monopole models as proposed by Riera et al. [*335*], with appropriate laboratory results. A monopole model of LFP generation is particularly attractive for the field model I am proposing. As the potential decays more slowly with distance for a monopole (1/r) than for a dipole (1/r²), the difference is significant in terms of the electrical coupling for synchrony. The 1/r decay is also consistent with a planar structure as opposed to volume conductance (1/r²), significant in the pumped AFP model (below). A monopole model was supported by experimental evidence despite an apparent violation of Kirchhoff's law of closed electrical circuits, the differences attributable to the differences in ion and electron mobilities. A neuron accumulates an electrochemical potential over a time span considerably longer than the ~0.5 ms duration of an action potential. During the initial phase of a spike Na$^+$ channels open, allowing a rapid influx of Na$^+$ ions with a subsequent rapid decrease in the neighboring ECS potential [*336*]. The discharge of the neuron's potential in single or bursts of action potentials can constitute an effective monopole, "injecting" potential (a decrease in voltage) into the LFP as a monopole. Bursts of monopole action potentials from multiple chattering cells [*153*] in layers II–III may drive the LFP even though the 400–800 Hz frequencies are well above the upper components of the LFP (see Supplemental Materials). Their receptive and distribution fields are large.

Spike timing and the LFP

There is a relationship between spike production, timing, and the LFP [*254, 337*]. The amplitude of the LFP affects the spike production. The phase of the LFP affects the spike timing. The timing of the spike production affects the LFP. Agarwal et al. [*338*] provide experimental evidence supporting their idea that the pattern of the LFP in the hippocampus can be as good an indicator of place encoding as spiking patterns. Mendoza-Halliday et al. [*339*] found a relationship between the middle temporal lobe (MTL) LFP and spike timing in the dorsal visual pathway, suggesting that local field potentials encode some spatial information while also modulating spike timing. beim Graben and Rodrigues [*330*] propose that the LFP forms a continuous electrical field in the cortex to which neurons may become coupled, thereby contributing to the LFP. Such a field would be the AFP. They used a dipole model of current generation in pyramidal neurons. Tiesinga et al. [*156*] report large-scale coherence of the firing of cortical neurons associated with the LFP. Activity of the elements (i.e., neurons) in a cortical aperture is correlated with the LFP [*156, 331*], specifically the rising phase of the negative component of the LFP (nLFP+) [*150, 336, 340*] (Figure 17). Tiesinga et al. found that maximal spiking occurs during the maximum rate of internal potential decline in the pyramidal. This is equivalent to the maximum rate of increase of the external potential, the nLFP+.

The specific potential of each neuron will partially determine its phasic relationship to the LFP. The LFP in the ECS changes the field gradient across the neural membrane and changes active membrane currents [*341*]. Reato et al. [*322*] demonstrated in vitro a relationship between electrical fields analogous to an LFP and spiking behavior. Anastassiou and colleagues [*254, 259, 341*] showed a direct relationship between an extracellular electrical field and entrainment of neuronal spikes. Anastassiou et al. [*259*] found that small electric fields across the neural membrane cause neurons that are near threshold potential to spike. "Despite their small size, these fields could strongly entrain action potentials, particularly for slow (<8 Hz) fluctuations of the extracellular field." Neuron activity could be entrained by potential changes as small as 0.5mv. This ephaptic coupling is independent of synapses, i.e., not due to molecular transmission at gap junctions. This is consistent with the model by Holt and Koch [*227*] of local electrical interactions of neurons. Direct electrical coupling is probably less important, as only a small fraction (~0.1%) of gap junctions is electrically conductive [*232, 233*]. Although network synchronization can occur with weak coupling, it is problematic that gap junction coupling alone would be adequate to establish synchrony. Presumably network synaptic activity produces a near-synchronous state of the aperture.

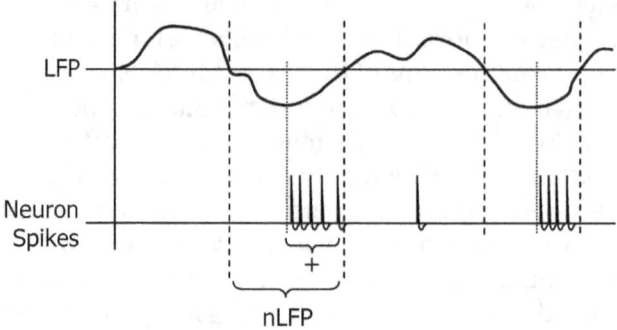

Figure 17. Sketch of LFP with spikes occuring mainly on the rising (+) phase of the nLFP.

There are two classes of spiking behavior: single and train [*336*]. As described above, a coherent aperture will have an LFP that is nearly uniform over its area. Single spikes will occur more synchronously, favoring the more rapidly rising LFP waveform, hence the higher frequency component. We can consider single strongly coherent spikes to relate to structure in time and space [*342*] (shape) in which synchronicity is particularly important [*343*], although Masquelier et al. [*344*] propose spike timing relative to the background LFP oscillation as providing "phase of firing coding." Individual spikes and spike train envelopes of subpopulations of neurons receiving coherent inputs will tend to cohere, as their f_ns (and electrical potentials) will increase together, usually to the point of synchrony reflected in their relationship with the LFP [*76, 156*]. The interactions of both the single spike and spike trains with the LFP will in turn shape the LFP. This behavior is consistent with the pumped LFP aperture model.

Spike trains have variable interspike intervals [*345, 346*]. A train results from a more strongly depolarized neuron. "Strongly" may relate to the size of the neuron and/or the depth of depolarization. A train may be considered a pulsatile discharge of the cumulative depolarization, perhaps linked to the LFP [*347*] and the moderating ion redistribution through the intracellular matrix. The spike interval is constrained by the gating window, the natural frequency of the neuron (f_n), and the amount of remaining charge. Initially the spike intervals will be uniform due to gating and natural frequency limitations, and will increase in time due to the decreasing proximity to threshold achieved during the refractory period. Such a model is outside the scope of this paper. It is consistent with the experimental results and analysis of spike trains by Nawrot [*346*]. The coherence of the aperture will cause multiple trains to tend to be synchronous, but the neuron dynamics may override this tendency. Remarkably, Gray and McCormick [*153*] report a class of "chattering" pyramidals in the infragranular layers that produce bursts (trains) of high frequency (up to 800 Hz) spikes, the bursts correlated with LFP depolarizations.

The pumped aperture field potential (AFP)

The state of the LFP reflects the probabilities of pyramidal spiking in space and time. Is the LFP a cause or effect of synchronous activity? It is both a cause and a result of coupling that provides the potential for a synchronous area, an endogenous electric field feedback proposed by Fröhlich and McCormick [*348*]. A pyramidal neuron may be considered an LFP point reinforcer or local "pump," ultimately contributing to an AFP coupling field. Neural spikes decrease electrical potential of the LFP [*332, 336*]. A decreased LFP increases the probability of a spike occurring in a neuron near threshold [*322*]. A pyramidal spikes according to its internal potential relative to its threshold, relative to the LFP. If a pyramidal is close to its threshold when the LFP phase is negative, it has an increased probability of spiking. If it spikes, such a pyramidal will reinforce the LFP, as its spiking produces a sharp influx of Na^+ ions, further decreasing the external potential, followed by a slower K+ repolarization with a lower potential shift: an effective brief monopole. Thus if a pyramidal is ready to spike, it reinforces the LFP by spiking, almost in phase, at its location: it "pumps" the LFP. Tight coherence of minicolumns within a macrocolumn (see below) could further increase a monopole effect.

The point-reinforced LFP interacts with other pyramidal points of activity. The activity of a pyramidal is reflected in its natural frequency, f_n, hence in the frequency with which it reinforces the AFP. A pyramidal at any given location may not spike every time a lower phase AFP occurs, but it may do so erratically, coincident with the low-potential phase of the LFP, according to its level of activation. Noise arising from the brainstem (e.g., locus coeruleus [349]) and other inputs to the aperture can increase the activation of pyramidals, hence their f_ns. An optimally activated pyramidal will spike at every LFP low potential. Over the aperture, increased activity will appear as noise—stochastic neural spikes—if uncorrelated. Local pyramidal activity may also increase activation of nearby pyramidals. Such reciprocal interaction could easily run away into uncontrolled oscillations over the entire aperture—and beyond. Inhibition is required to maintain stability [184, 350]. This is probably provided by the large basket cells that span macrocolumns [287, 288].

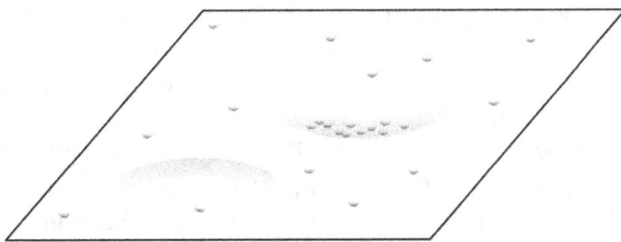

Figure 18. LFP "sheet" with pyramidally generated "dimples" concentrated in a voltage depression.

The LFP is distributed through an electrically conducting medium of ECS, gap junctions, and astrocytes. As discussed above, the probability of a pyramidal cell producing a spike is a function of the level of activation of the pyramidal and the phase of the LFP. The LFP may be envisioned as an electrical potential sheet stretched over the aperture. The potential is reflected in the local vertical displacement of the sheet. When the LFP is low, there is an increased probability that a pyramidal will spike. A spiking pyramidal causes a transient local "dimple" in the sheet as it reinforces the negative LFP, with no synaptic delay, decaying as $1/r$ [335]. The dimples are isolated events, not interacting unless proximal. In the aperture's disordered state, there are changing potential bumps and depressions in addition to overall changes in the LFP. With up to seven million minicolumns in an aperture, the complexity of the LFP sheet can be high. When the overall activation in the aperture is low, not every pyramidal will spike; consequently the location and occurrence of dimples will be random, although preferentially occurring in the local low potential depressions (Figure 18). As the level of activation increases, dimples will occur more frequently, having a combined effect on the local potential, depressing it further. This local effect is essentially a local coupling via the LFP. The LFP, being locally supported across an active cortical area, will increase the coherence path length at the LFP natural frequency (the aperture frequency, ω_a) and reciprocally in the unit activities. As the level of activity rises, perhaps due only to undifferentiated noise from brainstem structures of weakly cohered inputs, the natural frequencies of the pyramidals and their concurrent multiple couplings rise. The aperture concomitantly increases its nearness to synchronization. As the aperture grossly synchronizes, the LFP smoothes, the local bumps and depressions lessen. This is consistent with the results of Thiagarajan et al. [151] who found that separated neurons may be synchronous with the LFP even though intervening neurons were not. The LFP, when synchronous, creates a global (aperture) network, discussed further in Supplemental Materials.

The pumped LFP model is not a strict coupling model that would easily fit a Kuramoto network model [17] as the couplings cannot be described on a neuron-to-neuron basis, whether synaptic or electrotonic. beim Graben and Rodrigues [330] describe a model of coupling of a neural network to an LFP, but presume all elements in the network become coupled to the field. Not every pyramidal need be spiking to generate or maintain a synchronous LFP. A sufficient number of pyramidals, randomly spiking but coincident with the LFP negative phase (Figure 17), can collectively support a synchronous LFP over an aperture, reminiscent of the volley discharges in the inner ear [351] in which the individual neurons can only respond up to ~1 KHz, but the population will have coherent frequency-following up to ~5k Hz as the neurons randomly volley. As their numbers increase, the *scintillating* dimples spike in phase with the LFP like fireflies over a field on a summer evening (159). Such scintillating elements, not constituents of spike trains, will be considered disorded, although there may be a continuum from synchronous to fully disordered.

A hypothetical two-dimensional (flat) laser is roughly analogous to an LFP pumped aperture. In a laser, an energy source pumps elements to higher energy states such that they will randomly release photons (analogous to electric potential dimples) that are coherent with passing photons (i.e., an AFP phase). When the energy input is adequate, the system will become coherent over its length or area. Consistent with the LFP model, it is the activity into a pyramidal that influences its natural frequency (f_n) or, more appropriately, its probability of spiking at any given time. It appears that an LFP synchronized over an aperture creates an aperture-wide global network of weakly coupled elements (see Supplemental Materials). This model is also consistent with the results of Engel et al. [352] who found that each instance of aperture synchrony may have a different natural frequency, despite seemingly identical stimuli, as the history of the aperture will affect its response characteristics. A synchronized subpopulation, presumably instantiating information, will produce spike trains of varying length, that are in the gamma band (30–80 Hz). The randomly spiking pyramidals distributed throughout the aperture, particularly if near the critical point, will instantly respond to the LFP depressions, propagating the LFP. Macrocolumns are synchronization foci, increasing the LFP pumping. Levels of activity translate into natural frequencies for the neural (pyramidal) oscillators. These freely oscillating elements synchronize as they reciprocally interact with the LFP, pumping it in the process. An input may cause some minicolumns to respond with increased activity, increasing their natural frequencies. Responding minicolumns will result in coherent macrocolumns, creating a larger LFP dimple with a greater lateral spread. The reality of the LFP as a global (i.e., aperture-wide) synchronizer of those elements near spiking is unproven, although Thiagarajan et al.'s [151] finding of spatially separated elements phase locked to the LFP is consistent with this model; it conforms to Ockham's razor, the law of parsimony. This model differs from those in which some intervening pyramidals must be synchronous (i.e., networks). It is discussed further in Ch. 4 as comprising the coherence map

Aperture-scale dynamics

Nunez [81] developed a framework describing large-scale cortical dynamics and the EEG as hierarchical levels of spatial granularity, with more general principles applying as the scale increases, with no central synchronizing mechanism. Synchronization within and among apertures changes according to the state of wakefulness [353, 354] and awareness [355]. Sleep states can affect both thalamo-cortical and intracortical synchronization of the LFP [356, 357]. Synchronized activity may be increased by cortico-thalamic interactions during the normal sleep state [358, 359], being reduced during slow wave sleep [317]. As described above, synchronization may be generated by noise in the absence of stimuli [360]. Li et al. [361] find that the activity of a single cortical neuron may be sufficient to move LFP UP/DOWN oscillation of sleep into a persistent UP state of wakefulness. One must bear in mind that such a conclusion is based on sampling a very small number of neurons out of a larger population that could have distributed coherently active neurons.

The synchronous LFP is more than "local," as it appears to be synchronous, without loss, over considerable distances under some conditions [118, 150, 151, 319, 358], although Katzner et al. [362] disagree with this conclusion, finding LFPs quite localized, originating within 250 μm of the recording electrode. Thiagarajan et al. [151] used unanesthetized animals whereas Katzner et al. used anesthetized animals, which may explain the difference, as the synchronization of an area may involve the active participation of the thalamus (see Ch. 5): some anesthesias essentially silence synchronization in the thalamus. Petermann et al. [152] and Thiagarajan et al. [151] demonstrated the relationship between the LFP and the spiking of neurons distributed over an area. The signals of both the LFP and the spikes were detected with electrode arrays in the motor cortex of two macaque monkeys. The spikes were detected from proximal neurons, typically one per electrode. Thiagarajan et al. propose that the LFP is a "coherence potential" distributed over a cortical area at least 8 mm wide (the extent of the array), consistent with the results of Kajikawa and Schroeder [118]. When synchronous, the LFP had the same complex waveform and amplitude across the entire sampled area. The LFP at separated sites typically had correlations with time differences among them of less than 10ms (95%), heavily weighted around zero. The duration of cross-correlation of the LFP would often extend to hundreds of milliseconds. This coherence was not a function of the distance between neurons or of an oscillation. In other words, when the area was synchronous, the LFP was synchronous, hence uniform in form, across the entire cortical area under study. Not all neurons' activities within the electrode array were correlated with the LFP. Spikes of subpopulations of neurons were found to be phasically correlated with the negative excursions of the LFP. Intervening neurons did not necessarily spike with a correlation with the LFP or other neurons. They were disordered. The correlated neurons behaved coherently (by definition). The robust distribution (span) of the coherence potential (the LFP) rose in a sigmoidal fashion, reaching a limit when 30–40% of the neurons were coherent. Further increases in the number of neurons recruited produced no further increase of the LFP. This is consistent with the findings of Tiesinga et al. [156] who report large-scale coherence of the firing of cortical neurons associated with the LFP, and research by Okun et al. [363] who demonstrated a subthreshold relationship between the cortical LFP and neuronal firing.

Gray [4] found subpopulations of neurons with non-overlapping fields but similar feature (function) properties, with synchrony over distances of 2–7 mm that he associated with perceptual grouping and binding. This is discussed further in Ch. 4 as *coherence maps*. Volgushev et al. [364] found correlations among membrane potentials of both nearby and widely separated (13 mm) neurons at low frequencies (<5 Hz) during periods of slow oscillations that can be presumed to reflect the LFP. Frien and Eckhorn [319] found functional coupling in V1 of both LFP and multiple unit activity (MUA) recordings reflected in spectral coherences under a visual coaxial drifting grating stimulus for both low frequencies (0–12 Hz) and high frequencies (35–50 Hz) with significant effects of similarity of orientation as the distance between sites increased to 6 mm. Low frequency coherence was associated with locations of simultaneous stimulus elements whereas high frequency coherence reflected similarity of orientation of stimuli. Spectral coherence may not have the same information as waveform coherences; the two cannot be equated. Similarly, Jia et al. [162] found spiking with a gamma rhythm synchronous across separated sites with stimuli differing by the spatial frequency, high-to-short, low-to-long distances. Varela et al. [365] reviewed phase synchronization and large-scale integration of the brain, reporting synchrony in multiple frequency bands in columns with non-overlapping receptive fields in V1 separated by 2-7mm. Neurons that share similar feature properties tend to synchronize with the formation of dynamic links. This is consistent with the results of Thiagarajan et al. [*151*] that not all neurons need to participate in order for an extended LFP to occur. Further, synchrony is not constant [*366*]. Its existence is dependent upon the animal's state and external experience. Thus the LFP, when synchronous over distance, may be considered an indicator of the coherence of the underlying neuronal activity. El Bousanti and Destexhe [*3*] propose, with evidence, that coherence and spatial scale covary, with high dimensionality in asynchronous irregular states reflected in higher dimensional chaos at larger scales, and presumably higher information evident in higher information entropies. Thus there appears to be a potential state of coherence over an extended area that is referenced to the LFP, with only small time differences across the cortical span. Laminar differences in LFP, which are significant, are discussed below in the laminar coherence model. Although the significance of the LFP and concomitant neural activity is important, the relationships of activities among neurons are also important, as the pyramidals' activity comprises the output of an aperture. Attractor models are logical extensions of potential LFP-network activity relationships [*54, 169, 201, 367, 368, 369, 370, 371, 372*] as they result from the distributed properties of a network that are expressed in both network spiking behavior and the attendant electrical fields.

Although ostensibly a field model, the pumped LFP model includes aspects of both network and field models, as the minicolumns (and their pyramidals) are members of a coupled array—a network. In the pumped LFP aperture model, network synchronization is driven by coupling and the level of activity that is reflected in the natural frequency of the main neural elements, the pyramidal cells. Pyramidal cells are the primary source of the LFP, locally "pumping" the LFP at the low potential phase according to the activation of the neurons. A pyramidally reinforced (pumped) LFP may support synchrony among separated macrocolumns, hence generating a synchronous aperture that supports coherence. The Kuramoto model and the development of a synchronous LFP are discussed further in Supplemental Materials. The pumped LFP aperture model I suggest describes a field synchronization mechanism, but not a field information instantiation.

Caveat

The pumped AFP model, although having significant support from the literature, is nevertheless speculative. The model provides a process that may contribute to the synchronization of the aperture. Such a synchronization is already demonstrated through Kuramoto models. In either case, a population, or subpopulation, of synchronously spiking elements will produce a synchronous LFP across the aperture, an AFP. Therefore, references to an AFP do not necessarily require LFP pumping; it is the AFP that is ultimately significant. The AFP provides a phase reference for element activity and participation in, or exclusion from, a coherence map (Ch. 4). The pumped AFP model emerged as a logical extension to the exploration of the source of the LFP.

Laminar Coherence Model

An aperture that is synchronized without the coherence patterns of local phase differences and element participation may be considered as having a coherence propensity. This is a transition from disorder to coherence, a limited form of synchronization that may occur in the supragranular layers, as described in the laminar coherence model. Coherence finally arises as a coherent response to coherent inputs in a synchronized aperture, typically reciprocal with other structures. This model does not have robust support; it emerged through study of the laminar frequency differences, providing an explanation for the function of those differences.

The LFP is not uniform through the cortical plate. With the laminar model of aperture coherence, I suggest a path to a state of readiness (coherence propensity) that is transitional to coherence, required for the communication of coherent information among apertures. Although such a model is not necessary, it proves useful by separating the transitions to coherence. Cohesiveness (coupled elements) can lead to synchronization with a coherence propensity, and subsequently transition from synchrony to coherence (coherence of a subset of elements). Only the latter is associated with information in its various active instantiations. This model differs from that of Grossberg and Pearson [373] who ascribe different memory functions (spatial and non-spatial working memory) to different cortical lamina of a common circuit according to cortical location. Gray and McCormick [153] discuss chattering pyramidal cells in the SG layers that could form a synchronous substrate isolated from the local functions. The laminar coherence model suggests that an aperture-wide coherence propensity arises largely from the synchronization of the SG layers while coherent information instantiations are associated with infragranular layers.

Nobili [161] proposes that a subpopulation of neurons with intrinsic subthreshold oscillations can have their phases rapidly reset to synchronize and desynchronize by electrical activity through the extraneuronal medium, held in a state of readiness by inhibitory coupling. Synchronization of the EEG can occur in about 1 ms across the aperture, a small fraction attributable to any EEG frequency component. The result is an EEG (AFP) that can synchronize neurons across a population, allowing them to operate in parallel. This is consistent with the laminar cohering model, and further strengthens the hypothesis that the LFP is more than an epiphenomenon of neural activity. Nobili's proposal is supported by the model of Rosenblum et al. [374] of the synchronization of phase coupled oscillators.

Generally, supragranular activity is associated with transient states, infragranular with information instantiations. The supragranular synchronous gamma-band frequency increases in power with both adaptation (Hansen and Dragoi [131]) and attention (Buffalo et al. [130]). Masamizu et al. [375] found two differing layers of activity in the motor cortex. They associated supragranular activity with the coordination of activity among areas, infragranular activity as reflecting well-learned activities. Thus one may posit that learning was correlated with an increase in the coherence of specific subsets (CMs, Ch. 4) of infragranular, but not supragranular, neurons. Mizuseki et al. [376] found place field (e.g., information) correspondences in infragranular pyramidals' activity in CA1 (hippocampus), with high frequency bursts usually coincident with low frequency theta oscillations. It is suggested that supragranular synchronization is reflected in intercortical synchronization while infragranular activity reflects intracortical coherence. The hippocampus is not representative of the rest of the cortex; therefore the laminar frequency profiles found by Mizuseki et al. may be considered an exception. The high frequencies in infragranular bursts may be explained as high frequency components of the monopole source model of the LFP, and hence are artifacts not representative of the underlying lower frequency behavior.

Cortical aperture coherence is reflected in the cortical intralaminar coherences, lamina being differentiated by frequency bands. Bollimunta et al. [377], Buffalo et al. [130], Hansen and Dragoi [131], and Maier et al. [132], found distinctly different laminar frequency bands. Buzsáki and Schomburg [1] have also pointed out laminar differences in frequency spectra and interlaminar coherence. Activity in the supragranular layers is primarily in the gamma band (LFP$_s$, 30–100 Hz) while the activity within the infragranular layers is primarily low frequency (LFP$_i$, 5–20 Hz alpha, theta) with little coherence between the supra- and infragranular layers [130, 131, 132, 319], although such lack of coherence may reflect the relatively small subset of active elements in infragranular responses. Maier et al. determined the lack of coherence between the LFP of the supra- and infragranular layers based on a spectral analysis of the LFP versus a correlation analysis, which might have revealed coherence among layers in the fine structure of the LFP. This appears to be so in their figures. A correlation analysis, analogous to that performed by Engel et al. [352] may have been more informative. It is clear, however, that supra- and infragranular layers behave differently, particularly when the animal is attending [130]. A higher frequency gamma band LFP of synchrony in the supragranular lamina may provide an aperture synchronization, potentially incorporating local LFP pumping, which more limited subpopulations in the infragranular lamina may synchronize with at lower frequencies (<20Hz) in forming coherent maps (CMs) of activity (Ch. 4). As Burns et al. [*378*] note, although a supragranular "clock" is a convenient concept, it is a great simplification. "Gate" is a more appropriate concept, defining sampling windows through a cohering substrate, not requiring the information-bearing activity to have the same frequency as a putative strobe. While a gating frequency model is too rigid, as LFPs are usually not regular oscillations, it provides a sense of the relationship between a broad aperture synchrony and a coherent subpopulation. As an interesting parallel to the two-level laminar coherence model, Behrman et al. [*379*] propose a cohering network of essentially molecular quantum dots that are coupled in a substrate lattice through the equivalence of optical phonons. I am not proposing a quantum neural network (QNN).

The supragranular layers may drive the LFP$_s$ when the aperture is synchronized, with the infragranular pyramidals responding to that synchronization in the establishment of coherent subsets of elements that result from function activity, as described by coherence maps (CMs, Ch. 4). In addition to aperture-wide synchronization, the supragranular cells may provide top-down excitation to the infragranular layers [*380*]. Alternatively, high frequency supragranular trains may become phase locked, as groups, to a lower frequency LFP. Local synaptic circuits can be described both laterally, to include the local PCC and other synaptic couplings, and columnarly (e.g., Jones [*127*], Wang et al. [*381*]). Lateral and vertical synaptic circuits are not separate. PCCs may have laminar organizations (e.g., as roughly indicated in Figure 6). Buxhoeveden and Casanova [*120*] describe a "temporal column" as having variable vertical integration within a complex macrocolumn. Figure 6 is a schematic sketch of cortical lamina. No functions are suggested. Neural activity is projected into the granular layer (IV) where it is then connected to the supra- (I–III) and infra- (V–VI) granular lamina with output projected from the infragranular lamina. The apical dendrites and axons of pyramidal cells (P$_s$ is a supragranular pyramidal, P$_i$ is infragranular), oriented normal to the surface, span various layers, with interneurons (e.g., C) forming connections constrained within the cortical plate. Laminar differentiation includes the developmental laminar distribution of inhibitory chandelier cells [*285*] with a resulting laminar PCC.

The supragranular LFP$_s$ synchronization may arise from the actions of "chattering" pyramidal cells [*153*] that produce high frequency (300–800Hz) bursts of spikes in phase with the low potential level of the LFP (e.g., in Figure 17, if spikes are for only one neuron). As pyramidals in the supragranular layers have longer AISs [*382*] they are more sensitive to the LFP$_s$, potentially contributing significantly to the synchronization of the LFP$_s$ [*138*] through pumping and to the phase-modulated synchronization of the Kuramoto models [*187*]. The other, larger pyramidals in the infragranular layers would also contribute to the LFP as monopoles. Thus, in concert with the laminar coherence propensity model I am suggesting, the supragranular layers may be significant sources of synchronization, the infragranular layers acting with this synchronizing potential. The vertical orientation of the pyramidals may increase interlaminar synchronization through the LFP pump. This is generally consistent with the model proposed by Buxhoeveden et al. [*120*] of a mini-column that has both vertical (normal to the cortical plate) and laminar aspects with dynamically activated laminar components of varying vertical extents.

Laminar frequency differences and coherence need further testing, as the analytical methods used (e.g., Welch's average with a Hamming window on normalized filtered signals) may have emphasized such differences, since spectral power densities were often used for calculation of coherence, potentially losing signal cross-correlation information. The state of the animal, anesthetized, awake or asleep, may similarly cause differences. I propose that the supragranular lamina synchronize rapidly over the aperture through LFP$_s$ phase modulation, principally of the chattering pyramidals. This synchrony may, but not must, underlie coherence of subpopulations in the infragranular lamina, as the cortical minicolumns (hence macrocolumns) provide response functions to coherent inputs with resulting coherent infragranular output projections. The infragranular LFP$_i$ is more limited and reflects the activity of the CM (Ch. 4). The infragranular lamina is further enervated to cohere though thalamocortical reciprocal connections [377, 383]. Thus intracortical and projected synchrony may differ. This cortical laminar coherence hypothesis needs additional exploration, as the temporal relationships of activity within and between the lamina are important.

3.5. Results

Principles and Corollaries

Principles:

- The cortex is composed of functionally distinct areas—apertures (parcellation).
- A cortical area may synchronize.

Corollaries:

- Not all elements need to synchronize for the cortical area, or aperture, to synchronize.
- Synchronized elements may have local phase differences, i.e., they cohere.
- Aperture coherence may occur in response to coherent inputs.
- Aperture coherence may emerge.
- Concepts from radiant energy systems may be applicable.

Summary

The cohering aperture is a structure from which an integrated model of CNS operation flows. A cohering aperture model draws upon concepts from networks, fields, and radiant energy to provide a unified description of the behavior of a functionally distinct cortical area. Broadly speaking, an aperture, including a coherent multi-element array, is an opening through which radiant energy flows or in which it is instantiated or transformed. In the coherent apertures model, coherence instantiates information in the context of synchrony. Coherence has two components: the pattern of elements which participate in the synchrony, and the phase relationships among those elements relative to the synchronous behavior. Synchrony is evidenced by a local field potential (LFP) that is synchronous across the aperture, reflecting an essentially infinite coherence path length, with time delays among events across the aperture being considerably less than the neural propagation delays. The presence of synchronous cortical areas is well documented. From both experimental evidence and network descriptions, a functionally distinct cortical area which is capable of synchronizing is modeled as an aperture composed of a network of minicolumnar and macrocolumnar elements with strong slow local synaptic and weak fast extended electrical couplings. The transition from a disordered to a synchronous state is abrupt, occurring at a critical point. Three synchronization models are presented: i) a phase-modulated network, ii) a pyramidally pumped local field potential (LFP), and iii) a laminar coherence.

    i) **Phase-modulated network mathematical models** of the synchronization of the cortical array of elements coupled by synaptic connections and local electrical potentials, derived from the original work of Kuramoto, have been developed by others; I have combined and expanded them. The properties of cortex provide rapid distributed electrical—or electrotonic—coupling. Although electrical coupling is weaker than synapses, it produces a high level of coherence with short time differences, the LFP being a major agent affecting spike timing, allowing rapid synchronization.

ii) The **pumped AFP model** describes a pyramidal energetic phasic reinforcement of the LFP over the cortical aperture. As the population of random LFP pumping "dimples" increases, in response to noise and structured inputs, the coupling within the aperture increases. Nearing synchrony the aperture itself behaving as a node coupled to all elements.

iii) **Laminar coherence** arises from a laminar structure superimposed on the columnar elements, with coupled behavior between the major lamina groups. This gives rise to a model of orderly synchronization and cohering through interlaminar interactions.

Information can be distributed over the extent of an aperture, bound together coherently, including phase relationships. An aperture's ability to maintain coherence, as meaningful temporal (phase) relationships across its area, is reflected in the information-bearing capabilities of its projections to other apertures. Synaptic circuits of excitatory pyramidal and inhibitory chandelier cells produce narrow local gating windows that both quantize the cortical elements into synchronized and disordered populations, and support meaningful local phase differences. Not all elements within a cortical aperture necessarily participate in its coherence; considerable spaces among elements can occur.

As described in Supplemental Materials, within the limited model presented, a synchronous LFP will occur rapidly at some particular undefined frequency. This is not actually the case, as the CNS operates in multiple frequency bands, there being a dominant one under specific conditions. The lack of a frequency defining mechanism brings into question the validity or completeness of the frequency-dependent pumped LFP model to the extent that a specific frequency cannot be predicted. Buzsáki and Draguhn [384] have discussed cortical oscillations at length. The logarithmic progression of the bands may point to some underlying processes, or be the result of the manner in which these bands have been defined. Although nearby frequency bands tend to compete, the LFP can include multiple frequency components. Different frequency bands may be associated with different activated subnetworks of apertures and their tasks. The degree of central involvement may differ. The limitations of a single-frequency model notwithstanding, the role of the coherent aperture appears valid for instantiating and communicating information that is processed in CNS operations.

Conclusion

A functionally distinct cortical area, an aperture, may synchronize through the activity of a subpopulation of elements. Aperture synchronization is necessary for the coherence of a subpopulation of elements with relative phase differences, giving rise to the coherent aperture. It is not clear which of the three mechanisms is responsible for aperture synchrony and subsequent coherence: network, pumped LFP, or laminar synchronization. All may be, in a progression. The net result is a synchrony that may be modulated in element membership and timing across the aperture; hence the aperture is coherent, i.e., able to instantiate information. The network model has the strongest support, the laminar synchronization model the weakest. These models are not exclusive of each other, nor are all necessarily correct or required.

4. Information Instantiations in Apertures

4.1. Introduction

While computer models may be able to perform tasks such as image analysis, that does not mean the CNS operates on similar principles, particularly past the earliest stages. This chapter addresses the general instantiation of information in apertures. The synchronous cortical aperture forms a substrate for information instantiation. Information is an abstraction. Its instantiation is a physical realization of the abstraction that may have an undecipherable form, yet can be physically manipulated irrespective of its putative information. Not necessarily mutually exclusive expressions of information have been proposed. Encoding and subsequent underlying canonical circuits have not been found [385]. The rough canonical circuit in Figure 6 (Ch. 3) suggests a general form of cortical circuits not specified for any particular function, yet illustrative of a complexity that may support instantiations. I am not proposing a canonical circuit, but rather an underlying framework of some CNS processes based on synchronizing mechanisms and local (aperture-specific) functionals that perform transforms in progressions, divergences, and convergences, with information subsequently instantiated within and across apertures. As discussed in Ch. 3, not all elements in an aperture need be synchronized in a synchronized aperture. Information is proposed to be instantiated in a coherence map, a phase field,[11] a dynamically formed division of coherent and disordered subpopulations of minicolumns and macrocolumns. Direct synaptic coupling among mini- and macrocolumns is not required for them to become synchronous; therefore there can be synchronous islands within a synchronous aperture.

4.2. Information Models

Introduction

A test of an encoding system is the ability to predict, a priori, the instantiations of the information before a new unique stimulus is presented; otherwise it is a post hoc correlation. Yamins and DiCarlo [10] suggest such a priori predictions as tests of a model of CNS operation. I suspect such predictability not to be the case for all but the most rudimentary stimulus or behavior. Although instantiations are physical, I do not expect to deduce a "neural code" [103], if such a thing exists. I assume that information instantiations are not consistent across individual persons. Within an individual, active and stored instantiations must be closely related in the CNS, as one transforms into the other. With or without a code, information instantiations may be latent or active (e.g., within the cortical plate), and transient or persistent. A latent instantiation is an inactive form that may be transformed into action, projecting from one area to another over axons. I propose a model of the form of information instantiations and its entropy, based on coherence, that enables operations in the CNS.

Considerable work has been done by many attempting to deal with the issue of information in the CNS. Proposed methods of representing active information in the CNS range from the behaviors of individual neurons (and sub-neuronal structures) to the LFP. I will summarize some

---

[11] A field with elements in one of two or more states, e.g., an ocean with ice and water.

spike encoding and field models, as they illustrate issues. Frequently encoding and computation appear together in cortical models [*144*].

Spike Encoding Models

Neuronal coding

The current consensus is that neural spikes carry information in the CNS, in a manner as yet undeciphered. A neural code is elusive due to the very large number of neurons in the cortex, its history, noise, and, until recently, adequate simultaneous sample sizes. If such a code were to exist, deciphering it would be difficult. Some encoding concepts are fine-grained refinements of Hull's [*386, 387*] stimulus/response models of behavior, in which neurons inherently respond to specific inputs [*164*] with corresponding oscillatory outputs [*164, 193*], subject to modulation [*191, 388, 389*]. The outputs from retinal ganglion cells (RGCs) exhibit some of this S/R relationship, responding with spike trains projected to the lateral geniculate nucleus (LGN) that reflect the dynamics of retinal stimuli [*156, 390, 391*], activity incorporated in the patterns of activity from the LGN to the visual cortex [*392, 393, 394, 395, 396, 397*]. As retinal illumination will stimulate more than one photo-receptor, RGC outputs will inherently have some correlation [*398*]. Chichilnisky [*399*] developed a method to determine single RGC responses, using randomly changing light elements, that appears to mask the local coupling. The responses do not necessarily reflect reality, particularly the effects of coherent stimuli, as there is cross-coupling within the retina [*244, 250, 272, 400, 401, 402*]. Local correlations reduce the independence of neurons' activities and subsequently reduce the maximum potential information entropy of the population. Padmanabhan and Urban [*403*] find that intrinsic neural differences, acting as noise, result in temporal irregularities with a concomitant increase in neuronal independence, and a subsequent potential increase in information quantity and, one presumes, a corresponding increase in information entropy. As will be discussed below, the relationships among neurons present other opportunities for information representation in the absence of isosynchronous behavior.

Spike trains and ISI

Potential encoding includes the structure of inter-spike intervals (ISI) and spike numbers in spike trains, particularly from primary sensory systems, most notably vision. Spike timing neural behavior in response to stimuli is also seen elsewhere, e.g., the medial temporal lobe (MTL) in behaving monkeys [*342*], precise timing of pyramidal gating windows in the hippocampus [*313*], and phasic structure, also in the hippocampus [*404*]. In addition to visual pathways, entorhinal cortex grid (and associated hippocampal) cells also express some location specificity with temporally structured activity [*201, 367, 368, 405, 406, 407*]. Time-dependent sensory/response relationships appear supported by experimental evidence, but do not allow a priori prediction of a response to a novel stimulus outside the maintenance of coherent (typically synchronous) responses to temporal stimuli. There is no single apparent pattern of neural firing behavior, although dynamic responses to dynamic stimuli are common. Pyramidal cells, and their retinal equivalent retinal ganglion cells, produce spike trains in response to changes in stimuli. Kalluri et al. [*408*] describe multiple firing patterns in the vestibular nerve responding to constant depolarizing current steps, a condition that does not have a direct physiological correspondence; however, it is noteworthy that there are cell differences. Generally there appears be a decrease in an S/R relationship in higher levels of processing (abstraction).

Entropy in spike trains has been considered to reflect information [*409, 410*]. A spike train has three general features: onset, train, and correlations; such differentiation may prove useful in considerations of coherent apertures and networks. Nawrot [*346*] modeled a prototypical spike train as a rate-modulated spike sequence with a gamma process governing the probability of a spike, hence having an increasing inter-spike interval (ISI) after an initial low probability, with the time constant of decay characteristic to the specific neural type and history [*389*]. The ISIs will be influenced by external activity. There is no universal agreement that ISI timing provides an efficient temporal code relative to the information needs [*48, 411*]. Unitary event analysis [*345*] of the collective firing behavior of assemblies of neurons has revealed that there is significant correlated activity among such populations, indicating that the joint spiking behavior is information related. This is consistent with models of information distributed over populations of neurons. The importance of timing among ipsilateral cortical areas might be reflected in the maintenance of isochronal transit times in projections over different distances through an inverse speed-distance relationship [*412*], maintaining a consistent temporal structure. Hoppensteadt and Izhikevich [*48*] proposed frequency communication channels among cortical minicolumns with modulations of interspike intervals as phase encoding information. When an aperture is synchronous, separate frequency channels become problematic for information embodiments, although phase structures are of interest. The LFP of a synchronous cortical area is not composed of a large number of frequencies, limiting information channels, similar to the limitation in the LFP encoding proposed by Parameshwaran et al. [*150*]. Further, such separation of features into frequencies controlled by the thalamus implies some form of "tagging" of separable features within a cortical area.

Field Models

There are natural tensions among cortical neural field models, network array models, and biology. Ultimately biology (anatomy and physiology) rules. It is virtually impossible to have a meaningful understanding of the function of cortex based solely on biology; there is simply too much going on [*9*]. Array models and field models are attempts to reduce the complexity of the cortex to an understandable level. Field and array models each encapsulate different types of understanding, but are not incompatible. A neural field attempts to model the collective behavior of a population of elements, perhaps without the intermediary of a defined network [*70, 413, 414, 415, 416, 417, 418*], although some relationship between an array of elements is usually stated or implied [*54, 360*]. A cortical field can be characterized as a continuous function, mathematically manipulated, and subsequently sampled to model the behaviors of discrete elements in an array [*419, 420, 421*]. Active information may be expressed in the correlation of activity within an array, and of subpopulations of neurons within the array, a contiguous population response [*414, 422, 423*] implying a field, combining the abstractions of a coupled array and of a continuous field. Bressloff [*56, 63*] explored a neural field model with a homogeneous population of synaptically coupled neurons to study the developmental formation of functional units. Marder and Taylor's [*91*] comments on functional equivalence come to mind (homoplasy). Consistent with a field model, Yger [*424*] has proposed that topological consistency across a cortical patch reflects underlying distributed information more than specific neurons. The resistance of the CNS to noise, synaptic remodeling, and cell death can be modeled within the context of fields. While I may have may oversimplified the field concept, it provides a balance to neuron-specific models. Population models are conceptual bridges between field and network models. Graf et al. [*425*] found specific population responses to edge orientations by integrating over 1280 ms, obscuring any phasic characteristics. Their population response "decoder" performed better than one with more abstract parameters.

4.3. Information in Apertures: Dynamics

Introduction

Presumably information is instantiated within a cortical aperture. Here be dragons. Perhaps there are no reliable discernable neural codes at all but the most primitive sensory and motor levels.

Information in a network of neural elements (minicolumns), or an aperture, is not contained in an orderly pattern. The problem that appears insurmountable is the undecipherable complexity of the aperture's activity in all but the earliest levels of information flows, the pattern of activity becoming progressively less universal. Individuals will have unique functions and patterns in their cortices [426] as a result of genetic differences, random developmental processes, and differences in experiences: our uniqueness is expressed in our brains. We can, however, characterize information by its entropy.

Undecipherable Complexity

At increasingly higher cortical levels, information becomes increasingly undecipherable; it loses the characteristics of an encoding as information becomes progressively transformed through distribution among apertures with axonal divergence, receptive field convergence, and differing functions with their receptive fields (Ch. 5). I propose that *undecipherable complexity* describes the condition of an instantiation from which information cannot be extracted. Undecipherable complexity is a state that de Gennes [427] describes as a property of turbulent flows, e.g., fluid flows in porous media [428]. I use the term "undecipherable" since the information is not encoded in some fashion that would allow it to be deciphered. My handwriting might be indecipherable to most, but I can read (decipher) it. An undecipherable physical instantiation of information can be manipulated without regard to its content, altering the state of the aperture, as reflected in its entropy. In spite of undecipherable complexity, we must presume that for declarative entities there are invariants [429, 430, 431], a universal form of some set of features constant under certain transformations. An invariant is similar to a pattern or model. A musical pattern may be invariant under key changes. A square is a square irrespective of its color, edge, or size, all of which are transforms of a square. Faces and hands have universal structures. That is not to say that invariants in the CNS are pictures or templates. They are highly abstracted, may be distributed over multiple apertures, and may be undecipherably complex [430]. Wason [15] demonstrated that even orthonormal space is an invariant, the perception of Euclidean space occurring even when displays are affine transformations of orthonormal space–spatial structure scaled by its content. I demonstrated that an affine vector field is scaled by a separable scalar field; further, a dynamically sheared space would also be perceived as orthonormal. Cutting [432] had previously demonstrated that projecting images of moving objects on slanted screens (i.e., front row, aisle) destroyed neither the perception of rigidity nor orthonormal space as the images had undergone primarily affine transformations.

We can discuss functions within the CNS in terms of information instantiations, even if we cannot decipher them at higher levels. I address the roles of undecipherable complexity and information entropy in CNS integration in Ch. 6. Undecipherable complexity is a disquieting concept, but I have found no reasonable alternative.

Information Entropy

A malleable communication medium may incorporate information when the medium acquires a structure that incorporates a pattern, making the abstraction of information concrete, i.e., an instantiation. Information entropy can characterize information, but does not contain it. In spite of its complexity, an information instantiation may be described by its entropy: we might not know what it says, but we can have an idea of how much information there is. Shannon proposed [21, 22] information entropy as a statistical indicator of the potential for different patterns, i.e., the potential independence of the elements comprising a message: the greater the number and the greater their independence, the greater the Shannon information entropy, no matter what their configuration. According to Nemenman and Bialek [409], the relatively short duration (message) of spike trains makes estimation of entropy problematic, calling into question the use of Shannon's information entropy model for the behavior of individual neurons. Weaver [*433*] differentiates disorganized and organized complexity, the latter having a lower entropy for the same complexity, as the independence of the elements is reduced within a pattern. Organization can be considered *orderliness*. The Shannon and Weaver [*22*] entropy characterization of information can be applied in two (or more) dimensions. A Persian carpet has an orderliness above that of a randomly scattered deck of playing cards. In a model of a modest (120) network of neurons, Tkačik et al. [*434*] demonstrated that the entropy of patterns of activity in correlated neurons is related to information. Elements not participating in the pattern, if they can be excluded from it by a rule, may be excluded from the determination of the entropy.

Thiagarajan et al. [*151*] and Parameshwaran et al. [*150*] propose that information across a cortical area is encoded in the complexity of the LFP waveform distributed uniformly (thus synchronously) over an area. This has the advantage of a reduced sensitivity to noise and cell death. The reduction of the activities of a large population of neurons to a single, albeit complex, waveform represents a very large decrease in information-carrying capability, as activity at any point equates to activity at any other point. Given the low bandwidth and the uniformity of the waveform distribution over the area (hence a large number of neurons, or mini- or macrocolumns), the potential information density in an LFP is probably far too low to carry the proposed information quantity. Their demonstration of an extended "cohering potential" does provide compelling evidence for synchronous binding of information instantiated over an area, however.

Modulation

A function at a point in an aperture responds to a feature of the input, projecting to other apertures. If the system is to express information instantiations, those functions must be modifiable. New information must rest on modulation of features of the aperture. Such modulations will affect the functions within an aperture and memory. There are two general types of modulation: response strength and timing. Response strength is principally of synaptic origin, timing is of synaptic and non-neuronal influences, such as astrocytes, LFP, and ECS in the ECM. Modulation may be considered in three memory contexts: working, stabilized, and consolidated. Information in memory is complex: it needs to be considered in the operations of perception and access. Modulation is discussed further in Ch. 5 and Ch. 7.

4.4. Phase Structures

The coherent aperture model relates information to several dynamic phase relationships: global phase, aperture phase, microphase, phase differential, local phase, local phase vector, and group phase. A group of phasically related active elements or their collective outputs comprises a *phase structure*. When multiple apertures are individually coherent and form an essentially synchronous coherent network, the network activity has a *global phase*. (The term "synchronous" will mean "essentially synchronous" to include small local phase variations, unless specified otherwise.) The *aperture phase* of activity—the average phase usually evidenced in the LFP—is synchronous across a coherent aperture as the aperture frequency (ω_a). *Microphase* is the temporal phase relationship of an element's activity relative to the aperture phase. A phase differential is the phase (time) relationship between two elements that may or may not be contiguous. The *local phase* is the phase relationship relative to the average phase of nearby elements' activities. In this context it is a scalar. A local phase may have a direction of the maximum local phase difference, the vector sum of the local phase differentials (over an area, this is the gradient, a "slope map") [*435*]. In this context it is a vector referred to here as a *local phase vector*. "Local" does not have a discrete boundary in a quasi-continuous aperture. Cortical areas incorporate local phase vectors in their responses, for example, to edges and motion in V1. A series of spikes from an element has a *group phase* relative to the aperture phase. Although the phase reference of the series of spikes from an element is uncertain, the group phase may be considered as the phase of the leading spike relative to the aperture phase, with following spikes as microphase structures and/or indicative of amplitude, probably aperture-specific. Rather nicely, the aperture phase serves as a reference to align the microphases, with the result that microphases do not need to be accurately integrated over the aperture in order to establish a phase relationship (differential) among separated elements, areas, or subpopulations. This is evidenced in common fate (the perceptual grouping of separated elements moving in unison [13]) and visual perception of spatial structure though motion [15, 431]. Thus there is a relationship between the behaviors of neurons as elements and as participants in populations [436]. The power to instantiate the most information in the cortex is in how finely time (phase) can be reliably sliced; thus, the importance of the stability of the composition of the cerebrospinal fluid (CSF) and of the temperature of the brain, as evidenced by the pervasiveness of homeostatic drives.

4.5. Coherence Maps

Introduction

Concepts

In a synchronous subset within an aperture, information may be instantiated in temporal phase-related activity [344] resulting from the interference of multiple inputs [437] or other wave-like activity in the cortex [54, 360], perhaps resulting in an adaptive connection matrix among oscillators (e.g., neurons or columns) that incorporates memory [438], reminiscent of the coupling and connection matrices in the Kuramoto model [17] (see Ch. 3 and Supplemental Materials). In an alternative model, some have interpreted the responses to specific stimuli of single or sparse sets of cells—"Jennifer Aniston neurons" — that capture a single concept [439]. This has limited support, although the two models are not necessarily inconsistent, as they reflect a random stable subset of the whole that identifies, but does not incorporate, information, as discussed below.

In the cortical aperture, explicit connections among elements are not required, as it is the relationships among elements that are of interest. In a discussion of synchronization across an aperture, it must be borne in mind that this is not the synchrony of all elements but the coherence of a subset of neurons that reflects local functions responding to appropriate features, e.g., edges and orientation. A subpopulation of such specific responders may be interpenetrated in a larger population of essentially random or disordered elements. Scintillating elements (Ch. 3) are considered disordered. If that subpopulation is organized, information entropy decreases (i.e., Weaver [433]), and if the relationship among elements allows noise to be excluded or reduced, entropy also decreases.

Synchronous subpopulations of elements can support information instantiations in the coherent apertures model. Typically, neural net models result in synchrony or phase locking of all elements' activities, resulting in correlations among elements [11, 148, 159, 160, 192, 206, 379] that are considered to instantiate the information across the aperture, the synchronization or coherence itself embodying the processing or result (e.g., Parameshwaran et al. [150]). Small parts of a larger array may become synchronized [195, 440, 441]. In reviewing synchrony in neural networks, Uhlhaas et al. [5] include distributed patterns and in- and out-of-phase synchronization, which is coherence. Arenas et al. [17] and Strogatz [195] have reviewed synchronizing networks extensively, noting that a population of oscillators may split into two subpopulations: synchronized and disordered [442, 443]. There are mechanisms that can lead to likely subpopulation synchronization. Niyogi and English [444] developed a model in which a subpopulation of neurons self-develop into synchronized clusters during Hebbian learning in a network of coupled oscillators. Sompolinsky et al. [445]

propose that neurons with shared receptive fields (RFs) are strongly coupled (feedback) while those with different RFs are weakly coupled. Brunel [223] found that neurons were not equi-likely to be connected and those that had reciprocal coupling also had strong synaptic coupling.

Baran et al. [446] introduced a *coherence map* as a filter for interferograms. Areas with high coherence were less filtered for phase noise than areas with low coherence. The CM (below) in the aperture model extends this concept as a phase field [447, 448] with a sharp distinction between synchronized and disordered elements serving to filter out disordered elements, with the result that aperture entropy decreases through increased orderliness.[12] "Disordered" is defined as either not synchronized or silent. A CM as a whole over the aperture is the combination of both synchronized and disordered elements in which the synchronized elements have a disordered background. The term coherence refers to the information bearing potential within that context.

Organizing structures

Information is embodied and processed within subpopulations. Two models of subpopulations are useful: *motifs* and *maps*. Motifs and maps may have some underlying connectedness, sharing a "fitness for synchronization" [440]. The term "motifs" generally describes a subgraph of elements in a graph or network (e.g., the set of cortical apertures). Motifs have explicitly defined subgraph edges (tracts) connecting its nodes, the apertures. Typically motifs are patterns that may recur as building blocks within the network of cortical areas [449, 450]. Sporns and Kotter [451] use the term *motif* to describe both structural and functional patterns within the larger brain graph that may occur at several scales and may be interpenetrated. A *map* is constrained to refer to patterns of elements <u>within</u> a cortical area that might not have well-defined connections (edges). A map, such as the visual orientation map in V1 [119, 452, 453, 454, 455] or rat barrel cortex [129], is a description of relationships over a cortical area that does <u>not</u> require active participants to be explicitly connected, since a map may contain islands.

Correlation, coherence & synchrony

Models of information storage in a cortical area have generally focused on the strengths, polarity, and number of synaptic couplings among neurons to either represent information within circuits [34] or distributed across some area. These couplings are often Hebbian [33]. Clearly any mode of variable coupling may contribute to information embodiment, variations occurring at widely different time scales (e.g., LTP or LTD vs. synaptogenesis). Correlations may exist among a few local elements such as retinal ganglion cells [272] or, in the case of the cortical aperture, among many [4, 275, 352], organizing complexity, subsequently reducing aperture entropy. Coherence is the mutual-correlation among members of a population, and thus an expression of orderliness. A coherent population has stable phase relationships among its members; consequently, activities of elements that are coherent have the potential to carry either more information with the same entropy or less entropy for the same information [433] than independent elements.

Synchrony, as a form of coherence, has its elements operating in unison, sometimes referred to as *isochronal synchrony* (zero lag synchrony), such as found in a synchronous macrocolumn. I shall adopt the broad definition of coherence as elements having definable temporal relationships. An active map of coherent elements grossly pumps the AFP. For static inactive elements, it is logical connections that will, under the right conditions, cause the elements' activities to become synchronous upon proper stimulation. Such connections include the coupling mechanisms described in Ch. 3, largely phase modulation for long-term memory. For expedience and to reduce confusion, I have adopted the convention of referring to a subpopulation of elements that <u>can</u> be coherent, if

---

[12] A possible indicator of orderliness is the mutual correlation of spike trains, differing from the scintillation in the pumped AFP model.

synchronous, as coherent, and by extension, an aperture that can become synchronous and have meaningful internal phase differences as a "coherent aperture."

The Coherence Map

Introduction

Information in a cortical aperture is modeled as a coherence map (CM) composed of a synchronous, coherent, or potentially coherent (latent), distributed subpopulation of elements with local phase modulations, embedded in the complete aperture array that includes a background of disordered or silent elements (Figure 19). Such a model is consistent with the work of Kuramoto and Battogtokh [442] on the coexistence of synchronized and disordered phases in coupled phase oscillators, a proof extended from the theoretical work of Okuda and Kuramoto [443] on the synchronization of subpopulations of weakly coupled oscillators. The coherence of the CM mini-column elements, usually in synchronous macrocolumns, is a context for their phase relationships. An active CM is synchronous with, and the source of, the LFP (Ch. 3). For the sake of this discussion, elements that do not produce correlated activities but provide other information, such as color blobs in V1, are ignored. Presumably those elements provide information in the topological context [327] of the correlated activity.

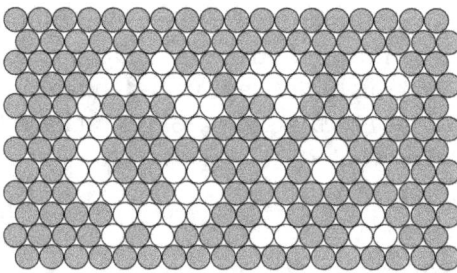

Figure 19. Coherence map of synchronous (white) and disordered (gray) elements.

Spike timing

The CM model presumes that amplitude (number of spikes and/or neurons responding) and timing (phase) have the potential for carrying information [344], with phase being the more efficient factor [456, 457]. Phase, being a continuous variable, supports non-discrete (e.g., non-digital) models of cortical operation that may, in the extreme case, have analog equivalents [41]. In models Yarrow et al. [458] found that stimuli that occur at the flanks of neurons' tuning (phase response) curves are prone to phase fluctuations, providing a slope along which phase is readily modulated, maximizing information encoding. The phase response curve is subject to cholinergic neuromodulation [180]. Biology uses the slippery slope to advantage. Nawrot [346] performed and reviewed analyses of spike intervals and numbers in spike trains from single neurons, concluding that variability had both noise and time-series components, the latter exposed in the distribution of interspike intervals, potentially giving rise to temporally coded information [392, 395]. Mazzoni et al. [343] studied the time course of neural activity in V1 of anesthetized macaques viewing naturalistic movies. Seeking timing codes, Mazzoni et al. found that object information ("what") is carried in sequences they called "spike patterns code," measured from single neurons with temporal structures with a resolution of about 10 milliseconds relative to the external stimulus. Episodic information ("when") was phase structured with respect to the low frequency LFP over a period of

100s of milliseconds; changes in responses corresponded to stimulus changes. These results potentially illustrate meaningful temporal responses at multiple scales.

Synaptic plasticity may both cause, and be caused by, spike timing, reflecting experience [49]. The degree of synaptic influence on timing is related to its location on the neuron [459]. Variability of spike timing can result from variation in the inhibition from chandelier cells on the AIS, as multiple chandeliers synapse in this critical area [320] (Ch. 3). Although these mechanisms tend to progressively constrain the timing of spikes, the diversity of neurons tends to decorrelate the timing, creating a dynamic balance [403] for phase-of-firing modulation, creating greater information content capabilities. The LFP can influence the timing of neural activity [254]; thus spike phase can be modeled as relative to a macroscopic LFP that Masquelier et al. [344] refer to as "phase-of-firing" coding. Neural activity can be phased locked to LFP oscillations, potentially converting rate activity into temporal phase activity [5, 344] in spike trains [156]. The pyramidal pumping LFP (Ch. 3) is a phenomenon triggered by the low potential reversal with little subsequent phase lag, but may not have an associated train if the pyramidal has a low probability of spiking (low oscillation frequency, f_n). The spikes within an individual pyramidal's spike train will coincide with the LFP low phase. The number of spikes in that train may reflect the intensity of the response. Theta-band entrainment [436, 460] and the PCC gating window tend to regulate spike timing [313], producing a spike train, analogous to a wave packet, from a neuron (pyramidal) [342] in which each spike suppresses the next spike [226], suggested to be via the PCC, until the pyramidal effectively discharges, serving to create a train linked to the gamma component of the LFP, with input amplitude evidenced in phase variation [347]. Temporal alignment exceeds that which would arise from the stimulus alone, reflecting a time window [461] arising from the behavior of interactions among neurons [462]. Small (<3ms) meaningful phase differences may occur within a synchronous macrocolumn.

Phased activity in the aperture

The activity of a pyramidal reflects the inputs over its receptive field, effectively modulating its natural frequency. The number of cycles over which a particular pyramidal maintains synchronization with the LFP reflects the strength of its activation. A neuron's spikes may be correlated with the activities of other neurons [163, 344]. Grün et al. [345] found that the joint spiking activity among neurons may organize them in functional groups. Even if separated, minicolumns that are stimulated appropriately at about the same LFP phase, e.g., synchronously, will have similarly increased natural frequencies and hence will tend to synchronize, binding them together. The frequency of synchronization does not necessarily reflect frequencies in the input. Synchrony of nearby pyramidals has been demonstrated [4], consistent with macrocolumns. A macrocolumn of minicolumns that are synchronous will provide a strong synchronization node in the aperture. If the aperture network is close to the critical point, coherent inputs will cause multiple macrocolumns to synchronize, increasing the probability of aperture LFP synchronization.

The temporal structure of the activity of groups of neurons is married to how information is instantiated within apertures. Groups of neurons in minicolumns and macrocolumns are elements in the aperture network that perform a function, with the activities of individual neurons, typically in the infragranular layers of the column, being the outputs of the element [110, 120, 453]. Chawanya et al. [55] model, with experimental evidence, synchronous activity among neurons as binding them together with phase and amplitude reflecting specific features. This is consistent with the pyramidally pumped LFP and synchronous macrocolumn models.

Active aperture gross synchrony is the result of proximity to the critical point and the coherence of the input(s). The laminar and pumped AFP models are convenient for envisioning the relationship between a coherence propensity and coherence. The readiness for coherent spiking in the infragranular pyramidals (layers V–VI) is influenced by the LFP and synchronous activity in the supragranular lamina, with the supragranular LFP$_s$ initially providing a state of readiness, essentially creating a near-critical point. An aperture near the critical point may experience neuronal avalanches in response to multiple inputs with a resulting changing LFP pattern [152]; thus a Fourier analysis of aperture activity may be difficult to interpret. The strength of local coupling in a near-critical network will spread over some distance from the site of local synchrony [162] as the local average is more narrowly defined in time, with a subsequent increase in coherence path length. As multiple locales produce areas with increasing coherence path lengths, the path lengths from separated locales that have a common synchronous source will overlap. Within distances of 2–7 mm, the LFP and subsequent activities—if they occur—of elements (minicolumns) will be nearly synchronous. At this point the local phase reference is essentially the same as an aperture phase reference, the LFP, causing those elements that are near threshold, having a significant probability of spiking, to spike. Within the aperture, only a subpopulation of often non-adjacent neurons spikes phasically with the AFP, based on some common feature or input characteristic [4] (Ch. 5). Grossly synchronous activity constitutes binding. Thiagarajan and colleagues [150, 151] also found such subpopulation synchronization. The AFP as an emergent clock [463] need not have clocklike [378] regularity. This is a chicken-and-egg situation: synchronization within an aperture near the critical point is realized as the result of a coherent input that results in neurons firing in response to the AFP that exists because the aperture is synchronous. This is a delicate balance that requires some form of feedback to prevent spontaneous oscillations [441]. The integration of the coherent aperture into a CNS network provides this control at a global level, as discussed in Ch. 6. Note the coherence of the input is not the same as the coherence (or synchrony) of the aperture, as discussed previously (Ch. 3).

Map

In a coherent network, and by extension a coherent aperture, information may be distributed over that network as interrelationships, supporting high information capacities over an aperture [422] with strong noise immunity. In 1998 Hoppensteadt and Izhikevich [328] proposed that weakly coupled thalamocortical columns within an aperture create (mini)columnar phase modulations, columns being linked through common frequencies, an FM principle. Such columns would be coherent, but the idea of a coherent or synchronous network or aperture was not broached. A coherence map (CM) is a phase field model [464] similar to the coherence field proposed by Sabatini et al. [465]. A phase field differentiates between two or more distinct phases: a liquid sea with solid ice floes, or a map of an archipelago of coherent hilly islands rising above a water level of disorder. This dichotomous model of locally isolated areas being coherent with a macroscopic map of AFP synchronization is generally consistent with the work of El Boustani and Destexhe [3], describing stochastic local areas coexisting within large scale synchrony. This phase quantified model is a more binary synchrony map than the continuously variable coherence map filter of Baran et al. [446]; however, in addition to the separation of the aperture into two states, providing order, the filter aspects limit the synchronous subpopulation by excluding disordered elements, significant with respect to entropy characterization. A phase field [447, 466] models microstructure with an *order parameter* that is similar to Weaver's [433] concept of orderliness. In a cortical aperture the coherence phase field order parameter will have values between +1 for synchronous activity and 0 for complete disorder. The subpopulation of synchronous elements which create a critical level (Ch. 3) will have an order parameter value greater than ~0.5. Thiagarajan et al. [151] found an average of 30%–40% of sites was all that was necessary for a synchronized cortical aperture.

Coherent elements, i.e., minicolumns and macrocolumns, <u>within</u> a cortical aperture CM may not necessarily be physically connected or contiguous [4, 162], being synchronously bound together [55] through a synchronizing AFP (Ch. 3 and its Supplemental Materials), although more direct coupling may also operate including through links to other apertures. Synchronously bound elements may share similar feature properties [*4, 5, 319, 440, 450*] such as the relationships among orientation-specific elements in V1 found by Bartfeld and Grinvald [*327*] and Ko et al. [*467*]. CMs will be dynamically assembled [*366*]; for example, as a visual light/dark edge moves over the retina it creates coherent activity in a small number of elements (minicolumns or macrocolumns) in V1 that may, or may not, be separated coherent (i.e., ordered) areas moving over the cortex. This models the phenomenon of feature binding in common fate in which moving visual elements perceptually group, separating from a background of stationary elements [*468*]. Estebanez et al. [*163*] describe distinct, and changing, subpopulations of correlated neurons in the rat barrel cortex, responding to whisker stimulation (perhaps macrocolumns of minicolumns).

The coherent CM subpopulation is grossly synchronous, phase locked to the aperture phase, with information instantiated only in the synchronized phase. Individual minicolumns within the synchronous portion of the phase field will have small variable local phase relationships to the AFP; thus the activity across an aperture can be modeled as a topographic map that is the product of a synchronous phase field of greater than 0 and a map of the local phase, preserving small local phase modulations of up to 6 ms. The small phase modulations within the PCC window may carry information, as might the length of gated spike trains. PCC gating (Ch. 3) and refractory times may cause in- and out-of-phase spikes to diverge toward two timing states [*469*], which has an analog in binary phase holograms [*470*].

CM classification has two factors: status and duration (Table 1), being active or latent and persistent or transient. An active CM has coherent activity in its synchronous minicolumns. A latent, or inactive, CM is embodied as local phase (or amplitude) modulators preferentially activated by an appropriate input of phase-structured activity; thus it embodies a low threshold to actively synchronize, e.g., a phoneme-specific CM [*471*]. As opposed to persistence, a transient CM dissipates, not supported by underlying durable physiological mechanisms. An active persistent CM continues to provide an output if it continues to receive an input it can respond to. A latent CM may be persistent or transient. Persistent CMs may form through preferential coupling modulation and homeostatic energy expenditure minimization processes, for example, ion concentration balancing that favors simultaneity through local phase modulation via glia [*237*], ephaptic coupling, gap junctions, and synaptic remodeling (see Ch. 3), subject to modulation. When appropriately stimulated, continuous, and noisy, phase modulations among individual cortical elements in a persistent CM can be reduced to limited phase values in an active CM through local binary phase quantization, recovering the initial information [*318*]. Long-term memory, as a latent persistent CM, may use the phase domain, responding to appropriate stimuli with activity in a phase continuum that is quantized into a binary phase relative to the LFP.

		Duration	
		Transient	Persistent
Status	Active	Variable activity	Consistent activity
	Latent	Temporary potential response	Consistent potential response

Table 1. CM states.

Multiple information structures may reside in the same aperture. A CM contains a bound distributed subpopulation that may be interpenetrated with other (latent) CM potentially synchronously bound subpopulations [*163, 448*], separable by phase quantization of small local phase differences. Shorter term persistent memory, active or latent, may use synaptic modulation as LTP and LTD, and transient glial responses, driving neural activity as Hebbian networks to accomplish coherence, which is potentially more energy costly than phase modulation. CMs have varying degrees of plasticity from formation through modulation and dissipation.

4.6. Information in Coherence Maps

Information Instantiation

Patterns

CMs instantiate information through the participation of minicolumns and the phase relationships of their spikes. Tsodyks et al. [*441*] modeled randomly connected (i.e., scale-free) networks of simulated excitatory and inhibitory neurons that could spontaneously self-organize to become synchronous across the network in response to different stimulus features, illustrating responsiveness to specific inputs but a lack of localization to a particular CM. The random connectedness of the "neurons," as opposed to an organized cortical array, may have prevented the emergence of a CM of activity in this model. A local phase shifting cortical model of Ermentrout and Kleinfeld [*360*] produced surface activity patterns across an array that changed from traveling waves in the absence of stimulation to synchronous activity during stimulation, similar to the emergence of a CM. These models do not indicate how information might be instantiated.

The instantiation of information is not well understood. There may be patterns within a particular animal, but there are no universally decipherable codes. Patterns of electrical fields in the visual cortex, presumably reflecting neural activity, might be the same during awake and sleeping states [*472*]; however; a specific representation of the experience is not evident. The behavior of an individual neuron in an animal, even under identical stimuli, is not identical in each measurement instance. It has often been proposed that information is distributed within populations, perhaps embodied in a structure of the local phases of spikes. Although simultaneous recording of the activities of a large number of nearby cells is now possible, Macke et al. [*473*] make the point that the relationship between the population activity and the predictions of theoretical models is not straightforward, and not successful to date. Bressloff [*63*] proposes a network model with a neural field specifying the state of the network. His model is limited to a homogeneous population of neurons. Further investigation is needed. A CM is a pattern of activity, at all but the earliest stages peculiar to the individual—to a degree, a statistical entity.

A CM is not an encoding. It is a physical entity that can be manipulated, as discussed in Ch. 7. The CM is an organizing structure for instantiating information, supporting processes that decrease information entropy through increased organization and decreased complexity [*474*], without defining how specific information is instantiated, allowing undecipherable complexity. Spatially structured instantiations are, of necessity, temporally structured; else the structure will lack coherence. Vision has to have the coherence of the perception to bind the components of an image together. A subset of all elements (macrocolumns and minicolumns) participates in the synchronized phase of a CM at any time and in response to the particular features of an input, although an individual element may participate in multiple ensembles (Ch. 6, iv), for example, different orientation-specific sets in V1 [*475*]. (An ensemble is a set of apertures that play together, much like a string quartet.)

Some CMs probably develop in response to experience [*476*]; for example, place and path information is represented in a network of hippocampal grid cells [*201, 367, 405, 476*] and nearby entorhinal cortex [*368*] through coupling patterns of apparent cortical synaptic spine modification developed early in learning. Tsanov and Manahan-Vaughan [*477*] considered synaptic plasticity in the hippocampus to underlie encoding of visospatial memory formation. We can presume the distributed subpopulations of neurons in the superior temporal gyrus that respond to different specific phonemes [*471*] are examples of persistent latent CMs that become active with appropriate inputs. This consistency of some subpopulations does not conclusively demonstrate information per se beyond a response to consistent stimulations, although multi-voxel pattern analysis also indicates consistent subpopulation responses to categories in visual perception and memory [478]. Some persistent CMs may be of genetic origin, e.g., the invariant of face structure [479].

Three facets of modulation

An active CM results from neurons' spiking behaviors that can be described by IF, WHEN and WINDOW. Three modulation processes interact in the participation of a neuron (or its minicolumn) in a CM (Figure 20):

> Activity: **IF** a neuron will spike, as a probability
>
> Phase: **WHEN** a neuron will spike; and
>
> Phase quantization: **WINDOW** of time within which a neuron's spike occurs.

IF is the probability a neuron will spike as a result of its level of activity, or natural frequency. The inhibitory and excitatory synaptic inputs to a neuron are the primary drives of its natural frequency. As described in Ch. 3, its natural frequency will influence its ability to synchronize and to pump the LFP. The preponderance of interneurons is inhibitory, so active excitatory synapses may form a pattern against this inhibited, or damping, background (admittedly a simplification). A cohered input projection will elicit a pattern of responses in a subpopulation that forms an initial CM if the natural frequencies in the subpopulation are sufficient to synchronize.

WHEN a neuron will spike reflects when it reaches its threshold. A strong excitatory input may cause a spike with little delay. In the absence of other influences, such as electrical interactions, inputs on the AIS do not have a significant effect on the activity level of a neuron, but on WHEN a spike may occur relative to a gating window, as discussed in Ch. 3. Electrical influences are significant modulators of WHEN a neuron will spike, its phase being relative to some reference, typically local activity (e.g., LFP). These are generally termed electrotonic effects, as electrical conductivity through the ECS and astrocytes, which form a syncytium through gap junctions. Astrocytes are rich dynamic modulators of the local environment, further affecting the threshold. See, for example, Simard and Nedergaard [257]. Generally the electrotonic effects tend to synchronize spikes to the negative, rising phase of the LFP. Homeostatic regulation of K^+ in the neuronal microenvironment will tend to synchronize nearby neurons [480]. Longer term homeostatic processes may result in changes in the connectivity and conductivity of astrocytes and the ECS, stabilizing phase differences to reduce active transports of ions. This is beyond the scope of this chapter.

A **WINDOW** of time is circumscribed by the PCC (Ch. 3). The window is small relative to the period of the natural frequency of the neuron; thus, small phase modulations can bring a spike within or outside of the window. As phase modulations are relative to the AFP, in a coherent state, neuron's spikes will be within or outside an aperture's common window. This quantifies the phases of spikes into two states: synchronized and disordered. Thus the activity levels of neurons (IF), modulation of phases (WHEN), and quantization of the phases into two states (WINDOW) can produce an active CM. As the level V–VI pyramidals are considered here to represent the minicolumn, the participation of minicolumns in CMs can be defined by these neurons alone, the activities of other neurons being considered internal processes.

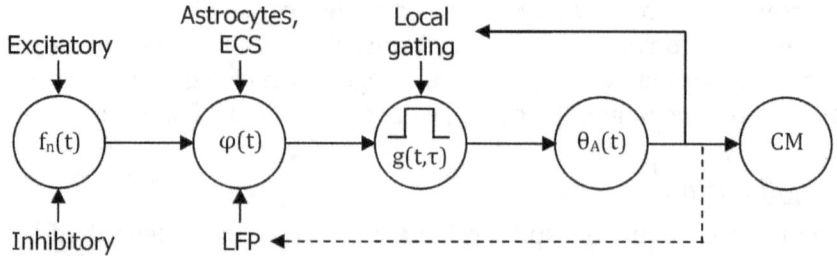

Figure 20. Evolution of element participation in a CM. $f_n(t)$ = natural frequency of neuron, $\varphi(t)$ = phase activity, $g(t,\tau)$ = local gating, τ = gating window, $\theta_A(t)$ = coherence phase of aperture, CM = coherence map, LFP = local field potential.

Local Phase: Information in Coherence Maps

Introduction

If <u>all</u> the elements within an array are (iso)synchronous, behaving the same, the information content is low, as each element carries no differentiated information. Information is instantiated in a CM in two ways: first, inclusion of macrocolumns and minicolumns in the synchronous phase (the coherence map), and second, local phase structures in the synchronous CM phase. As noted previously, the synchronous portions of a CM may be discontinuous with only AFP synchrony binding them together. A local phase structure is composed of the activities of coherent elements within a locale, characterized by the relative phases among those elements. A synchronized macro-column may constitute a locale for its minicolumns. Further research is needed. Local phase structures arise in part from the spread (overlap) through divergence and divergence of input activity that can be modeled by a point spread function, with abrupt transitions at aperture borders [*138*], illustrating that although apertures may not have anatomically distinguishable margins, they are functional units.

A cortical aperture may preserve temporal precision derived from the features of its inputs as evidenced by neural activity correlated with various superposed LFP rhythms [*343, 156*]; thus there will be a significant correlation of activity among a distributed population of neurons [*462*]. In the work of Wang et al. [*357*], the vibrissa of anesthetized rats were stimulated, differentially adapting the thalamus and cortex with adaptation in the cortex degrading deflection detection while differential velocity detection increased, presumably increasing temporal sensitivity. This change was not observed in the thalamus, indicating the cortical performance was modified by experience. As thalamic activity is often modified by anesthesia, care must be taken in interpreting these results [*481, 482*], as rats lack inhibitory interneurons in all but the LGN in the thalamus [*483*]. Estebanez et al. [*163*] found subpopulations of cortical neurons behaving as coherent subfields only during coherent (whisker) stimulation when nearby whiskers were stimulated within a short (a few milliseconds) time frame. Subpopulations of retinal ganglion cells rapidly respond to stimuli with relative time structures [*357, 391*], active transient CMs. The temporal precision in spikes from visual stimuli is preserved through the thalamus to the visual cortex [*164, 390*].

A CM's cohered phase is a phase structure, no matter its state. Multiple processes are responsible for the generation—and storage—of the local phase structures. The synchronous elements within a CM are coherent by definition, with local microphase relationships that are constrained to within a few milliseconds [*144*] (Ch. 3), reflecting the individual temporal responses of the minicolumns. The precise timing of neural spikes resulting from coherent sensory inputs and internally generated activity is phase-linked to the AFP's negative values (i.e., via pumping, Ch. 3) [*151, 259, 336*] creating synchrony with a de facto gating of activity [*461*] of an assembly (sub-population). This ensures that timing precision is maintained to within a few milliseconds [*342*],

upper limit on the noise of the synchronized minicolumns. The delineation and quantization maximize the Fisher information and information entropy [*505*]. As the information in an aperture is instantiated in the CM, the entropy of the CM is an indication of that information; its Fisher information is the information versus its instantiation. Wu et al. [*422*] extrapolate correlated neural activity across an area with a neural field model, finding the Fisher information decreases to a saturation limit as the width of correlation increases, then increases again <u>without limit</u> in proportion to the number of coupled neurons, numerosity overcoming noise.

4.7. Coherent Information Structure (CIS)

The activity of the CM projects to other structures, principally other apertures and the thalamus. There is a direct relationship between the CM and its projection from an aperture. The map of the ordered phase that is a subset of the aperture's minicolumns, and their coherent phase structure, constitute instantiated information. The phase-structured coherent activities of layers V–VI pyramidal cells create the minicolumns' outputs that Grün et al. [*345*] describe as the "temporal structure of their joint spiking activity." I shall use Lizier et al.'s [*165*] term "coherent information structure" (CIS) from elementary cellular automata [*166, 506*] as a noun for the *projection* from an active CM containing a coherent phase. This is similar to the concept of the "population response structure" described by Cunningham and Yu [*469*] in their review of dimensionality reduction. An active CM will continue to produce CISs in response to the appropriate synchronous inputs, being persistent and modulated such that its entropy can decrease.

4.8. Results

Principle and Corollary

Principle:
- A synchronous aperture can produce a Coherence Map (CM), a phase field of a pattern of synchronized and disordered elements.

Corollary:
- A CM may be characterized by its information entropy.

Summary

Information can be instantiated in the coherent apertures model such that it may persist or project within the CNS. Information in cortical apertures can be described as instantiated in phase fields, more appropriately called "coherence maps" (CMs) to avoid confusion. Not every minicolumn in an aperture participates as a member of the synchronous phase in a CM. A CM instantiates information in a distributed form in a subset of coherent minicolumns, with local phase relationships among them, analogous to a phase hologram. The minicolumns in the CM subset are bound together synchronously through the AFP. A CM may be active or latent, transient or persistent. The information in an active CM is related to its information entropy, even if the CM itself is undecipherably complex. The entropy over an aperture, the "aperture entropy," emphasizes the importance of the CM organization in entropy minimization. In the latent forms, the phase relationships among minicolumns are retained as synaptic and electrical phase modulators among nearby elements, particularly within a macrocolumn. Incoming patterns of activity, either as single or interfering projections, may cause a latent CM to become active by matching, whole or in part, the latent phase structure, subsequently projecting its coherent information structures (CISs) to other apertures, receiving similar structures from them.

5. The Aperture Operator

5.1. Introduction

How does a cortical aperture *operate* on inputs? One might say, "It's the response, stupid." I am not presenting a model of detectors and encoding but of responses and transforms. As examples I shall refer primarily to the visual system. An in-depth discussion of visual processing is beyond the scope of this paper (e.g., see Callaway [507], Bullier [508], Felleman [52]). There is ongoing work in understanding specific cortical functions. I shall not summarize such work but instead address the mechanisms by which functions can be realized and, most significantly for the current work, modulated. I shall consider an *object* to be a data structure that may consist of some thing or an area of interest such as a scene. An utterance may be considered an object. A *feature* is a distinguishing characteristic of an object (Figure 21). A feature may be described by its constituent *facets*, which are feature subspaces. Although this might seem to be splitting hairs, it is important to recognize that to which an aperture responds. Its local *functions* respond to facets of features. A *functional* is composed of a set of elements in an aperture with similar functions. An aperture may have multiple functionals, comprising its *operator*. A function is local; a functional is aperture-wide. Functions may be bound together, producing responses to features. One can consider a functional as effectively performing a transform on the activity it receives across the aperture, passing the result on to other apertures in exchanges.

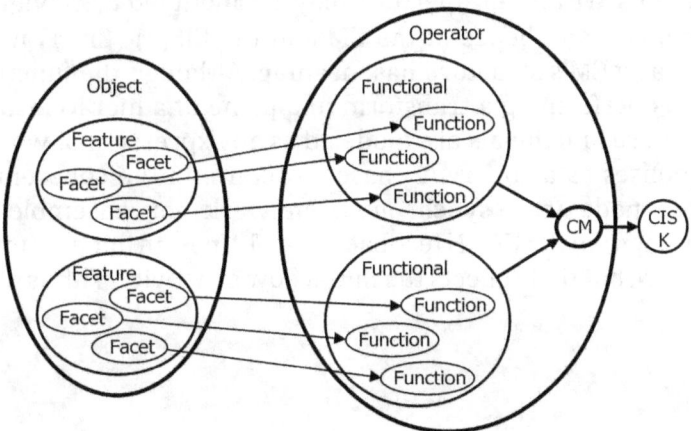

Figure 21. Logical relationships among concepts.

5.2. Functions and Functionals

Properties

The visual cortex operates on the image details in a retinal photoreceptor mosaic [509] through functions' responses to the constituent facets of features of the image. Functions are produced by minicolumns that may be bound into macrocolumns. A function is a relationship between a single input and a single output, although the input may comprise a receptive field as its single input, e.g., a retinal ganglion cell's receptive field. A receptive field may arise from input convergences (many-to-one) and divergences (one-to-many) to a function (Figure 22). There may be no direct correlation between the input of a single axon and the output over a single axon. There does not need be a 1:1 ratio of inputs to outputs over an aperture; thus apertures do not need to be of similar sizes. A function may be the response pattern of a macrocolumn composed of minicolumns, for example, an edge orientation; thus the "single output" has a meaning which may extend beyond a single minicolumn. Function responses comprise the feature inputs for functions in other apertures. At higher levels of abstraction, features include faces, hands, and phonemes, which are invariants. Individual functions may be modulated over a range of times, with a range of persistences, for entropy changes and memory.

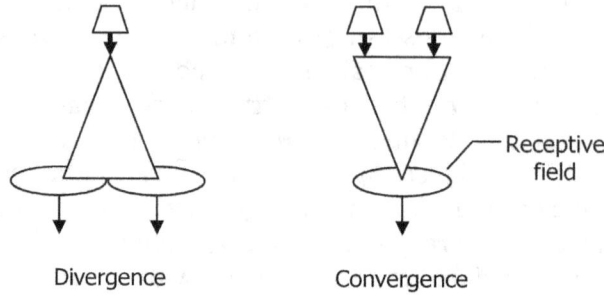

Figure 22. Divergence and convergence.

A functional is a set of closely related functions distributed over an aperture, which may have multiple functionals that are interpenetrated, their functions able to share areas of their receptive fields [510]. The elements within an aperture may be modulated, changing the responses of functions, their functionals, and hence of the CM and the CIS(s) (Ch. 4) it produces. It is not a population code [144], as a CM's structure has meaning. Although the functionals of an aperture might be considered as performing a transform in a plane of a hierarchical convolution neural network (HCNN) [10], here functionals are modeled as an expression of what their functions can produce in their responses as a coherent phase structure. In the coherent apertures model, a cortical aperture is a node in a synchronous network—or ensemble—with bidirectional communication, different from the HCNN feedback model. Upon an initial stimulation, the ensemble may appear to be a HCNN, but that reflects the initial flow of activity. This is discussed more fully in Ch. 6 and Ch. 7.

Functions

Basic characteristics

The functions provide the responses to the inputs, building the aperture's operator. A neural spike is just a spike. Element (minicolumn or neuron) activity—its function response—has a context of location in the aperture and response classification. Location is retained by the element's topological location, e.g., within a retinotopic mapping [511], that is maintained at higher tier cortical levels, although there is an expansion of functions' receptive fields [512] (Figure 22) in successive apertures. The function of a minicolumn is its response to a specific facet, typically responding to differentials with respect to one or more variables (e.g., space, time, intensity). There are two main types of retinal ganglion cells (RGCs) in the retina: magnocellular cells (M) with high luminance and temporal frequency, and low color and spatial frequency sensitivities, and parvocellular cells (P) that have the converse sensitivities with low luminance and temporal frequency, and high color and spatial frequency sensitivities. The retina is a specialization of cortex, performing functions [*513, 514*] responding to four spatial frequency bands in octaves [*515, 516, 517*], which can be considered four functionals of the aperture projecting to the LGN in the thalamus. The spatial frequency bands result from the RGCs that have circular ON and OFF center-surround receptive field (RF) functions [*513, 518*] in four octaves of diameters (Figure 23), with moderately overlapping tuning curves. This is an example of interpenetrated distributed functionals. Widely differing spatial frequency bands with no frequency overlaps are independent, as illustrated by Figure 24, the two image regions failing to bind together in a single stable percept. Conceptually, functionals may provide a mapping between a discretely elemental network and a continuous aperture field that may be modeled mathematically (not discussed here).

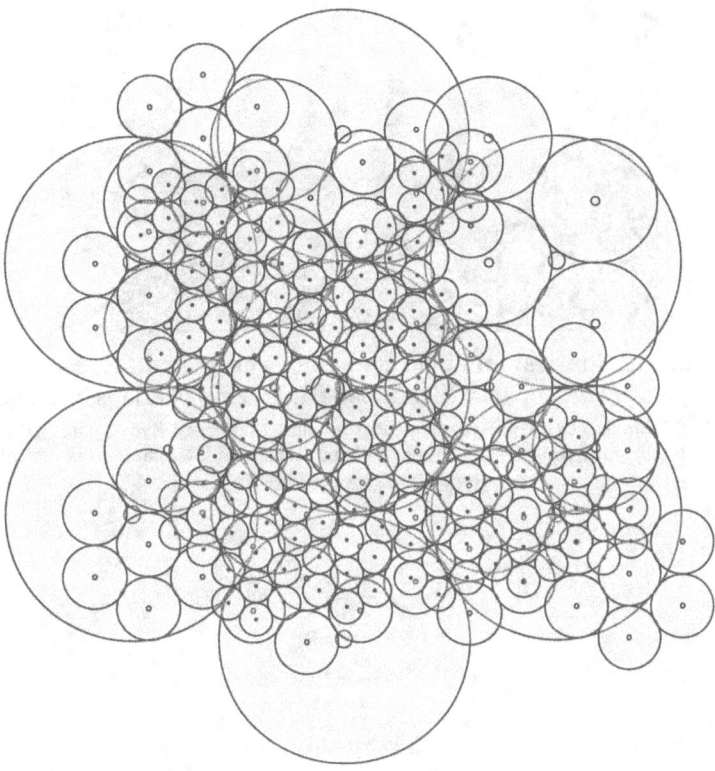

Figure 23. Aperture with four coincident functionals of octave spatial frequencies (partially populated to show different sizes of RGC RFs). RFs for each octave are circular and conformally mapped to the retina. The mapping reflects local photoreceptor density.

The early visual system has functions responsive to the four overlapping spatial frequency bands, with their phases, edges, and orientations. A set of orientation-sensitive V1 functions may differentially draw upon a population of inputs to implement their respective functions [*280*]. An edge is an oriented line demarcated by a difference in brightness [*519*]. The visual system responds to an edge with oriented spatial frequency functions formed from combinations of the local differentials of the ON- and OFF-center RGCs. Pollen and Ronner [*475*] found spatial frequency specific adjacent simple cells in VI with orthogonal orientation tunings that responded to drifting sinusoidal gratings of varying orientations with 90^0 phase differences, as paired sine and cosine filters. This is overtly inconsistent with the work of Hubel and Wiesel [*119, 280*] who found a regular progression of optimal angles in adjacent cortical ocular dominance minicolumns [*520*] arranged in macrocolumnar orientation pinwheels [*327*]. Perhaps the distinction between individual neurons and cortical minicolumns is significant, as the response of a minicolumn, as a function, may be formed from the responses of multiple connected neurons, although it is the single output that defines the element as a function.

Figure 24. Illustration of effects of interpenetrated RGCs spatial frequency functionals with no spatial frequency overlaps in image.
Akiyoshi Kitaoka (Dept. of Psychology, Ritsumeikan University, Kyoto, Japan), Out of Focus (illusion), copyright 2001, www.ritsumei.ac.jp/~akitaoka/motion-e.html. Reproduced by permission.

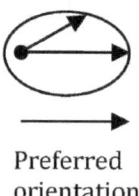

Preferred
orientation

Figure 25. Response characteristics of a mini-column to varying stimulus orientations.

The thalamic LGN projection spread within V1 produces the equivalent of a point spread function [*138, 512, 521, 522*], thus providing inputs to multiple functions. A minicolumn in V1 may respond to an edge [*519, 523*] with a specific orientation in a visual stimulus [*119, 453, 524*] (Figure 25), with multiple minicolumns having overlapping receptive fields [*510*] that are responsive to brightness changes [*525*]. Orientation minicolumns with radially differing orientation responses are bound into direction pinwheels [*452, 454, 455, 526*] that are hypercolumns (V1 macrocolumns) performing higher level functions. The response of a pinwheel is the collective of responses of its minicolumns, a vector normal to the edge orientation. The orientation responses of minicolumns are simple functions; their pinwheel arrangement provides a context for the simple functions. Viewed collectively over the aperture as a continuous functional, the result is a gradient, a continuous vector field. A field is consistent with the modeling by Yger et al. [*424*] that topological invariance does not depend on the details of connectivity within a synchronized network of excitatory and inhibitory neurons.

Phasic

Time windows vary in the CNS. Activity within the cortex has "what" information embodied in short-span spike patterns with resolutions of a few milliseconds, normally associated with the ventral stream, whereas "where" processing is embodied in longer scale (hundreds of milliseconds) phase-structured patterns of activity [343]. Both of these patterns could be associated with the AFP during aperture synchronization, as discussed in Ch. 3. Function behavior includes temporal phase modulations responsive to the stimulus features [156] creating local phase structures. Specific subsets of neurons in the LGN respond reliably to different visual stimuli of black and white patterns with different specific time-structured activities [393, 394] that are relayed to the visual cortex. Feature-specific responses are preserved within local populations of cortical neurons spanning shared receptive fields, while specific neurons may participate in multiple cell assemblies [510] that are active transient CMs. These responses will change abruptly at feature boundaries, e.g., borders, with shifts in timing. Hung et al. [523] found a distinct, although not large (<15ms), temporal shift in relative spike timing across brightness edges with brightness modulation of images. This is consistent with the phase window models. Cortical responses have usually been explored with moving stimuli resulting in stimulus movement across the retina, hence the cortex. Feature discontinuities are demarcated by phase borders of the stimuli to which the cortex is sensitive [523, 527]. Onat et al. [528] decomposed stimulus location from stimulus motion by using stationary stimulus boundaries demarcated by moving gratings, demonstrating that directional responses in patches could be maintained in spite of an overall stationary structure, further illustrating the existence of local phase structures as components of the synchronous phases of active persistent CMs. Such stationary moving gratings may be defined as sinusoidal rotating phase structures to which the cortex responds according to orientation and spatial frequency.

Henriksson et al. [525] found that the congruence of spatial phase coherence across frequencies produced the strongest responses in multiple cortical areas. This illustrates the coexistence of multiple functionals (i.e., spatial frequency bands) within an aperture that may be aligned retinotopically at earlier levels and more distributed at higher levels as spatial phase structures rather than as spatial geometric structures, despite foveal magnification in V1 [511]. This is consistent with the findings of Wang et al. [381] of tight spatial phase alignment within a minicolumn's lamina, which have considerable spatial frequency specificity. Different minicolumns with different principal spatial frequency maximum responses may reinforce edge responses across spatial frequencies, also illustrating multiple functionals (e.g., spatial frequency bands) within an aperture. Tootell et al. [512] found an expansion of retinotopic response location at increasingly higher cortical levels while achieving more robust responses at higher levels. Expansion of this topic is outside of the scope of this paper.

Neurophysiology

Marder and Taylor [91] propose that a relatively small number of modifiable classes of circuits can be used to model the cortex. Although there is no general agreement [385] that there is one common canonical cortical microcircuit [140, 141], it has been proposed that there is a basic pattern that appears across the neocortex [143, 284]. Many subpopulations are composed of stable circuits [467, 529, 530, 531, 532], often established developmentally [402, 533, 534, 535, 536] with surprising flexibility [94, 95, 537, 538] and correspondence across modalities [91], supporting Marder and Taylor's contention. Functions comprise the responses of local circuits and the expansion of input projections resulting from convergence, divergence (Figure 22), and asymmetries. The circuits can be modified, e.g., synaptic remodeling, and modulated, e.g., coupling characteristics, changing the timing of responses relative to other functions within (and among) the functional(s) (Figure 21). A small change in delay of an element (i.e., neuron or minicolumn) relative to some other local element will produce a local phase structure. These modulations can extend over the entire aperture within that functional, with a resulting phase structure for the functionals' response. A change in delay might cause a potential spike to fall outside of a PCC gating window (e.g., Swadlow [308]), the element thereby either not producing a spike or producing one at a later time. As a consequence, a modulated function may have elements that do not respond synchronously with others or participate in spike trains in different manners. Additionally, functions may or may not be coupled to each other, depending either on direct coupling or overlapping responses to stimuli. The absence of the latter is evident in Figure 24.

Functions may form dynamically, reflecting temporal characteristics of input projections [57, 437] and thalamic interactions linking minicolumns [328]. They may reflect the perceptual task [539]. Complex cell functions may be formed from simple functions such as a cross-orientation suppressive response due to the interactions of dominant orientation excitatory responses and suppressive response subunits [540]. A detailed discussion of cortical function realizations is beyond the scope of this paper (see Shepherd [140]). Figure 6 (Ch. 3) is intended to give a glimpse of the complexity with orderliness of cortical microcircuits, providing a skeleton for considering the implementation of functions. Although the circuit will differ regionally, these differences might be considered modifications to a general plan. This cortical microcircuit is consistent with the laminar coherence model.

Neurons are leaky integrate-and-fire elements, activating when a threshold is reached in a particular neural structure, e.g., in the distal region of the AIS of the pyramidal cell. The larger the neuron, the more ions must cross the neural membrane in the proper direction to reach this threshold. Therefore, a neuron's behavior is a function of synaptic activity, time, LFP, and the local environment, as discussed elsewhere. Although Figure 6 does not depict many synapses, which may number over 1,000 per neuron, the figure can be considered as including synaptic weighting. A single cortical minicolumn will have 80–90 neurons rather than the few depicted. Some neurons may synapse principally within the minicolumn, others may extend well beyond it. Some may be interminicolumnar and some intermacrocolumnar. More neuronal types occur than are shown. Intersynaptic interactions are not shown. No glial cells or other extra-neuronal modulators are included, as important as they are. Projections into a cortical aperture diverge. Lateral coupling analogous to a point spread function [522] has been described in the macaque primary visual cortex [138]. Coupling asymmetries could produce a function responsive to an edge's orientation [523, 540]. Computer models may be able to perform tasks such as image analysis, but higher level processes are still a mystery. I am not proposing a specific canonical circuit but an underlying framework based on a cohering mechanism and local (aperture-specific) functions that perform divergent transforms as functionals in progressions, with information subsequently distributed within and across apertures. A deep stack hierarchical convolutional neural network (HCNN) has captured this concept of successive functionals, but in a more straight line fashion with internal and reciprocal projections [10].

5.3. Coherent Functions, Functionals, and Coherence Maps

Conceptually, the coherent response of an aperture is the product of the response functions in each functional class and the coherence map. The terms "functional" and "function," types of operators, imply linearity,[15] which is not strictly the case here, as there are underlying cohering-related interactions. Synchronously bound functions within an aperture comprise the synchronous portion

---

[15] $f(A + B) = f(A) + f(B)$.

of the CM (Ch. 4) creating the subsequent CISs. Thus a CM, and its CISs, may contain elements from multiple functionals. A particular functional may, or may not, participate in the generation of all CISs.

Coherent Functions and Functionals

Instantiating information would be difficult if <u>all</u> the neurons within an aperture were to fire synchronously, creating a single information state. A minicolumn is logically independent. A minicolumn's response behavior is the result of its local circuit, expressed in the output by its pyramidals (layers V–VI). Specific functions within an aperture respond to specific facets; thus not all elements are responsive to the same facet, nor are all that are responsive to a facet stimulated equally. Within limits, the fewer neurons responsive to an input, providing the aperture's output [*198*], the more efficient (lower) the information entropy. For example, for visual orientation functions, not all neurons (or minicolumns) within a pinwheel need fire, although nature abhors a silent neuron, so firing rates or train lengths may differ in an orderly (and synchronous) fashion within the pinwheel. An input that is coherent over an aperture raises the activation of a subpopulation of neurons together and hence raises their natural frequencies. This drives the synchronization of a subpopulation of responsive functions, limiting and organizing the subpopulation into local phase structures. For example, if activity over the retina is coherent, the activity of some subpopulation of functions within the V1 aperture will be synchronous if the critical point of activity (hence coupling) is reached. It is likely that there is an increased probability of synchronization of a subset of functions within a functional class, as functions within the same functional class have an increased probability of receiving coherent stimuli (CISs), e.g., an edge will have similar spatial frequency components along its length. The overlap of tuning curves of adjacent functional octaves [*541*] will produce some coherence among functional classes. The synchronization within the subpopulation is gated and further aligned by the underlying PCC system, electrical coupling, and the AFP (Ch. 3) and/or from thalamocortical interactions [*79, 83, 308, 353, 358, 377*]. The coherence of the stimulus <u>itself</u> (a CIS) is an initial filter for selecting those elements that will cohere. The gross (gamma) self-synchronization of an aperture that is near the critical point is thus both the cause and the result of the synchronization of a subset of functions in response to the coherence of the input projections. It is noteworthy that at any given moment a different subpopulation of minicolumns within a macrocolumn may be active [*120*], potentially performing a *local function*; thus a minicolumn will participate in a local function and in synchronization.

Coherent macrocolumn function

The coherent aperture is reflected in the interactions of functions of minicolumns. Macrocolumnar synchronization through a pyramidal-chandelier circuit (PCC) and the LFP forms an attractive model, with strong experimental support, for binding the neural activity of functionally related neurons, minicolumns. For example, a macrocolumn of a pinwheel of orientation response minicolumns in V1 collectively produces a vector of an edge and its motion [*275, 452, 520*]. An analogous macrocolumn of minicolumns in the rat barrel cortex (i.e., a segregate column) produces a net deflection vector of direction and force [*121*]. The overlapping systems of orientation and ocular dominance columns [*280, 327*] call for deeper analysis [*533*].

The synchronization of minicolumns reflects the stimulus that spans elements. The review of the temporal correlation hypothesis by Gray [*4*] is particularly relevant. To summarize: if minicolumns have overlapping receptive fields, irrespective of orientation specificity, they will tend to synchronize. Minicolumns with non-overlapping fields will tend to synchronize relative to the stimulus coherence (e.g., common motion). Thus if the stimulus has the same orientation across separated minicolumns, those columns will synchronize, a result demonstrated in previous research by Engel et al. [*352*]. Minicolumns with overlapping receptive fields are likely to be within the same macrocolumn. Engel et al. demonstrated that local coupling can result in complete macrocolumnar coherence that will be synchronous under the right stimulus circumstances. Light bars moving over the cat retina were the stimuli, tuned by length, speed, and orientation, while neurons and LFP were studied for best response. Oscillation occurred in the stimulated neuron. Orientation minicolumns in the visual cortex of the cat became synchronized in the gamma frequency band, no matter what the orientation of the stimulus, if located within 2mm, with small phase differences consistent with a "tight" coherence model. More separated points (7mm) had coherence only if both minicolumns were stimulated by optimally orientated stimuli or by stimuli oriented midway between the two optima. Minicolumns at this spacing that were not simultaneously and nearly appropriately stimulated did not become synchronized (hence not coherent), although Gray and Singer [*275*] previously reported coherence confined to stimulus-orientation-specific minicolumns. The LFP follows similar responses, although LFP oscillation synchrony was found at separated distances without simultaneous stimulation.

Engel et al. propose that oscillations with nearly zero time lags occur among independently oscillatory neurons through unknown horizontal non-synaptic mechanisms, although they open the possibility for polysynaptic pathways, similar to network synchrony described by Kuramoto [*11*] and others [*5, 17, 48, 64, 66, 187*]. Chen et al. [*540*] report nondominant subunits in visual cortex that have large receptive fields, the effects of which on more widely spaced minicolumns and on the AFP are not known relative to the study by Engel et al. These results may be interpreted to mean that macrocolumnar coherence is a local phenomenon, consistent with the span of the chandelier axon field. The PCC synchronizes minicolumns within a macrocolumn to the degree that they are appropriately stimulated. The PCC is not driven by a particular function [*220*], serving instead to synchronize a macrocolumn and provide binary phase quantization. In addition to coupling, the PCC provides the local mechanism for pyramidals within a minicolumn to synchronize, and for some or all minicolumns within a macrocolumn to synchronize. Additionally, gap junctions produce a local macrocolumnar network among cortical astrocytes [*235*]. Building on local synchronization, the AFP supports the emergence of aperture-wide CMs, with a synchronous phase derived from functionally associated subsets of quasi-independent elements embedded in the larger network (Ch. 4). Further, the magnitude of response (as a number of spikes in a train) of a minicolumn may reflect specific feature(s) such as orientation and direction of movement of a visual stimulus (Figure 25). The partial independence among minicolumns is significant from the standpoint of an information instantiation. Fine-grained modulation of minicolumns' phases *within* gated windows may occur, providing local phase modulation within a map (Ch. 3).

An analysis of cortical activity must take into account the nature of the inputs from the primary receptors, as they provide experimentally manipulable probes, and subsequently shape our understanding. Some coherence could be of retinal origin, as coupling within the retina has been reported [*240, 250, 272, 400, 401, 518, 542, 543*]. Some early-stage processing, such as direction sensitivity, occurs in the retina [*544*], as does sharpening [*545*]. Spatial tunings may be sharpened by intervening inhibition within the retina [*529, 546*]. The retina, a specialized form of cortex, has a nonuniform array of photoreceptor elements [*120, 124*] that serve the arrays of receptive fields of the ganglion cells [*453, 509*], functionally subapertures. The receptive field sizes are subject to change with light intensity [*547*]. The retinal ganglion cells, as relayed through the thalamus (LGN), are conformally mapped from the curved retinal space to a flat space of uniform spacing in the cortex, resulting in a retinal non-uniform spatial frequency response [*548*] over the four octave spatial-frequency bands [*515*]. The visual system is sensitive to the spatial phase relationships among spatial frequencies [*525*]; thus the specific location of retinal stimulus probes must be controlled to ensure cortically mapped equivalence (the specifics of the retinal projections are beyond the scope of this paper). The visual system is also sensitive to temporal patterns [*398, 549*].

Thalamocortical relationships can influence synchronization over distances [*308, 353, 358, 550, 551*]. The thalamocortical relationship is discussed further in Ch. 6.

Synchronous Functions, CMs, and CISs

The activities of separated functions responding as minicolumns and macrocolumns to coherent input projections from other apertures and the thalamus can dynamically assemble as a synchronous subpopulation [*360*] through receptive field feature-specific tuning [*57*] and aperture synchronization. An aperture may have one active CM at a time,[16] even if transient, binding together its not necessarily contiguous responsive functions by synchrony. As described previously, such a synchronized subpopulation may extend over distances larger than the individual receptive fields of its constituents [*4*], forming the synchronous phase of a CM, generating an output CIS. The image of a structure with a large visual angle, moving across the retina, will result in a corresponding large moving synchronized phase of a CM—with functions responding synchronously in V1— propagating farther into other visual centers. Thus the PCC, AFP, coherence propensity, and the coherent functions in concert may create the synchronous phase of a CM. Persistent latent CMs act as rapid response filters, operating on potentially low signal-to-noise ratio inputs, affected by higher level projections of expectations (potential outcomes, see Ch. 7). An aperture may contain multiple latent CMs, each with a set of functions that respond to patterned coherent input which preferentially stimulates many of the elements within a CM. The stimulated elements recruit other members into the synchronous phase of CM through a decreased threshold due to a proximity to the critical point and coupling effects within the CM, resulting in an active CM. CMs responsive to faces [*478, 479, 531, 552*] in the fusiform face area and hands [*150*] exist in higher level cortical apertures. CMs in the human superior temporal gyrus (STG) may respond to specific phonemes [*471*].

Functions over Multiple Apertures

A cortical aperture responds to incoming projections, producing outgoing projections (CISs). Even though the input projections largely target layer IV, as a simplification one can consider the major projections to both enter and exit the aperture from the inside of the cortex. This can be conceptualized as a thick mirror (Figure 26) with functions, projecting an output to (at a minimum) the source of the input in a manner analogous to a phase conjugate mirror [*553, 554, 555, 556, 557*], while not reflecting the same pattern but a pattern resulting from the aperture's functionals [*558*] and multiple inputs. The mirror is nonlinear since, among other things, there can be interactions between inputs in producing outputs, or a local phase relationship may change with intensity. The output ("reflection") can be expressed as functions' outputs over the projection. At anything other than the primary cortical levels, a relationship of activity to information would be undecipherably complex (Ch. 4), although some topographic information would be retained with decreasing fidelity and with convolutions as the activity propagates among cortical apertures. Transforms may result from functionals in a set of apertures. Multiple diverging progressive transformations may be formed as sequences of (partial) derivatives on different bases (functionals). For example, the interpolation among the four overlapping spatial frequency octaves forms a de facto coarse Fourier transform that can then be incorporated into more abstract—and progressively less retinotopic— embodiments of an object's characteristics. Development of the transforms for invariants is a subject for future research.

---

[16] Subject to future research.

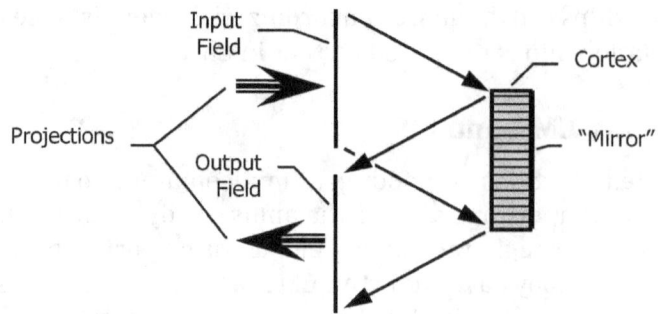

Figure 26. Thick mirror approximation of cortex.

Fuzziness and Quantization

The behavior of individual elements is not consistent in time or space, having some statistical distribution. This is a fundamental property of the CNS. A repetition of the same input—if that were possible—would produce different spike responses in a set of functions. The fuzziness of the functions' responses is increased by the fuzziness of the input CISs. The discreteness of the neural spike creates quantization noise that is offset by the repeated measures in the large numbers of elements involved, the different functionals [559], and the time course over which events are aggregated. Composed of large distributed instantiations, CMs are relatively robust in the face of noise, sharpened by binary phase quantization by the PCCs (Ch. 3). Noise and modulation may cause small changes in the timing of neural potentials with subsequent phase sensitivity relative to a "window of excitability" [308, 560] or phase coherence [486] for action potential generation in the AIS (Ch. 3). These processes create subsequent phase segregation into one of three states: i) synchronous; ii) delayed until subsequent appropriate LFP synchronizing level; or iii) disordered (if at all). Thus a small continuous phase modulation is converted into binary coherence phase states (Ch. 4) of synchronous and disordered, the first two states behaving synchronously. As the receptive fields of functions are finite—and sometimes large—exact activity patterns are not required to create useful responses, since receptive fields aggregate inputs. Conceivably, functions could be modulated through "thick mirror" reflections to match inputs' stochastic distributions.

5.4. Modulation

Evidence

Short-term perceptual effects illustrate modulation of functions and CMs. McCollough contingent aftereffects (for color and orientation [561]) cause short-term adaptation to colors in oriented gratings, producing the perception of the complementary colors in subsequent black-and-white gratings, specific to the spatial frequency and orientation of the adapting and test gratings, even with significant time spans between presentations. McManus et al. [539] found that the shape of an expected image affected the retinotopic responses to contours in V1, and that shape selectivity could be modified by the perceptual task, indicating higher level modulation of the responsiveness of an aperture to its inputs. Memory is evidence of modulation, although, with difficulty, one could maintain that modulation does not affect functions. It would be a semantic argument as the model here relates to the function as a relationship between an input receptive field and the resulting output.

What Is Modulation?

As discussed in Ch. 4, information, an abstraction, is realized in an instantiation, however diffuse that might be, there being no direct relationship between a bit of information and a synapse, or subset of synapses. The CNS has physical processes operating on physical information-bearing entities. It is beautiful in that regard. The brain changes over the life of the individual, reflecting experience and learning [562]. At all time scales, physical attributes within the brain are modulated by physical processes to enable perception, instantiation, retention, recognition, and recall. Instantiations must provide compatibility among these operations, employing plasticity in multiple forms [563]. I shall use the term "modulation" to embody both an alteration and its result, as the two are tightly bound experimentally. Thus modulation of attributes results in the modulation of local functions across the aperture. Retention (e.g., memory) requires a modulation of attributes, no matter the form in which information is instantiated. Recall requires a reconstruction of an instantiation, either the same as the initial instantiation or based on the invariants [564, 565] that are derived from the instantiation. A reconstruction, rather than recovery, model of recall is favored, consistent with a MultiVariant P Analysis (MVPA) of visual cortex that reveals that sensory processes use the same structures as does working memory [566]. Instantiation results from operations on some form of input activity, such as sensory input, to produce an information instantiation with functions capable of producing the coherent outputs of a CM; therefore, local modulations of the functions change the CM's CISs, reflecting the transfer function of the aperture. The aperture as a whole is modulated by incoming projections (CISs) from central (e.g., thalamic) and other cortical areas, and by internal processes such as homeostasis [265, 567]. The coherence of input(s) might modulate stimulus-related activity, perhaps through phase-locking to aperture-wide rhythms [156, 176] in which the statistical distribution of events with respect to the input topology is more salient than the local organization of the cortex [424]. Modulations resulting from interference patterns arising from the interaction of activity from multiple sources [437, 489, 568, 569] within the aperture may produce transient CMs. This will be particularly true of non-linear interactions. As a general rule, one may presume that the potential for modulation is greater the farther information has propagated from the sensory system. In the earlier stages of perception, functions within an aperture are structures dominantly stable on time scales of weeks once past development [94, 95], providing fast reliable responses that can be projected to other cortical apertures. Modulation with respect to operations is discussed in Ch. 7.

Mechanisms of Modulation

Introduction

One cannot say there is a single modulation process in the CNS that controls the retention and recall necessary for even the most fundamental learning. Although there are multiple modulation mechanisms [563], the inability to separate them does not affect the underlying model, as the same result may be achieved through different mechanisms. Multiple mechanisms may work in concert. Generally, modulation progresses from a strength of coupling (weighting) to a timing (phase) mode, although the modulation mechanisms may overlap. Hebbian-based structures may perform an interim distributed representation that slowly converts to a distributed phase modulation form [498]. El Boustani et al. [570] propose spike-timing plasticity rules that predict homeostasis in synaptic weights, work supported by Turrigiano and Nelson [571] and Moldakarimov et al. [567], with some experimental evidence. Neurons change with experience: dendritic inhibition [200], intrinsic resonant frequency [191], and localization of ion channels [296]. Ion channel localization is particularly sensitive in the AIS [262, 309, 572]. Gap junctions among neurons provide cell-to-cell communication of ions, molecules, and electrical signals [210]. Neurons and glia are created and die (apoptosis) over the life span of the animal. The AIS is a particularly sensitive site for extracellular influences (Ch. 3).

Synaptic

Synaptic modulation has short- and long-term mechanisms [573]. Synaptic plasticity [218, 477, 574, 575, 576] is the most evident modulation, affected by frequency of activity [388, 577], spike coincidence [578], the activity across the neural membrane [579, 580], and various molecules [180, 266] and ions, most particularly Ca++ [572], although the effects of backpropagating potentials is not clear [581]. Long-term plasticity [577, 582, 583] results in long-term depression (LTD) [584] and long term potentiation (LTP) [585], which have been studied extensively. Changes in synaptic coupling (number and characteristics) are frequently subsumed within models of Hebbian synaptic weightings [33]. The location of synapses on the receiving neuron can also affect the type of plasticity exhibited, with subsequent modulation sensitivity [200, 538, 586]. Coincident synaptic inputs create synaptic potentiation [587]. Ectopic neurotransmission [224] outside the synapse would appear to provide a sensitive system for short-term modulation. In addition to modulation, the neural circuits may be remodeled through changes in the numbers [588] and locations of synapses [535, 577]. Glia figure dominantly in this remodeling [537], which may occur over the life of the animal [269, 477].

Glial

Glia are now considered major components of CNS function [589], comprising over 50% of the cortical volume and outnumbering neurons [590] by a factor of at least 10 [591]. Of interest, Einstein's brain had significantly more glia per neuron in area 39 than normal [592]. Glia may modulate over both long and short terms. Neuron-glial interactions have been well established [593, 594], including involvement in the development, function, and death of synapses [500, 537, 542, 588, 595]. Astrocytes, restricted to the cortical plate, are of particular interest. Lee et al. [237] found that astrocytes play a significant role in recognition memory. Astrocytes have been implicated in spike-timing dependent LTP [501]. Han et al. [229] engrafted human glial progenitor cells into mice, with a significant increase in learning and LTP over controls of the same strain. As the immunodeficient strain had a lower than normal (compared to wild type) learning capability, the results are not unambiguous. Astrocytes may perform phase modulation. Glia provide other potentially modulatory actions through gap junctions between neurons and glia, and between glia and glia [210], through the interchange and redistribution of K+, Ca++ [549, 596], H+, glutamate [212, 597, 598], and GABA [599]. Glia are involved in morphological changes in synapses [269]. They appear to participate directly in the dynamic formation of the local organization of neurons [600, 601].

Extracellular space

The extracellular space (ECS), including both the fluid space [257] and the extracellular matrix (ECM) [265], may change, directly affecting the behavior of synapses [266]. Of interest, the shape (tortuosity) and volume of the (ECS) may change dynamically, modulating the diffusion of ions and molecules [242, 243], and presumably modulating electrical currents, subsequently modulating the behavior of adjacent neurons. Given that action potentials are electrical events triggered at thresholds, the electrical environment produces a significant short-term effect on the behavior of neurons [415]. Phase responses are modulated by LFPs [322], small changes in the presynaptic potential [388], and cholinergic effects [180]. In addition to the exchange of messengers, gap junctions provide direct electrical communication among cells [210], modulating their behavior.

Homeostatic drive

Homeostatically driven modulation will occur at multiple time scales. The homeostatic tendency may be revealed in the decrease of a response to an unchanging external stimulus. A system with a dynamic input does not have a steady state minimum entropy. Noise and input projections provide a level of readiness by approaching the critical point necessary for synchronization across the aperture. A proximity to the critical point will be constantly changing as a balance of energy minimization and activity. Cooperative modulation among processes may be driven homeo-statically, attempting to stabilize the local cerebrospinal fluid (CSF) among other things, presumably reducing entropy in the process, with an increase in information instantiation efficiency. Despite the system's attempts to maintain a constant composition of the CSF, there are local changes. Consider a pattern of activity in which there is a net redistribution of ions in the ECS over time. The cortex may attempt to restabilize a balance through active transport of ions, such as the glial redistribution of ions [241, 257] (e.g., K^+ in the retina by Müeller fibers [599]), change in ECS volume or tortuosity [242], change in the extracellular matrix (ECM)[265, 602], rapid synaptic plasticity [603], and change in neuronal activity [480, 604]. Intracellular ion concentrations may be regulated through homeostatic modulation of chloride [217] and K+ [605] channels, which may subsequently modulate thresholds on a short-term basis. Longer term homeostatic mechanisms may be implicated in memory, particularly through phase modulation, potentially creating distributed interpenetrated phase structures, analogous to phase holograms [456], which are more energy efficient than amplitude modulated (i.e., through Hebbian synapses) distributed structures. Thus, homeostasis may contribute to memory formation.

5.5. Results

Principle and Corollaries

Principle:

- The responses of the cohered functions within the CM produce one or more Coherent Information Structures (CISs) that are embedded in the projections of activity from an aperture.

Corollaries:

- A functional of an aperture is composed of elements performing some essentially common functions.

- The CIS instantiates some aspect of information in the synchronized portion of the aperture's projection.

Summary

The aperture in the coherent apertures model is a cortical area with a de facto operator that responds to inputs, producing outputs that are projected to other apertures. As such, it can be considered a structure composed of small elements, much as an antenna array is composed of elements. The elements are not independent, their collective effect resulting from, and being affected by, their coherence. Modulation of these elements can affect the characteristics of the operator. Functions are the local responses in the aperture, embodied in minicolumn and macro-column elements. A functional is a class of functions in an aperture which respond to a particular facet or feature of input projections. An aperture may have more than one functional class. Many mechanisms can modulate the functions. A rough canonical circuit, derived from the work of others, was considered as the basis for the local functions that are sensitive to the phases of inhibitory and excitatory input projections, although no specific functions were modeled. Functions may be bound together through synchronization and overlapping facet responses. A subpopulation of synchronous local functions comprises an active coherence map (CM), responding with an output of a coherent information structure (CIS) with a synchronous phase. As apertures receiving inputs from other apertures always project back to their sources, an aperture of coherent modulatable functions is analogous to a thick nonlinear phase-conjugate-like mirror, a model not inconsistent with the local LFP reinforcement model of pyramidal behavior. Functions, hence functionals, may be modulated and the specific phases of element responses captured; thus the thick mirror may have internal modulations responsive to particular inputs. Such an aperture can respond to coherent inputs with coherent outputs. Complex functionals may occur over multiple apertures as divergent progressive transformations that maintain some aspects of topological consistency.

6. Integration of Coherent Cortical Apertures

6.1. Introduction

In the coherent apertures model, the CNS produces an internal experience by integrating cortical areas that cohere (Ch. 3), contain coherence maps (CMs) that instantiate information (Ch. 4), and perform modulatable functions (Ch. 5) on inputs. Coherent apertures become bound in a synchronized subgraph (ensemble) through the thalamus and thalamic reticular nucleus (TRN) to bidirectionally exchange coherent information structures (CISs) projected from the coherence maps. Relevant apertures are self-selected for participation through co-modulation, driving inappropriate apertures to disorder, removing them from the functional subgraph [451]. The information is instantiated with increasing efficiency through entropy minimization.

CNS operations have been considered from many perspectives including computation, networks, and dynamics. The Adaptive Plastic Scalable Electronics (SyNAPSE [606]) initiative [25, 606, 607] with IBM, Stanford University, the University of Wisconsin-Madison, Cornell University, Columbia University Medical Center, and the University of California-Merced, is attempting to create an artificial intelligence, or cognitive computing, brain model for massive data analysis, one of several such efforts (e.g., Blue Brain [École Polytechnique Fédérale de Lausanne], C2 [IBM], Neurogrid [Stanford], IFAT 4G [Johns Hopkins], Brainscales [EU]). Just as computer science is exploring the limits of the classic Turing machines [36, 37], neuroscience is moving beyond the circuit-based computational model [31] of neural networks proposed by Hopfield and Tank [34] in 1986. The central processing unit "computer" model of a brain has progressed to less well-defined processes that occur in multiple spaces, such as the global workspace of Dehaene et al. [32]. Indeed, some go so far as to suggest analog [41] or quantum [42, 608, 609] computing models. I do not propose a quantum model, although the concepts of coherence and decoherence reviewed by Hepp [38] may have some relevance.[17]

The CNS is a world of connected constituents. While knowledge of how the brain is interconnected is essential, it does not necessarily lead to an understanding of how the brain works. The US BRAIN initiative [610] intends to provide innovative tools to map brain structure and function, not to "solve" the brain problem. The ambitious Human Connectome project, described by Sporns and colleagues [98, 99, 100], is developing a "connectome," a "structural description of the network of elements and connections forming the human brain." The project itself does not attempt to describe how the brain works, but to provide a comprehensive map of connections. The project of Modha and Singh [59] at IBM (CoCoMac) parallels the connectome library in the macaque monkey.

Much of the connectome is probably structured developmentally [611, 612]. Yap et al. [101] have summarized the functional connections (using fMRI) according to the connectome models, and have summarized some applications of this knowledge in modeling. The Brain Activity Map project (Alivisatos et al. [62]) is developing a model of the functional relationships among brain regions using fMRI, providing an insight into the relationship between the connectome and brain function. De Garis et al. [30] have surveyed the artificial brain projects such as that of Eliasmith et al. [24] to produce a large-scale model of the functioning brain. The TrueNorth project of Preissl [69] is a massively parallel functional simulator that builds on the CoCoMac [59] data to model the macaque brain's activity with neurosynaptic cores. Although its objectives are changing, the EU

---

[17] The term "decohere" is not used in order to avoid confusion with its use in quantum physics. "Disorder" is used.

Human Brain Project [*613*] was an attempt to create a digital simulation of the human brain to achieve a better—and useful—understanding of how it functions. The relationship between connectionist models and human brain operation is open to question, as McClelland et al. [*614*] point out. They have attempted to bridge the gap between the connectome map and cognition through non-computational explanations which focus on mechanisms within networked (connectionist) and dynamic systems that give rise to cognition. Liquid computing, or a liquid state machine [*40, 615*], maps a continuous input stream into a continuous output stream through a functional or operator. Liquid computing has been suggested by Grzyb et al. [*616*] as a model of operation in the CNS, with a model for a hypercolumn in the visual system as a starting point.

6.2. Oscillations, Synchrony, Micro- and Macrocoherence

Introduction

In the coherent apertures model, oscillations underlie and are expressions of the integration of cortical apertures into a functioning network. Activity within and among cortical apertures may be characterized by coherence, synchrony, and oscillations of the LFP [*155*], indicating participation in a common task. As discussed in Ch. 3, the LFP is a product of neural activity. Neural activity and columnar activity are not unrelated; when an aperture is coherent there is an overall nearly synchronous AFP against which are small local phase shifts of neural (and by extension mini-columnar) activity [*5*]. A set (or subgraph) of functionally related coherent apertures will have synchronized AFPs, as evidenced in the EEG [*1*], which is produced by AFPs. The AFPs, and hence the EEG, have roughly oscillatory frequency bands.

Oscillations

Frequencies and general locales

In neuroscience, the term *oscillation* applies to both regular periodic and irregular patterns. The important feature is a changing state with some pattern of regularity that can typically be decomposed into frequency components. Extracellular measured oscillations usually refer to voltages (LFPs and their gross resultant EEGs). Frequently the sources are difficult to specify either because the source of an oscillation is not accessible through the measurement technique or the source of the oscillation is distributed over some set or population, the oscillation emerging from the group's collective behavior.

As reviewed by Uhlhaas et al. [*5*], synchronized oscillations among CNS structures are well known. Fourier analysis reveals that the AFPs (usually as EEGs [*332*]) are more complex than simple oscillations [*617*]. Buzsáki and Draguhn [*384*] have reviewed oscillations in cortical networks, providing a useful illustration of the frequency bands and potential functional significance. Tiesinga et al. [*156*] provide generally agreed upon definitions of the frequency bands: delta, 0.5–4 Hz; theta, 4–8 Hz; alpha, 8–12 Hz;, beta, 12–30 Hz; gamma, 30–80 Hz. Scaled correlation is sometimes used to separate higher frequency oscillations from lower frequency ranges in analysis [*618*]. It would be simplistic to ascribe particular frequency bands to specific brain structures; however, there are general patterns. Bearing in mind that the site of detection is not necessarily the same as the source, generally the cortex is the apparent source of higher frequencies (beta and gamma) [*153, 340, 619*]. The thalamus is generally associated with lower frequencies (delta, theta, and alpha) [*620*]. Bollimunta et al. [*377*] report alpha current generators in all cortical layers, typically cohered with the thalamus. It is clear that no structure generates its activity in isolation (e.g., Destexhe et al. [*358*], Jones [*621*], Timofeev et al. [*353*]).

Oscillation may be considered of network origin, rather than from a single source or dyad. Wang [*155*] in a 2010 review paper discussed neural synchrony in cortical networks, particularly gamma and theta rhythms. Frien and Eckhorn [319] found that during processing, synchronous gamma oscillations in the cortex correlate to the binding of cortical areas. Such functional assemblies result in consciousness, attention [622, 623], and working memory [624, 625]. Synchronization in the beta–gamma range among assemblies of specific cortical areas [1] is associated by context [5] such as the planning of actions [340]. Liebe et al. [625] report that theta synchronization of the LFP and single unit activity in V4 and the prefrontal cortex predict short-term visual memory performance. One must bear in mind potential differences among experimental species.

Sources

The source of the LFP oscillation has been ascribed to neurons and to the thalamus, thalamo-TRN, cortical, and thalamo-cortical structures (Ch. 3). Determining the source(s) of zero-lag synchronization is difficult. In order to oscillate together, all participating structures must have similar natural frequencies [64]; therefore it is not surprising that under appropriate conditions each will exhibit similar natural frequency oscillations in isolation. Ultimately the source of quasi-periodic activity (e.g., oscillations) in the CNS results from neural activity, although neurons are not the only cellular mechanisms involved; Lee et al. [237] have demonstrated an astrocytic contribution to gamma oscillations. In their review, Buzsáki and Wang [619] propose that although the effect is cortical and collective, the source of gamma oscillations is neural, tied to perisomatic inhibition. Koch and Crick [626] proposed that the natural frequencies of pyramidals give rise to periodic behavior—the coupled result creating binding over multiple areas. Li et al. [361] report that UP/DOWN oscillations may be modulated into persistent states by bursts of single cortical neurons (probably an oversimplification). Kalluri et al. [406] found that some vestibular neurons had regular spiking patterns, although the relevance to cortical activity is not demonstrated.

The point reinforcement of the LFP, as pumping "dimples" in the pumped AFP model (Ch. 3), is a dynamic semi-random interaction between cortical neurons, most probably pyramidals, and the LFP, locally maintaining the AFP. Increased activation of neurons will increase their natural frequencies, increasing AFP pumping. When the number and strength of the LFP pumping sites are high enough, the AFP reaches a critical coupling level, synchronizing the aperture. The source is thus distributed over the aperture, although foci may organize the behavior. Although not specific in their model, Crick and Koch [627] propose synchronized phase-locked gamma oscillations of neurons, presumably cortical, in the binding of populations of neurons, resulting in working memory.

Bollimunta et al. [377] found the pyramidals in the infragranular cortical layers are local pacemakers capable of alpha oscillation, although with potential interaction with the lateral geniculate nucleus (LGN) in the thalamus. Within a cortical aperture there are laminar frequency distinctions [132], with higher frequencies (gamma) more predominant in the supragranular (SG) cortical layers (I–III), and lower frequencies (theta, alpha, beta) in the infragranular (IG) layers (V–VI). This may be compatible with the supragranular gamma oscillation in the laminar coherence model (Ch. 3), implicating the "chattering" neurons that were reported by Gray and McCormick [153]. The supragranular gamma oscillations can be linked to the pyramidal-chandelier circuit implicated in PCC phase window gating (Ch. 3). Surprisingly high frequency bursts of up to 800 Hz in supragranular chattering pyramidals occur within the negative phases of beta and gamma oscillations [153]. The laminar frequency differences are consistent with a laminar coherence model (Ch. 3), with the higher frequencies potentially providing a gating function for lower frequencies, akin to sampling. Thus Buxhoeveden and Casanova's [120] temporal minicolumn may have frequency differences over its depth but must have coherence of some frequency components. The hippocampus, which is not a representative cortical structure, can produce theta band oscillations [436, 628, 629, 630, 631], although Mizuseki et al. [376] report higher frequencies in the deeper layers.

The thalamus has been considered a source of oscillations. Local circuits within the thalamus can generate alpha and theta rhythms [569]. Destexhe et al. [358] demonstrated that with the "massive feedback projections" from cortical areas to respective areas in the thalamus, synchronization of thalamic alpha oscillations among areas was disrupted with the removal of cortico-thalamic projections, although disruption of intracortical tracts did not disrupt thalamic synchrony. During different sleep states, cortical-thalamic-cortical loops served to synchronize oscillations among spatially separated cortical areas. Funke and Götter [164] found that the coherent activity of the cortex may originate with the coherent activity in the thalamus (LGN) which, in turn, reflects the coherence of the peripheral source. Although the thalamus could be the source of oscillations, it does not operate in isolation from the cortex [213]. Alonso et al. [632] found that cortical areas could not sustain activity without thalamic input—thalamic and cortical activity being tightly bound. Bal et al. [620] modeled coupled thalamocortical loops, finding they could generate oscillations in the delta, theta, and alpha bands responsive to cortical feedback. Suppressing thalamic activity causes immediate and reversible unconsciousness [481, 482]. Contreras et al. [79] found that although local theta oscillations may occur in the thalamus of decorticate cats, long-range synchronization would occur with the presence of the cortex, indicating synchronization as due to corticothalamic projections, a finding consistent with my intermediating model based on the work of Fischer et al. [633] (see iv, below).

Bal and McCormick [634] proposed the thalamic reticular nucleus (TRN) as a source oscillator producing different frequency bands (7–12 Hz, 30–40 Hz, or 30–60 Hz), tested by imposing differing experimental conditions on guinea pigs in vivo with various states of awareness. Long et al. [635] suggested that electrical gap junctions among small clusters of neurons in the TRN may generate alpha band oscillations in the thalamocortical system of juvenile rats, a finding indicating the importance of the developmental age of the test animals. Destexhe et al. [89] produced synchronous oscillations in a model of inhibitorily coupled neurons in an isolated TRN. Jones [621] demonstrated that corticothalamic activity can control the TRN activity and subsequently control the oscillatory synchronization of the thalamocortical network as high or low frequency; Bollimunta et al. [377] demonstrated this effect in lower frequencies. This is consistent with the cortical feedback modeling by Bal et al. [620].

Synchronous Associations

Operational associations

The self-synchronizing models considered here employ multiple elements [17, 11, 187, 174, 182, 184, 636], binding both within apertures and in multi-aperture networks. Neurons have been extensively modeled as oscillators [193] that can form coherent networks that Galán [160] characterizes with two activities: synchronization and phase locking. Frank et al. [637] characterize this as the behavior of a system of coupled oscillators under external drives. Synchronization across an aperture (Ch. 3) is the result of the coherence of only a subset of neurons within that aperture. The responses of those neurons to the current input projections (Ch. 5) reflect the local functions, e.g., edge detection or orientation sensitivity. The responses are a component of an information instantiation (Ch. 4).

A coherent aperture may be considered an oscillator itself, with natural frequencies in (at least) the gamma band. A coupled group of such apertures, with similar natural frequencies, will synchronize [64, 179, 463] to the driving forces of the emerging oscillation in the network [172], while being sensitive to phase modulations of those individual sources [188, 638]. This zero-lagged synchronized network of apertures, an *ensemble*, need not contain all members of the larger network in which it is embedded. More broadly, a weakly coupled population of synchronous elements may be embedded in a larger network, much of which is either not synchronous or not synchronous with that particular sub-network. Wang [155] recognized this division, observing that despite the irregular activity of a single neuron, the LFP oscillations reflect the synchronous activity of the synchronous subpopulation interpenetrated in a larger population of essentially randomly or non-coherent (i.e., disordered) responders.

Consciousness is a difficult problem. The EEG and AFP oscillations are associated with ongoing brain processes (e.g., Ehm et al. [639], Hipp et al. [167]). At the grossest level, the presence—or absence—of synchronized oscillations is associated with consciousness, although experimental evidence varies. As noted above, suppression of the thalamus suppresses synchronized thalamic and cortical activity [481], although Barrett et al. [640] found what appears to contradict this finding: induced anesthesia induces cortex-wide synchrony. A few localized cortical EEGs were analyzed relative to consciousness; data did not include thalamic activity, so synchrony still may have involved an active thalamus, as the anesthesia propofol may not have fully involved the thalamus [641]. Huupponen et al. [355] found differing EEG frequency bands for wakefulness (alpha) versus deep sleep (delta). Consciousness and awareness are not synonymous. Zhang et al. [642] report temporally synchronous variations among regions in conscious animals. Melloni et al. [643] report widespread cortical (EEG) synchronization associated with awareness, with only localized coordination without awareness of short stimuli despite wakefulness. In the model of El Boustani and Destexhe [3] the degree of coherence in global (i.e., EEG) versus local activity may reflect locally regular (phase) activity occurring during sleep and locally more irregular activity during wakefulness; thus the local irregularity of wakefulness during active processing may appear at odds with global synchronous behavior during sleep. Perhaps a more appropriate model of active processing is the synchrony of a dynamically recruited functional set of a limited number of cortical areas, with the suppression of the inclusion of many apertures, as discussed below.

Perception involves multi-aperture synchronization that proceeds from, or is in coordination with, synchronization in the apertures. Within a cortical aperture, Frien and Eckhorn [319] found stronger stimulus-related neural synchronization at high (gamma) versus lower (delta, theta, and alpha) frequency bands for visual gratings of different orientations, and for field overlaps at larger cortical distances, than simple coaxial receptive fields would predict. Attention may result in a decrease in V1 alpha activity, with subsequent measurements favoring the gamma band (377), perhaps indicating a change of the EEG due to a local asynchronous irregular condition [3]. Changes in EEG gamma activity may precede changes in the visual perception of ambiguous figures by 200 ms [639]. From theoretical and experimental standpoints, Dehaene and Changeux [86] considered the relationship of sensory stimuli to cortico-cortical synchronization, finding beta and gamma synchronization among multiple cortical areas during consciousness. A visual stimulus, projecting from the retina to the thalamus, produces rapid responses in multiple cortical areas [78], with coherence in the cortex being responsive to the coherence of the stimulus in the thalamus [164]. Attention to images of faces, in contrast to images of houses, enhances gamma synchrony between the fusiform face area versus the parahippocampal place area [479]. This attention can produce relative phase shifts in the 8–20 ms range while maintaining gamma band synchronization of the separate cortical areas [479, 623], differentiating between synchronization and isochronal synchronization, a topic for discussion below.

Models of processing ensembles

During processing in the CNS, cortical areas may be dynamically organized into functional networks [644, 645]. Frostig et al. [117] found large-scale sensorimotor organization in rats, raising the potential for synchrony over multiple areas. Englert et al. [646] illustrated that a "reciprocal pair," such as cortical areas [647], could achieve zero-lag synchrony without a common driver (discussed further below). Hepp [38] proposes a coherence of reciprocally connected cortical areas that results in a global neural workspace, concomitant with consciousness. Zero-lag synchrony with complex phase activity has parallels in optical oscillators formed by multiple phase conjugate mirrors [648, 649, 650, 651], capable of storing multiple images. The activity across a coherent aperture will have local phase differences within the CM, a complex phase-structured activity, described previously (Ch. 4). Many models that effectively [*177, 192, 652*] or explicitly [*176*] refer to the Kuramoto model of synchronizing many peer-connected elements (Ch. 3) consider the cortex as the site of feature binding [*55*] and computation, with areas bound together with oscillations, and with segmentation of cortical areas into functional ensembles. Baars et al. [*39*] consider the formation of a global workspace as resulting from a narrow band theta/gamma or alpha/gamma phase coupling that binds cortical areas into a functional hub (although a rather extended model of a hub, similar to what I call an ensemble). Fries considered the gamma band to be primary in this functional synchronization [*560, 653*]. The oscillations may be concomitant with, not just responsible for, the elements' synchronization. The synchronization of cortical areas may involve other structures such as the thalamus (see above). Min [*80*] proposed the thalamic reticular nucleus (TRN) as central to consciousness perception, the TRN providing multi-cortical synchronization through maintenance of a thalamo-cortical synchronization. I propose that the topological overlap of cortical areas in the TRN provides the potential for binding among cortical areas [*635*], consistent with Hoppensteadt and Izhikevich's [*328*] more local thalamically mediated coupling among cortical columns. A relationship among the thalamus, TRN, and cortex has been described and modeled by Destexhe and colleagues [*89, 213*]; I have formalized this below as a composite node.

6.3. Apertures in Cortical Integration

Introduction

How do apertures become integrated into a synchronized network? Previous sections have presented a model of a cortical area as an aperture that may become coherent across its area as it is synchronized (Ch. 3). Information in such an aperture is embodied in latent or active, persistent or transient, CMs that, when active, produce CISs projected over axons to other structures, including apertures (Ch. 4). Information in an aperture, possibly being undecipherable, may be described by its entropy. Within an aperture, functionals with component modulatable functions respond to input projections, under the appropriate conditions producing active CMs (Ch. 5). A set of coherent apertures, by virtue of their activities, can become bound together into a synchronized functional network with a corresponding network oscillation, often in the gamma band.

Development of the Internal Experience

As some cortical oscillations relate to consciousness, we may ask how the synchronous network relates to the internal experience. The internal experience is an emergent result of the integration of coherent apertures into a synchronous subgraph, an ensemble, with a subsequent net reduction in information entropy by aperture comodulation through bidirectional causation. I shall explain the model of cortical integration for early stage sensory processing through a logical ordering of what are actually simultaneous processes. I define four types of groups of connected elements:

Graph: a connected set of nodes, often described as vertices connected by edges. All nodes do not need to be connected to all other nodes.

Subgraph: a connected subset of nodes within a larger graph.

Motif: a preferentially connected subgraph, usually of apertures, reflecting some functionality. It may be active or latent.

Ensemble: an active synchronous subgraph of apertures.

These all differ from the CM in which each columnar element in the aperture's population is a member of one of a few phases (e.g., synchronous, disordered) but does not need to be interconnected within that phase. Cortical integration is the formation and persistence of ensembles controlled by ongoing interacting processes that I have logically separated:

i) Coherent Apertures Produce CISs with Keys.
ii) Coherent Apertures are Nodes.
iii) CIS Exchanges are Enabled by Aperture Coherence States.
iv) Ensembles Form through Bidirectional Exchanges of CISs.
v) Bidirectional Exchanges Regulate the Composition of Ensembles.
vi) An Ensemble Seeks a Low Entropy State.
vii) The Internal Experience Emerges.

i) Coherent Apertures Produce CISs with Keys.

A *natural key* is an identifier derived from some natural attributes of a data source, typically simpler than the source itself. It is a surrogate for the entity. An email address may be used as an identifier for a person, e.g., the "user name." As used here, a natural key [654, 655] (K) is a concept more that a strict reality, a unique identifier for a unique pattern in a cortical aperture, or of its projections over axons. K is a sample of the whole that designates and characterizes the whole. It is composed of the states of a randomly distributed subset of the whole, its pattern identifying an information instantiation but not necessarily incorporating any information. It is nearly meaningless. For the sake of this discussion, a K is formed from a stable set of topologically randomly distributed points which create a de facto hash code [656] algorithm, and hence, in the case of the early visual system, retains some retinotopic organization [657]. Some of such a quasi-random sampling set might be derived developmentally and experientially, reflecting frequent intercortical associations. A complete set of coherent and disordered (including silent) points in the random sampling constitute a K. Although such keys may not actually be present, they could have a significant value in computer modeling.

K expresses the idea that a smaller set can suffice to uniquely tag a larger CM or CIS, and should not be taken too literally. K is a convenient shorthand for referring to the power of the subset of a whole. The efficiency of a natural key is its size relative to the entire population. Consider an aperture with 100,000 elements (minicolumns) and a natural key formed from 10% (10,000) of those. If we presume that the elements have binary states, then 2×10^{3010} ($2^{10,000}$) unique Ks are potentially available to identify CMs, an absurdly large number that illustrates that even a subset of the whole can uniquely identify the whole and provide a substantially unique pattern within an aperture. K is intended to capture that idea.

A K may be active or latent. A K is derived from a CM. An active CM will project a CIS that contains an active K (Figure 27), the two being only conceptually separable. A K_i, a subpopulation of CIS_i, matches a subpopulation K_j' within a latent CM_j in node N_j, increasing its likelihood of becoming active and projecting a CIS_j with its K_j from an aperture to other apertures and the thalamus. It should not require a complete activation of all points of a latent CM to activate it. An active K may activate a CM by matching, to some degree, the latent K it contains. If the pattern of incoming projections contains an active K that substantially matches a latent K, the CM associated with that K will be preferentially stimulated, as the K is distributed over the same aperture and will have local phase relationships that match the CM's latent phase couplings, the K being a subset of the CM after all. As discussed in Ch. 5, a latent CM has a modulated functional, embodied in the couplings among elements, and thus will preferentially respond to a match to that functional. In order to match, a natural key, being part of a CIS, must have a corresponding functional in the receiving aperture that responds to it. Local couplings will affect the phase of the aperture's functions' responses, if any, to phase-structured input and other activity (e.g., noise). Phase quantization (Ch. 3) will separate elements into cohered and disordered states, subsequently producing a CIS with an active K. An aperture's high propensity to cohere (Ch. 3) will increase the effects of the incoming K on the appropriate local K and its CM. CMs and CISs, and consequently their Ks, have significant noise components. Repeated stimuli do not produce the same result. This will be reflected in large "fuzzy" Ks [*658*]. Even with the fuzziness of Ks, CMs, and CISs, the potential for unique IDs is still quite large, sufficient to unambiguously identify—and robust enough to activate—different CMs. The large size of the natural key provides substantial redundancy that will resolve potential overlaps of Ks, and corresponding CMs. The same natural key, or a substantial portion of it, may be projected to more than one aperture. As a K subset is topologically stable, it may activate CMs through their functionals in multiple connected apertures that all have partially common K subsets. This can cause multiple apertures to synchronize.

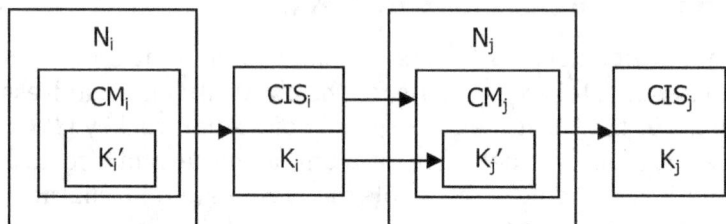

Figure 27. Persistent CMs with K′ subset maps in aperture nodes N that produce active CISs and Ks. Ks may activate or facilitate subsequent CM activation through their embedded K's. A K extends over the entire aperture.

The natural key hash code model has limitations. Membership is not well defined. The nature of phase encoding is not defined. Fuzziness is not well defined. Consider it shorthand for the state of a statistically stable random subset of the whole that can identify, and potentially elicit, the whole. The issue of place independence (or correspondence) raises similar issues. An image can be recognized even if it appears on different retinal locations or with different viewpoints or scales. How? How can a natural key work under the same changes? One must presume that such comparisons are made at higher levels in which the image—or other representations—is abstracted. Such entities are probably instantiated in multiple apertures, as discussed below in Ch. 7. More exploration of this issue is in order.

If one characterizes the aperture entropy by the degree of organization or orderliness of the activity in the aperture (Ch. 4), then a change in the entropy of a random sample of points over that aperture should correspond to a change in the entropy of the entire aperture. If a K serves as a surrogate for an active CM's entire information structure, then its entropy provides a surrogate for the aperture entropy.

ii) Coherent Apertures are Nodes.

Aperture cohering summary

Through application of the van Cittert-Zernike theorem, a concept in statistical optics, a coherent aperture may be considered a node in a graph or subgraph, supporting entropy minimization and the internal experience. As discussed above (Ch. 3), cortical apertures may become coherent over their area when a critical point of coupling exists either directly [*272, 306, 319, 340, 436*] or through the LFP [*151, 177, 348, 360, 619*], with models often derived from Kuramoto's 1975 work [*11*] and other models of coupled networks [*17, 148, 176, 187, 484*]. An aperture may have a coherence propensity state described by a laminar model [130, 131, 132, 153, 376, 377] and by a critically pumped AFP, as discussed in Ch. 3. The resulting aperture-wide synchronous AFP allows a distributed subpopulation of elements (neurons or minicolumns) [55, 487] to form a coherence map (CM, Ch. 4) [154, 414] that projects coherent information structures (CISs) to other apertures as a node in a graph.

Node described via the VCZ

Synchrony among a subgroup of cortical apertures is generally accepted as concomitant with integrated cortical processing, therefore in order to form a synchronous network of apertures, each cortical aperture must be reduced to a single unit or node. Under the van Cittert-Zernike theorem (VCZ) [147, 659, 660, 661], activity generated by a synchronous aperture [105, 203] can be treated as emanating from a point, hence a vertex or node in a graph. Coherent source and recipient apertures can be modeled as point nodes exchanging information, or information units, through bidirectional couplings, consistent with Mesulam's [662] concept that each cortical area is a nexus of convergence and divergence, which are aspects of deep stacks in hierarchical convolutional neural network models (HCNN) [10]. I have used the term "bidirectional" rather than "reciprocal" to reflect the bidirectional causation to be described below.

The VCZ describes the relationship between two complementary Fourier transform pairs [659] for radiating photons. Although the theorem is broadly formulated for randomly radiating photons, it may be applied to coherent radiation, which is a simpler case. The radiation projected from a *point source* in one aperture will coherently cover the area of another. As propagation in classical optics is reversible, synchronous (coherent) radiation over an aperture projects as a point [20, 498, 663] if there is no disruption of the coherence by the intervening medium. Thus a coherent node can both project and receive as a point (a VCZ simplification). A coherent aperture may be referred to as a *VCZ node*. The disordered elements do not participate in the information exchange; hence they are disregarded in the face of coherent activity (see PCC, Ch. 3) and thus do not interfere with the VCZ status.

The laminar coherence, pumped AFP, and Kuramoto-based models together (and singly) (Ch. 3) can describe an aperture at its critical point of coupling, thus potentially a VCZ node, as it has a level of internal coherence that allows it to accept and produce CISs. An active CM may produce small phase timing variations across the aperture [342, 404, 523], including local interference patterns [569] (Ch. 5), while maintaining an aperture VCZ node status. Such a CM produces a subsequent microphase-modulated CIS that can carry significant information [20, 105, 165] within the CNS [5, 156, 328, 336, 343, 344, 488]. The VCZ point-node approximation is appropriate, as the microphase span is small compared to the average synchronous activity period, even though the synchronous activity does not have a regular periodic structure, as coherence is maintained.

Aperture cohering

As a general model, I shall consider all apertures as peers. An aperture achieves coherence in response to inputs and its internal state (Ch. 3–5), the coherence of inputs separating the ordering from the disordering effects. A CM is a state of orderliness. As suggested in Figure 28, which disregards input CIS interactions, responses of an aperture to inputs either increase or decrease its orderliness, combining to determine its state, which may produce a CM if the aperture becomes coherent. The appropriateness of a particular input is determined by its coherence and a match to a functional in the aperture. Cortical gating circuits function as filters for coherent inputs. If from a synchronized or coherent source, some inputs may have CISs with Ks, whether of cortical or thalamic origin. Appropriateness, through the response of an aperture's functional, may increase the LFP and/or synaptic coupling toward the critical point with a subsequent propensity to cohere and an increased orderliness. If orderliness increases sufficiently, a CM will emerge as the aperture synchronizes. A sufficiently strong CIS, projected into an aperture, if appropriate to a functional within the aperture, will cause it to cohere, producing or activating a CM, unless excessive disorder from other sources overwhelms it. If an appropriate CIS is projected into an aperture that is already coherent, if it is not in conflict with the existing CM, it will increase the strength or sharpen the CM. This coherence results in a VCZ node, enabling the aperture to participate in a subgraph of multiple apertures. If the aperture is not able to attain or maintain a VCZ node state, it cannot participate in such subgraphs.

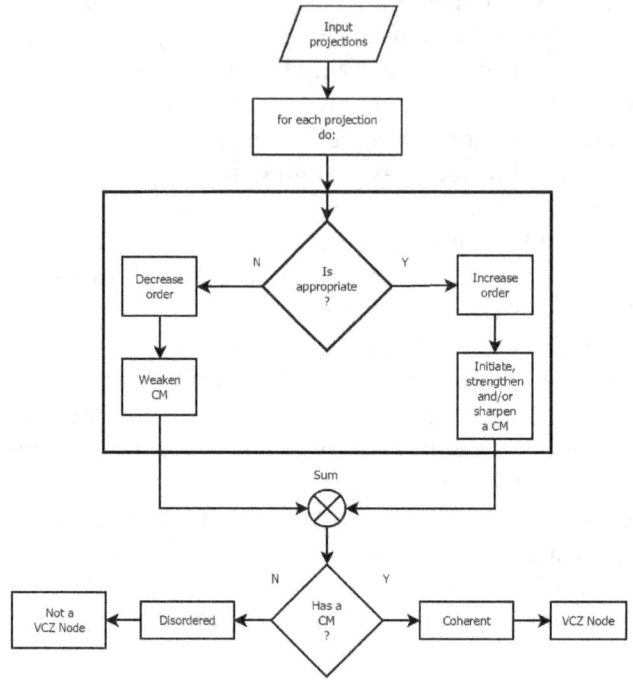

Figure 28. Aperture cohering process, increasing or decreasing orderliness for each input projection.

Noise is a double-edged sword, playing differing roles in cohering. Diffuse low level noise, primarily from brainstem sources (e.g., the locus coeruleus [349]), raises the LFP pumping and synaptic activity, with their coupling, toward the critical point for coherence. Noise input to the aperture via layer IV with specific neural targets, being disordered and thus inappropriate, will increase the disorderliness of the aperture. Targeted disordered noise will probably indirectly excite the inhibitory cells such as basket cells, decreasing the activity of pyramidals and subsequently decreasing the LFP pumping. Thus an aperture can be "weeded out" of a subgraph.

iii) CIS Exchanges are Enabled by Aperture Coherence States.

Communication is the exchange of information. Communication among apertures is accomplished through coherence. Weaver [22] describes communication of information as having three stages:

1) creation
2) transformed transmission
3) extraction.

Accordingly, 1) functions in CMs in coherent apertures (Ch. 5) produce information-instantiated CISs that are 2) projected over axonal tracts to other apertures. The point spread function (Figure 22) and interference describe activity that is projected into and out of a cortical aperture, analogous to an optical system. 3) The coherence of the projected CIS and proximity to the critical point of the receiving aperture form a filter set enabling the exchange information in the face of noise. The coherence, or near-coherence, of apertures (as VCZ nodes) serves as a filter for exchanging CISs, discriminating the CIS from background element activity that is noise relative to the CIS, simultaneously creating a coherent subgraph (ensemble).

The PCC gating window described in Ch. 3 (see Figure 13) will preferentially select spiking activity that is coherent (Figure 29). Gating is constrained to a temporal window of δ duration. Input gating will occur primarily in layer IV (granular), output gating in infragranular layers V and VI. Elements within a macrocolumn can be coherent if its functions respond to an input. If the aperture is synchronous, the coherent activity will be phasically locked to the AFP, extending the effective gating window over the aperture. Together the PCC-gated output (projection) of one aperture and the PCC-gated input of another form a coherent filter set if they are—or become—synchronous, the filter effectively isolating and enhancing the coherent components of the CISs exchanged.

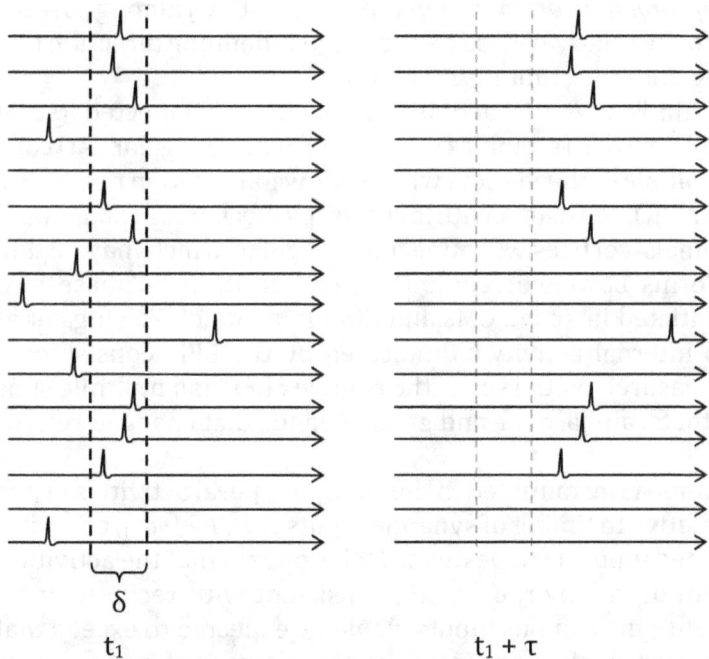

Figure 29. Neuron activity timelines with PCC gating filter. At time t_1 activity is PCC gated for an interval, δ, excluding (potential) activity at a later time, $t_1 + \tau$, that was not within the active gating window. The gating filter can apply to both projecting (output, V, VI) and input (IV) activity.

The process of cohering instantiates information both within the aperture's CMs and in projections of CISs with Ks among structures; thus the process of aperture cohering occurs within the larger context of cortical integration. The coherence of an aperture is concomitant with its participation as a VCZ node in the emergent ensemble, including a coherent simultaneous bidirectional exchange. An aperture will most commonly receive inputs (CISs) from more than one source. A (mis)matching of a pair of apertures' functionals supports cohering (or disordering) of one or both of the apertures [*433*], with disorder suppressing participation in the network. Communicating apertures have some topological correspondences. Ks reflect the topology of the apertures; thus the alignment of Ks will reflect their spatial correlation, particularly if the K populations are largely stable. The correlation of an internal active CM (if it exists) and input CISs (or their surrogate Ks) will affect the stability of the aperture's coherence. The net effect of the CISs and the current state of an aperture may result in subsequent modulations of the functions' responses. The modulations may result in an increase in the orderliness of the response, decreasing the aperture entropy (Ch. 4).

The problem of information in the CNS appears frequently: there isn't any. Information is physically instantiated, and that is what gets processed. It is convenient, however, to speak of information when actually referring to its instantiation. Oscillations appear intimately tied to the active instantiation, transfer, and processing of information. Buzsáki and Draguhn [*384*] have provided a review of neuronal oscillations in cortical networks. Buzsáki and Schomburg [*1*] propose that the synchrony of cortical areas comprises a network with inter-area communications. Buzsáki and Diba [*664*] propose that oscillatory activity underlies bidirectional information transfer among cortical structures, including the hippocampus, enabling processing and storage in the brain. Fries [*560*] proposed communication of neuronal groups through coherence as dictated by simultaneously open windows in synchronous apertures. More specifically, Tiesinga and Sejnowski [*486*] propose a *principle of communication through coherence* (CTC) among cortical areas in the gamma frequency range, with the relative phase between areas defining direction of communication, thus having phase shifting underlie stimulus selection.

Destexhe [*84*] finds that the activity of randomly connected (scale-free) networks of integrate-and-fire (IF) neurons resembles the asynchronous irregular (AI) cortical activity in awake animals processing complex information, whereas low information results in more regular activity (synchronous regular, SR), consistent with low entropy measures of Shannon and Weaver [*22*]; thus activity in awake animals' cortices will appear as irregular, which may be difficult to differentiate from complex waveforms. Lower correlations may reflect the local phase variations [*469*] inherent in information instantiated in active CMs and CISs. In a model, Tiesinga et al. [*156*] found phase-locking of spikes to internal activity as indicated by the LFP, consistent with sensitivity to a coherent input. As measured on the scalp, the complex LFP had multiple superposed frequencies including the delta, theta, alpha, beta, and gamma bands that changed relative to each other over time.

Active information is instantiated in the spatiotemporal activities of populations of neurons. Neural activity is sensitive to timing of synaptic inputs [*103, 164, 665*], consistent with Ising models responsive to correlated inputs [*11*]. Destexhe [*76*] proposes that the activities of the cortex may be more modulated than driven by the senses, consistent with receptivity to sensory input. The population response to synchronous inputs would be expected to exceed that predicted from the common modulation by a single input [*460, 461*], the network response enhancing synchronicity. The model of Renart et al. [*666*] indicated that a shared input resulted in an asynchronous state. They found a near-zero mean correlation among cells in the rodent cortex despite substantially shared inputs. Excitatory (e.g., pyramidal) and inhibitory (e.g., chandelier) populations appeared to balance each other, producing a near zero mean population correlation. Combining the two in a mean population determination may have limited validity. Separating the two populations would be, at best, difficult. A CM is composed of coherent and disordered members; therefore a random sample of points (K) is apt to show relatively low correlation. One would not expect most functions in an aperture to respond to any given input. As Yamins and DiCarlo [*10*] note, in the HCNN model, even though the functions across a cortical area may be relatively uniform within a functional, the pattern of the responses is what is significant. It is not a population encoding (or instantiation). A specific local circuit is not necessary for coherence (Ch. 3). Yger et al. [*424*] illustrated that a coherent response of balanced networks to the synchrony of external drives is invariant with respect to interconnection patterns.

An aperture's output response is characterized by its functionals' alignment with appropriate input(s). An aperture structured for receiving specific types of coherent inputs may respond coherently only to spatio-temporally structured coherent inputs, for example, structured versus irregular patterns of whisker stimulation in the rat [163]. Neural groups may form spontaneously in response to structured inputs [499]. A CM within an aperture that is near the critical point will cohere when it receives appropriate coherent activity [484, 485]; therefore, the response of a subset of nodes—the cohered subpopulation in a CM—in the receiving aperture will correspond to the projected information, the process selecting the information-bearing activity based on its coherence. A high coherence propensity, as abstracted in the laminar coherence [130, 132] or pumped AFP models, would increase the likelihood of a coherent response (Figure 28).

iv) Ensembles Form through Bidirectional Exchanges of CISs.

Introduction

During a process such as perception, a group of apertures becomes bound into a coherent subgraph, an *ensemble*, with an underlying frequency. Ensembles self-cohere in a manner comparable to aperture cohering. An ensemble can be characterized as having a coherence field that may be evident in EEG analysis [81]. The binding occurs through the nearly synchronous exchange of CISs, which can be characterized by the correlation of exchanged Ks. An aperture and its corresponding thalamic nucleus and TRN sector, form a composite node (CN). The development and maintenance of an ensemble incorporates couplings both within and across CNs. The processes of developing and maintaining ensembles and, most importantly, accomplishing something with them, draws on three models in which the CN clarifies the roles of its constituents. The three models, illustrated by solid state laser experiments, provide stages in the evolution of the ensemble, its resolution toward a solution, the outcome, and the internal experience. Noise is significant in several contexts, including CM sharpening, modulating ensemble membership, and homeostatic effects. It may also have a destructive effect, reducing the threshold for epilepsy.

Models of cortical integration

McClelland et al. [614] propose that cognitive processes emerge dynamically from a large number of simple non-cognitive processes, a non-computational model. This model separates processes from their network substrate, although there is a presumption that the processes occur within the substrate; "where" is not defined. "Where" must be considered. Although the connectome [100] provides a structure of connections among brain regions, a network architecture [59] that Sporns [43] points out, this still does not solve the problem of how the network relates to cognition. Destexhe [76] proposes that the brain network is dynamically modulated and organized into functional networks (i.e., subgraphs) by sensory inputs, since functional organization requires some but not all of the network, with inputs modifying the internal dynamics, consistent with the functional relationships of the Brain Activity Map [62]. Dynamically emerging oscillations and waves extend as a spatial structure, incorporating multiple areas with apparent local resonances that are both externally driven and self-sustaining [83]. The integration of a functional network of cortical areas is, to some degree, a network of peers in which activity flows bidirectionally among areas, rather than as a "waterfall" (sequential) model of processing. To illustrate, a circular afterimage generated by a bright light will be perceived to have a size appropriate to the distance of a surface on which the afterimage is "projected," greater distances producing greater perceived size. The physical size of the V1 afterimage mapping will change generally appropriately to perceived size [*667*], similar to the size mapping in dreaming [*472*], illustrating that perception is not simply a unidirectional construction, but a network phenomenon.

How are images to be perceived, remembered, and recalled as a unified whole if one presumes that image processing separates features within a set of cortical areas? This is a "binding problem (*6*)." Perception, memory, and subsequent recall require that the appropriate instantiations in the appropriate cortical areas be assembled [*668*] and "bound" together. Synchronous activity among dynamically assembled sets of specific cortical areas has been associated with attention and perception [*167, 479, 625, 639*]. This behavior, associated with the binding of cortical areas and their functions, has been modeled as synchronization [*5, 560, 622, 626, 627, 653, 669, 670*], built on the general model of coupled oscillators [*17, 187*]. Multiple coherent cortical apertures may dynamically combine to form a synchronized functional network [*502, 645, 653*], focusing on a task—perception, memory, recognition, recall—when the appropriate apertures combine, with other—disordered—apertures suppressed.

The overarching activity of the brain may be considered in terms of the large scale dynamics and stability of the various frequency bands (e.g., gamma, theta, alpha, delta) as described by Wright et al. [*671*]. Wang [*155*] has provided an extensive review of the role of cortical rhythms at multiple levels, including among cortical areas. Uhlhaas et al. [*5*] address key issues in their review of the state of knowledge of neural synchrony in cortical networks with three major themes: 1) binding of separated areas; 2) communications; and 3) communications encoding. The authors offer two cooperating solutions: i) a fixed backbone of connections; and ii) dynamically self-organizing patterns within and among cortical areas, which they focus on. They presume that dynamically self-organized patterns among cortical areas must be constrained by the anatomically fixed connections. I agree. Their resulting hypothesis is that synchrony among cortical areas is directly involved in the three issues, there being considerable evidence of synchrony among cortical areas associated with perception and experience. Uhlhaas et al. find that cortical circuits that comprise the connectome develop early in life, coincident with the rise of gamma oscillations, maturing into early adulthood. They find short distance synchronization occurs in the gamma band, with larger distances manifested in lower frequency ranges (beta, theta, and alpha). Long range synchronization can occur with short time lags [*623, 672*], perhaps with synchronization and transfer mediated through thalamocortical projections [*673*], the hippocampus playing an important role [*631, 664, 674*]. These topics are discussed elsewhere in this paper (Ch. 3). Underlying preferential organizations (motifs) may expedite binding synchronization [*440, 450, 451, 644, 647*].

As noted earlier, Crick and Koch [*627*] propose that "neural computations" are the result of phase-locked gamma oscillations creating a "searchlight" of visual awareness, essentially presaging the current models of synchrony among cortical areas as being the determinant of the internal experience [*675*]. A single frequency may not characterize all synchronizations. Huupponen et al. [*355*] provide evidence that, beyond gamma oscillations, consciousness is associated with lower frequencies (theta and alpha). El Bousanti and Destexhe [*3*] propose that spatially scaled frequency hierarchies reflect information at different levels of complexity. Conceivably there can be more than one subgraph at a time, or nested subgraphs may have differing underlying frequencies [*617*]. Watrous et al. [*676*] propose that specific frequencies demarcate particular network connectivities, finding multiple spectral components in play for spatial and episodic memory retrieval. Formation-specific patterns may be directed by specific structures, e.g., object recognition coordinated by the inferior frontal junction [*479*]. The gamma and theta bands of frequencies appear to underlie the coherence [*155, 156, 329, 625*] that is typically associated with the internal experience and working memory [*624*]. Performing spectral analyses, Tiesinga et al. [*156*] found that EEGs exhibited multiple superposed frequencies from delta to gamma ranges, with corresponding phase relationships of the phase-locked spiking. Sehatpout et al. [*677*] found long-range coherence among cortical areas during visual object processing, but at a lower beta band frequency. In contrast, Ramsey [*678*] found high gamma (80–150Hz) synchronization in different areas during speech tasks.

Definitions: graphs, subgraphs, motifs, and ensembles

The cortical system is a *graph* from which functional *subgraphs* (ensembles) are bound through the coherent, and nearly synchronous, activity of multiple coherent apertures, i.e., as VCZ nodes. Ensembles may incorporate pre-existing subgraphs—motifs—that perform as functional units, having a proclivity to form. Sporns [679], Bullmore and Sporns [324], and He and Evans [617] have used the term *graphs* from graph theory [680] to refer to a network of cortical areas as vertices (apertures) connected by edges of axonal tracts. The human connectome [98, 99, 100, 101], and its macaque counterpart CoCoMac [59], comprises a mapping of these connections, all of which are bidirectional among cortical areas, and between cortical areas and the thalamus. The degree of connection—the number of other nodes a node is connected to—is not uniform for all nodes, there being a "small world" [681] of apertures coupled though "rich club" hubs [682, 683, 684, 685, 686]; for example, the visual system has several separate tracts [687] that may be reflected in a hierarchy of synchrony frequencies [215]. There appears to be a bias toward increased connectivity via a high capacity "backbone" over efficiency in connectivity, despite increased cost [683, 688], as the level of separation between any two apertures potentially degrades communication of information between them [689]; therefore, increased connectivity increases the communication efficiency.

A *subgraph* is a connected subset of VCZ nodes within the entire cortical graph. The structure of a subgraph is constrained to the incompletely connected graph (Figure 30), which may diverge and converge, subsequently limiting the internal connectivity of the subgraph. As cortical areas are organized hierarchically [59, 81, 690, 691], subgraphs will be organized hierarchically [52, 692], a property seen in modular networks [215]. Sporns [43] refers to the role of hubs in the structure of such functional subgraphs. Although cortical processes are widely distributed, areas are recruited into a subgraph as appropriate [566]. Many apertures may operate simultaneously on different tasks, but not necessarily in completely overlapping subsets, as the various apertures' functionals may be used in multiple processes. For example, the visual system alone has some 32 identified areas, each with different operating characteristics [52], that have functionally significant connections forming dorsal and ventral "streams" with a predominance of adjacent and super-adjacent (next-door-plus-one) connections [691].

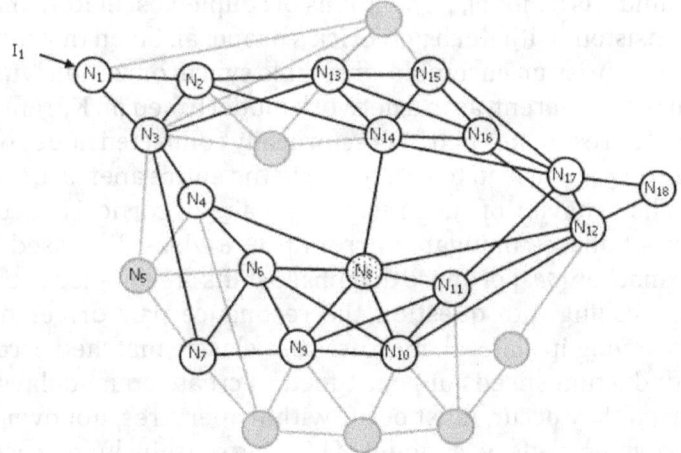

Figure 30. Functional subgraph (ensemble) of coherent (clear) aperture nodes with divergence and convergence. Non-coherent apertures in gray. MTL aperture stippled.

Sporns and colleagues [451, 683] denote *motif* as a subgraph structured around specific functions [451] or modalities, such as visospatial processing [687]. Some cortical areas are devoted to particular functions such as faces or scenes [479, 552] or phonetic structures [471]. Power et al. [97] have also identified such recurrently activated functional networks. Motifs alone do not provide a process [693]: they can be components in a larger subgraph that does execute a process. Moreno et al. [440] model motifs as latent functional subgraphs with a proclivity for synchronization based on the motif's pre-existing interconnectedness and complexity, ready for inclusion in larger subgraphs.

An ensemble is composed of a set of VCZ nodes that are coherently bound [5, 17, 693], operating on a common task. What makes an ensemble in the CNS more than a set of apertures with their connections and frequencies? The synchronization of the subgraph elements, VCZ nodes, causes them to act as a unit, an ensemble. We must ask not only what binds the apertures together, but what separates them from the rest of the apertures. An aperture must be in, or able to achieve, a VCZ node state in order to participate in the ensemble (Figure 31). On a coarser scale, an ensemble is analogous to the synchronized phase of the CM, with apertures that are synchronous (coherent) and those that are not. Different processes will spawn different, but overlapping, ensembles that serve sensory selection [460], perceptual disambiguation [167], and memory [502]. Visual processing is reflected in transient gamma-linked ensembles. Attention to a stimulus during visual perception has a concomitant increase in gamma synchrony in two widely separated cortical areas, the frontal eye field and V4 in the ventral stream [623], presumably components of an ensemble. Axmacher et al. [694] found increases in the strengths of bidirectional gamma coupling among multiple cortical regions during visual working memory with increasing perceptual load, with decreased numbers of voxels (volume units) within the areas. This decrease may indicate a decrease in apertures' entropies through organization. Ehm et al. [639] found dissolution and formation of gamma synchrony among cortical areas associated with perceptual reversals of a spatially ambiguous stimulus, indicating the concomitant dissolution and formation of ensembles necessary for a percept. It is evident that ensembles are bound through synchrony. The absence of persistent synchronized subgraphs may indicate a pathological condition such as autism [695] or schizophrenia [155].

Binding: CISs and Ks

How is an ensemble bound? As a model, populations of coupled oscillators may result in synchronized behavior [670], consistent with Koch and Crick's proposal. Given that a substantial number of apertures [52, 114] may comprise an ensemble, ensemble synchronization within a network of VCZ nodes, as oscillators, may be inherent, as in a network model based on Kuramoto's work [17]. Gollo et al. [647] propose that if a "resonance pair" of reciprocally connected nodes within an aperture set exhibits zero-lag synchrony, it will lead to resonance in the entire aperture set, whether or not the pattern is oscillatory. The behavior of such reciprocal pairs of cortical apertures is analogous to oscillations in a system of phase-conjugate mirrors [648, 649], as discussed later. One must ask, which pair? Fell and Axmacher [502] propose that phase shifts are entailed in the synchronization of different brain regions, calling into question the resonance pair driver model for the entire ensemble. Time delays among ipsilateral apertures are closely matched, irrespective of physical distances, as axonal conduction speeds appear tuned to create equal delays (0.57 ±0.37 ms, rat [412]); thus phase shifts, if they occur, must occur within apertures, not over the axons.

An ensemble is a dynamic entity. A model of binding must include processes for the recruitment and exclusion of apertures, resolution of purpose, and transition out of synchronization. I am proposing a model for forming an ensemble that can evolve in membership, function, and result through the exchange of information instantiations (CISs) among apertures that are near, or have achieved, VCZ node status. Their coherence status (see section iii, above) acts as a filter for both CIS exchanges, and for the aperture subsequently being recruited into, or excluded from, an ensemble. Ensemble development is progressive, co-occurring with the cohering of the apertures. The flowchart in Figure 31 illustrates the overall ensemble formation from VCZ nodes, N, that are recruited from a larger population. An aperture, A(i), receiving a CIS(ji) from another aperture, A(j), may be, or may become, coherent in the process (Figure 28), thus becoming a VCZ node, N(i), able to participate in an ensemble. If it does not become coherent, it will not participate. If there is more than one communicating VCZ node, e.g., N(i) and N(j), indicated as Sum > 1, then the nodes can form an ensemble. Given the connectivity required for CIS exchanges, a VCZ node must be connected to at least one member of an ensemble to participate in it. At a gross level, a subgraph is synchronous (gamma frequency band)—at least initially—and is coherent with local phase differences within apertures and CISs. Unanswered is the question of whether an aperture can be coherent (a VCZ node) in the presence of other VCZ nodes and not be part of an ensemble.

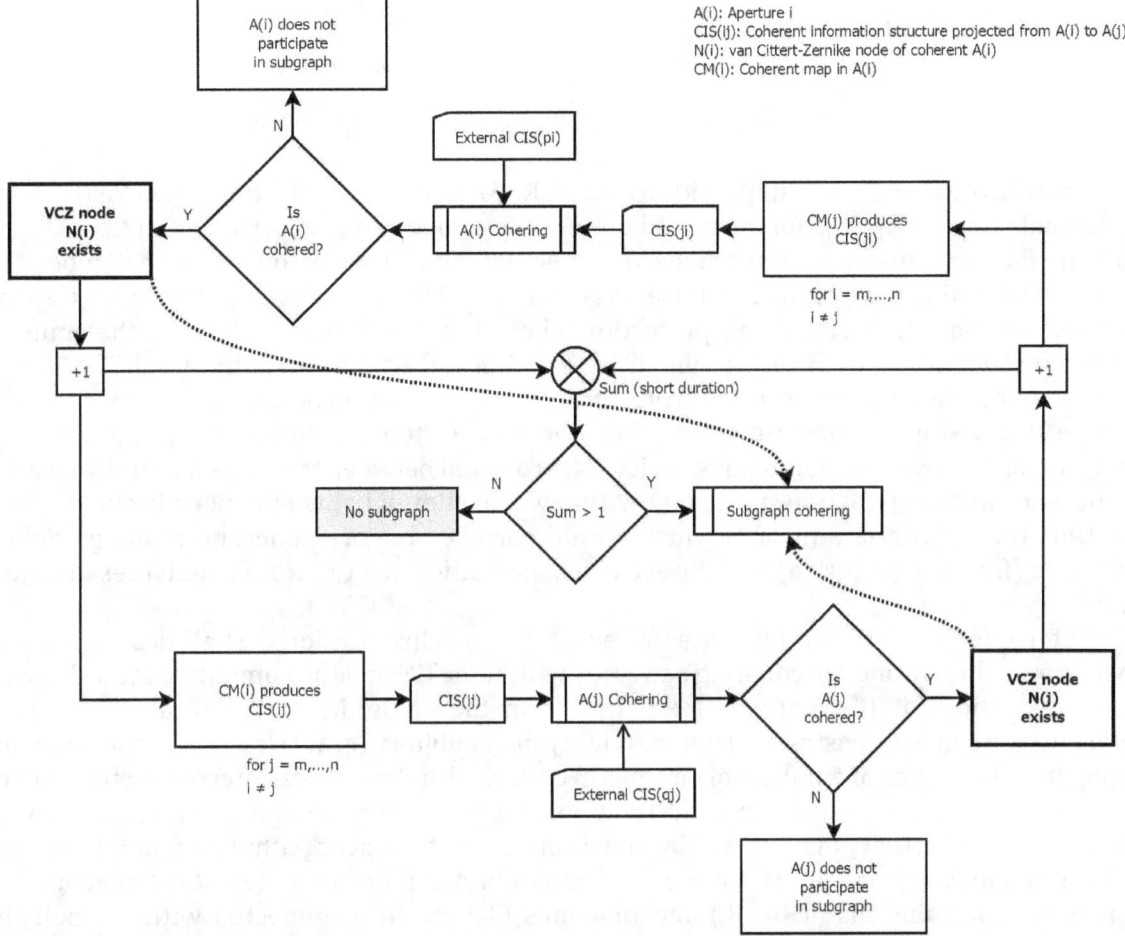

Figure 31. Logic of ensemble formation for i = 1,…,n, j = 1,…,n apertures, i ≠ j.

Composite node

It is useful to introduce a *composite node* (CN, Figure 33), an abstraction that is an outgrowth of the VCZ node, that enables the ensemble cohering process (Figure 34), incorporating both vertical (radial) and lateral communications that support bi- or multi-directional information (instantiation) exchanges among VCZ nodes. A CN is composed of radially corresponding layers of a cortical aperture (A_i), a sector of the thalamic reticular nucleus (TRN_i), and a thalamic nucleus (T_i). The three layers can become coherent [83]. A coherent aperture is a VCZ node. There is coherence within a composite node; thus both an aperture and a composite node can be a node in a coherent network or graph framework. The CN is a shorthand description of the relationships of structures and may have not physical reality as a unit. The thalamus and TRN participate in the connections in the cortical graph, hence the utility of the CN.

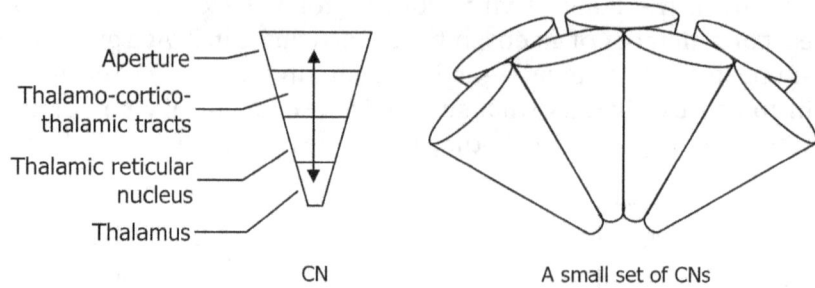

Figure 32. Composite node (CN) structure and relationships.

There are active relationships within the CN. Recurrent communication between the cortex and the thalamus is required for sustained temporally structured cortical activity [*482, 632*]. This recurrent flow modulates and sharpens thalamic activity. Transfer over the thalamocortical projections is modulated by synaptic noise levels generated in the cortex [*397*], providing feedback to the sensory input. Corticothalamic projections sharpen the receptive fields in the thalamus and increase the effective gain of sensory signals both within and across modalities [*696*]. The TRN, on the other hand, receives projections from the cortex and subsequently inhibits the thalamic response [*80*] and modulates attention [*697*]. Coherent (often synchronous) retinally sourced activity, delimited by the ordered phase of its CM, projected between the thalamus and V1, retains its coherence in both [*308*], the synchrony within visual stimuli being enhanced in the thalamus [*357*]; thus the appropriate thalamic nucleus and cortical area are coherent, although Fourier transforms (frequency spectra) of their activities may differ due to internal activities and other inputs.

Before delving into the mysteries of ensemble synchronization, I shall describe the CN activity flows, illustrating the cohering activities within the CN and the communications between CNs. In the schematic of Figure 33 (extended from the model by Ferrarelli & Tononi [*697*]) (meta-)neuron symbols illustrate excitatory (clear) and inhibitory (gray) classes, not specific neural components. The ratios of numbers of neurons in each level and their local interconnections are not illustrated, the intent being to describe the relationships among layers within and among composite nodes. Patterns of activity may be considered as *units* that flow over pathways among elements. Units can embody aspects of instantiated information. A composite node's (CN_i) aperture (A_j), thalamic reticular nucleus (TRN, R_i) and thalamus (T_i) are interconnected with a topological registration (2,3,6,7 in Figure 33) [*80, 697, 698, 699, 700*].

The CN flowchart in Figure 33 illustrates functionality but should not be considered as literal. Although unit flows in a composite node must be described in a linear fashion, units flow along multiple paths simultaneously. The composite node has cross-inhibited loops. The thalamus maintains the coherence and synchrony in the input, I_1, (e.g., I_{1a}, 1 in Figure 33) which is inherent in sensory input, e.g., changes in an image on the retina through saccades and movement. The units flow into thalamic "relay" excitatory elements (LGN) that project (2) excitatory units of spatio-temporal information to topologically equivalent excitatory elements in the corresponding cortical aperture [*701*], A_1. The thalamic elements also send collateral projections (4) to inhibitory elements in the TRN as the tracts pass through it [*697*]. These inhibitory elements project back to the thalamus (7) in the locale of the source elements with a spread function, creating lateral inhibition around them, a cross-inhibition circuit familiar in the CNS, causing the input from I_1 to have enhanced edges—the image is sharpened, or noise is suppressed [696], a TRN-thalamic sharpening suggested by Crick [702] and Turk-Browne [566]. A_1 projects topologically registered excitatory units back to the thalamic source element (3) [79, 690] (with an effective point spread function), presumably having performed some local function on the incoming unit [703]. Collaterals from the cortico-thalamic projection (5) [697] project to the inhibitory elements in the TRN [704], directly inhibiting the source element (6).

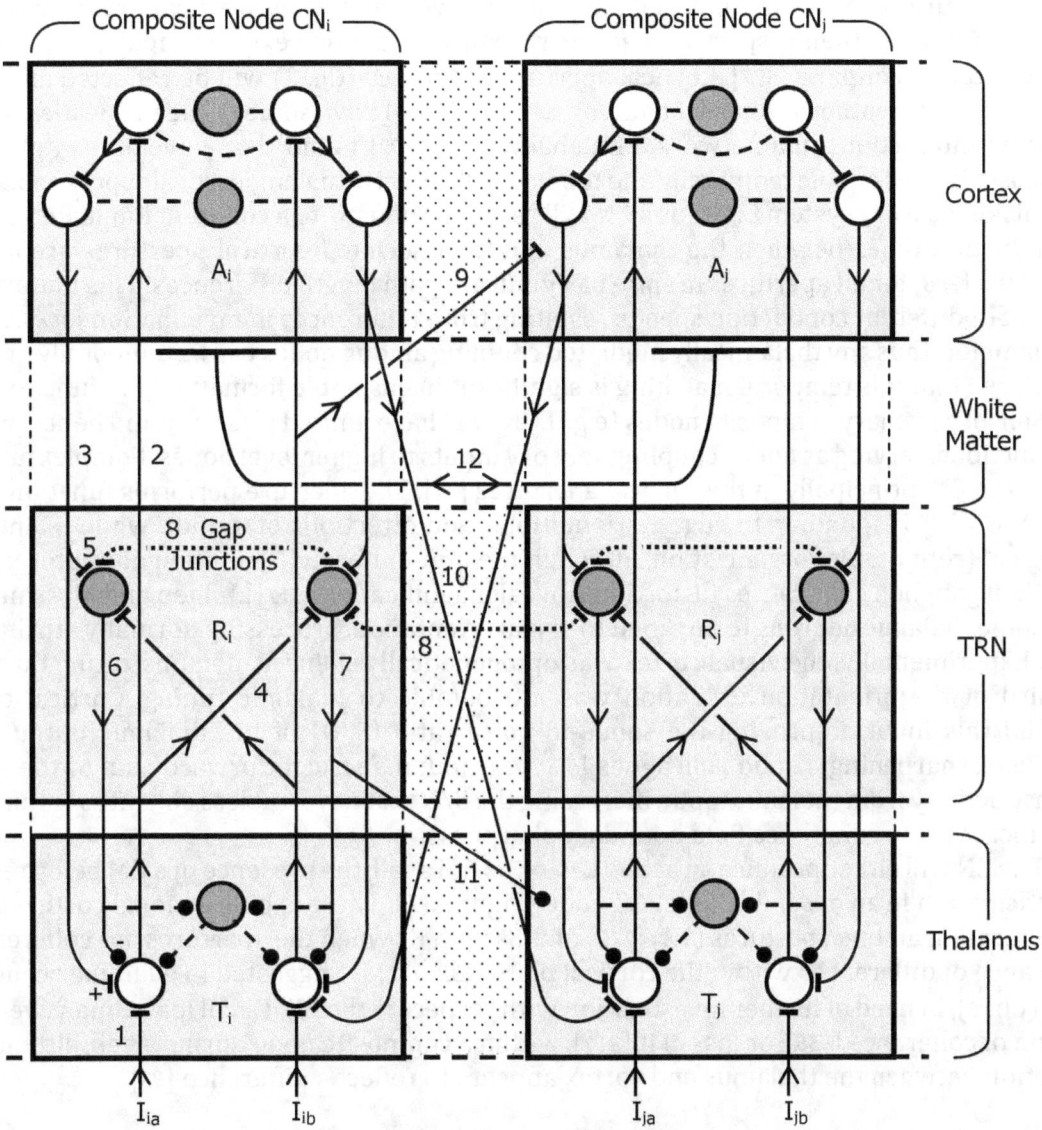

Figure 33. Composite node schematic with unit flows. White = excitatory, gray = inhibitory neurons. Cortical structure is suggestive. CNi & CNj are composite nodes. Ai and Aj are cohesive (potentially coherent) apertures. Ri and Rj are sectors in the TRN. Ti and Tj are nuclei in the thalamus. Bidirectional excitation between apertures through the white matter is indicated. Ii and Ij are peripheral inputs. Higher order thalamic nuclei will not have such inputs. Numbers label information flows. See text. Extended from Ferrarelli and Tononi (2011) [697].

The aperture's projections to the thalamus are both directly excitatory and indirectly inhibitory [213]. Inhibitory interneurons, diffusely inhibited from the TRN [80, 698, 705], further modulate the thalamic elements and their coupling. The elements in the TRN are all inhibitory (GABAergic) but mutually excite each other weakly through fast dendritic electrical gap junctions (8) [248, 706]. This fast weak coupling enhances the potential for synchronous activity within a sector of the TRN as a network [89] (see review paper by Arenas et al. [17] for relevant network analyses). This is further supported by Destexhe and colleagues' [79, 358] simulations and experimental evidence of coherent oscillations attributable to cortico-thalamic loops. The corresponding cortical aperture, TRN sector, and thalamic nucleus will tend to become coherent (and nearly synchronous) across their respective spaces in response to coherent external inputs, internal states [84], and internal coupling [627]. Cortical aperture coherence (Ch. 3) will be reflected among the three structures creating a coherent composite node. It will have some of the properties of a VCZ node in its interactions in a network. Minlebaev et al. [707] found that, developmentally, early gamma oscillations enable temporally and topologically precise thalamocortical synchronization in a rat whisker sensory system CN; thus a CN will respond robustly to a coherent input. Remarkably, the conduction times between the thalamus and its associated cortical apertures are uniform within 2 ms [708] for all apertures in spite of significant path length differences. This uniformity is the result of different conduction speeds resulting from differences in myelination [308, 709] and axon diameter; thus any thalamically mediated coupling among nodes will be temporally matched. We shall see that this temporal matching is significant in ensemble formation and function.

Sensory primary composite nodes (e.g., LGN–V1) have limited plasticity and hence perform basic functions, serving as nodes coupling sensory inputs to higher level nodes. Complex functions occur in the CN, principally in the cortical apertures [77]. The aperture performs functions on its input(s) in V1,[18] responding to edges, orientations, and directions of motion while maintaining topological (retinotopic) organization, although the non-uniform distribution of photoreceptors and retinal ganglion cells causes distortion (foveal magnification [511]).[19] Lien and Massimo [710] found some thalamic neurons to be tuned to visual orientation, processing normally attributed to cortex. Experimentally, the visual cortex was optogenetically silenced, demonstrating that about one third of the orientation excitation was attributable to thalamic tuning. Cortical circuits amplified this input, improving the signal-to-noise ratio (S/N) of the thalamic outputs with subsequent sharpening. Li and colleagues [550, 711] found a straightforward gain of the feature response activity from thalamic input, increasing the S/N that nevertheless enhances features that would increase the coherence field boundary sharpness.

The CN will either produce an active CM, or no CM at all, the presence of a CM being required for participation in an ensemble as a VCZ node (Figure 31). There is bidirectional cortico-cortico communication among apertures (12) [712] of CISs with Ks when the apertures are coherent. The significance of differences within the cortical plate [132, 377] is suggested in a laminar coherence model (Ch. 3), in need of further investigation with respect to the CN. I shall leave unanswered the question of coherence [358], or loss of it [317], within a composite node during sleep, although the interaction between the thalamus and cortex appears to reflect a difference [83].

CNs in ensembles

I have restricted this discussion to the CNs, omitting the systems governing affect and behavior, largely of brainstem and limbic system origin. The cortical activity is strongly influenced (perhaps driven) by these systems. They are important components, omitted here, and must be studied in the future. The CN helps in explaining ensemble formation. The central role of the thalamus, hence the CN, is a recurrent theme [358]. Crick [702] proposed that attention is perhaps more focused than

---

[18] Intensity and color perception are omitted here.
[19] Probably a conformal mapping from the retina to V1, which may be of computational utility.

emergent, an internal searchlight of focused attention (Treisman and Gelade [713]) of synchronized neurons [627] controlled by the thalamic reticular nucleus (TRN), directed by the thalamus. Synchronization of cortical areas characterizes binding and subsequent intercortical communication [1]. A CN has a characteristic period (and thus a natural frequency) and hence may be considered an oscillator. The TRN [89, 634] and/or thalamus [569] appear to be the principal sites of the characteristic period, subject to variation, although the complete thalamocortical loop is implicated [79, 353]. A CN is typically internally damped such that it does not oscillate in isolation. Although CNs have interconnections at all levels, aperture coherence requires the thalamus [83, 317, 714]. Disabling the thalamus causes immediate loss of consciousness [481]. This is not proof of its necessity in synchrony, but is consistent with this role. More investigation is needed in this area. The cross coupling from the cortical aperture in a composite node CN_i (Figure 33) may map to some equivalent in the thalamic node T_j of a composite node CN_j. As unlikely as it may seem that such a mapping might occur, this developmental targeting and pruning is not uncommon in the developing CNS [715], including the thalamus [716]. Such a mapping is essentially fixed with respect to the time scale of synchronization. Information instantiations may be limited in the exchanges, which serve mainly as ensemble control. A sparse set of interconnections may support synchrony. Subsets of CISs and Ks may project from one aperture to other apertures if the apertures have cross-connections through the thalamic-TRN system (Figure 33). Not all apertures must necessarily be cross connected, but sufficient cross connections must exist within the synchronizing network of CNs for a CN to participate.

Ensemble formation

Ensemble synchrony occurs though communications at multiple levels that can be conceptualized as communication between coherent—or cohering—apertures, simplified in Figure 34. One would expect the connections between CNs to be parallel at all levels: A_i to A_j, TRN_i to TRN_j and T_i to T_j. This occurs. An aperture must be, or become, cohered to be included in an ensemble. There are two interacting flows, within and between CNs. The specific mechanisms have been discussed previously (e.g., Figure 33). Subgraph formation is illustrated with two CNs, CN_i and CN_j, representing a larger number of sparsely connected CNs. The model has similarities with the coupled thalamocortical model proposed by Drover et al. [88], although binding synchronization is not included in their work.

The CN incorporates a cortico-thalamo-cortical loop. An input to an aperture, e.g., I_i, results in the projection of a CIS_I and its K_i' from a thalamic nucleus (T_i) to the aperture (A_i), passing through the TRN (TRN_i), where collaterals stimulate inhibitory neurons. The inhibitory neurons, receiving multiple inputs, modify CIS_I. CIS_I will contribute to the cohering of A_i, as described previously. The aperture projects some form of activity, here represented as a K_i, back to T_i, modifying its behavior. As K_i passes through the TRN_i it projects over collaterals to the inhibitory neurons. T_i projects a new K_i' and CSI_I back to A_i, closing the loop. Simultaneous with the flows within a CN, there are exchanges between A_i and A_j that may drive or impede coherence in each aperture (Figure 28). A_i projects a CIS_i with its K_i appropriate to the target A_j over axons through synapses, bearing in mind that K_i is an abstraction composed of the outputs of a consistent subset of neurons or minicolumns. Given the topological registration[20] between directly connected apertures [511], a stable subset of points in each (probably selected developmentally) for the Ks provides a surrogate for the correlation between apertures. Each of the points in K has some degree of indeterminacy in time and occurrence; thus each K is a bit "fuzzy" (noisy). This fuzziness enables progression toward a correlation maximization, as exact matches are not required during the resolution.

Apertures exchange information structures if they are both coherent (gating filter, Figure 29). Activity is also exchanged between thalamic nuclei in a reduced form, here denoted as K. These

---

[20] An isomorphism.

exchanges cross-modulate the CNs. Although there are delays, the exchange can raise the cohered apertures toward synchronization in a manner analogous to aperture synchronization described in Ch. 3. The direct electrical coupling of the TRN inhibitory neurons promotes the synchronization within and among CNs. All neurons in the TRN are inhibitory, sharpening the responses of neurons in the thalamus (696, 702) and perform sensory gating (697), thus contributing to the timing in the corticothalamic loop as discussed above. A synchronized aperture will project synchronous activity back through the TRN to the thalamus, with collateral excitatory synapses on the TRN's inhibitory neurons, causing their activity to reflect the synchronization of the projection, a synchronization enhanced by the gap junction coupling, further reflected in the synchronization of gating in the thalamus. The gap junction coupling extends beyond the CN to adjacent CNs. If another CN is also coherent, internally synchronized, the gap junctions will promote synchrony between the CNs. This will be reflected in the development of the synchronization of the cortical apertures.

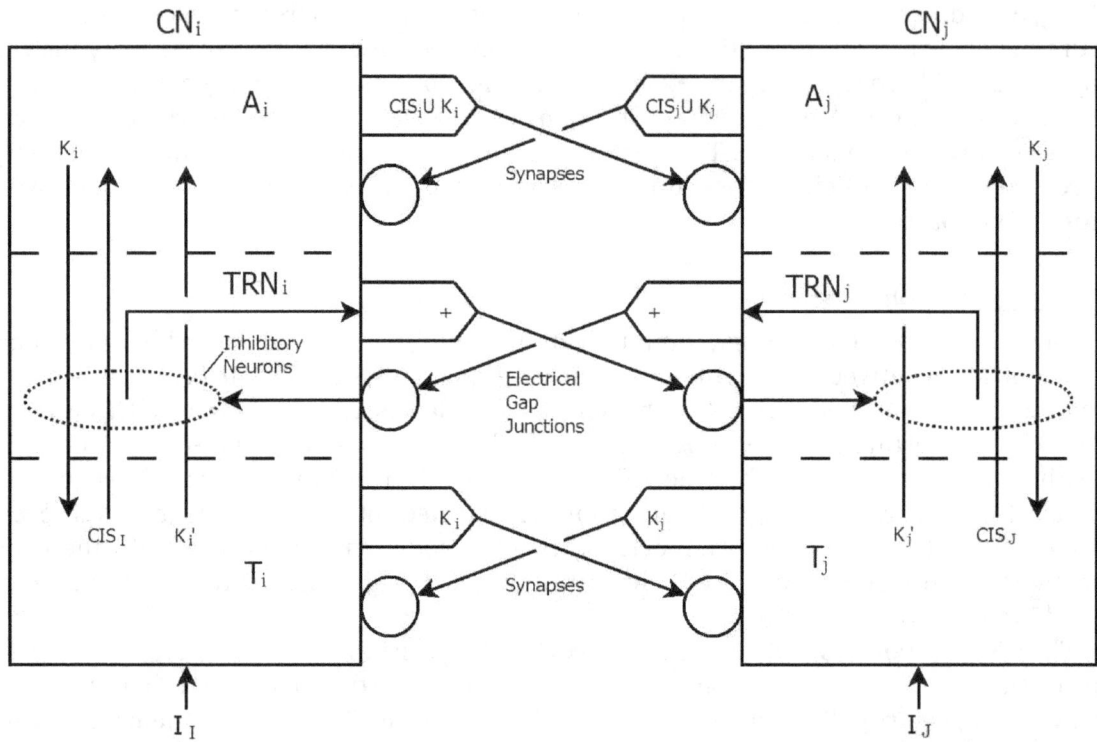

Figure 34. Inter-composite node synchronizing. i ≠ j. I = sensory input. CSI and K' are derived from K and I. See text for TRN electrical gap junction coupling. See Figure 33.

As an aperture progresses toward a steady state condition, if it maintains its VCZ state, its orderliness will stabilize, reflected in a decrease in change in orderliness, with a lessening rate of decrease in entropy for that aperture. These processes are ongoing, with apertures that cohere becoming VCZ nodes. Non-VCZ apertures diverge from potential synchronization as the mismatches increase as the natural keys diverge. The processes of comodulation, discussed below, will contribute to this convergent/divergent participation dichotomy, converging toward a net minimal entropy state for the ensemble with each CN retaining some disorderliness.

Three modes of synchronization

Composite nodes must dynamically assemble, based on a commonality of task, into a synchronous transient network [5], an ensemble of coherent apertures. How is this accomplished? Composite nodes do not form a regular lattice, nor do they form a completely connected network. Most cortical apertures are not connected to many other apertures, a network of low degree. Some apertures (nodes) have higher degrees of connection forming clusters [717] or motifs [451, 718]. Referring to Figure 33, there are several forms of neural coupling among composite nodes (8, 9, 10, 11, 12) [248, 673, 706, 719]. Thalamic sectors of CNs are coupled; TRN sectors are coupled. Thus the T-TRN complex of a CN can be considered a unit that is coupled to other T-TRN units, exchanging little information. Aperture synchronization is one of the proposed functions of the T-TRN system. Some researchers have focused attention on the TRN alone. All areas of the cortex and thalamus are directly connected to the thalamic reticular nucleus (TRN) [720]. Drover et al. [88] developed a model with the TRN synchronizing separated cortical areas. Min [80] has proposed that the TRN plays a central role in the thalamocortical synchronization necessary for conscious awareness, filling both synchronizing and filtering roles. He includes the cross-modal topographic mapping in the TRN as "associated with cross-modal or unitary consciousness." This model may overload the capacity of the TRN. Saalmann et al. [721] proposed the pulvinar in the thalamus alone as the regulator of cortical synchrony. Crick [702] modeled the T-TRN system as synchronizing cortical areas.

Three modes of synchronization of apertures into an ensemble are considered here (Figure 35), the modes flowing one into another in stages, involving the apertures' entire CNs.

1) **Intermediated synchronization**: cortical apertures form intercommunicating networks progressing from the synchronous exchange of coherent activity through an intermediate structure, then

2) **Common drive synchronization**: exchange CISs while being driven by a common structure, and subsequently

3) **Peer bidirectional synchronization**: synchronize with direct exchanges among apertures.

Each mode supports the exchange of information instantiations, at progressively more efficient levels.

The synchronization modes are consistent with the application of radiant energy concepts. Ensemble synchronicity is composed of apertures exchanging information bidirectionally. Given the coherence of the VCZ nodes, the coherence of the information they exchange, and the need for synchronicity among nodes in order to exchange information, models based on solid state lasers provide useful analogies. Apertures that become coherent before or during the subgraph operation are VCZ nodes, equivalent to solid state lasers. A solid state laser produces a monochromatic light beam with a moderately long coherence path length, with individual photons locked together coherently. The output amplitude is not constant, but varies randomly as noise. The noise profile constitutes the equivalent of information. The exchange of noisy (i. e., information laden) beams between two lasers is analogous to one cortical aperture projecting a CIS, the activity of a CM, to another aperture that can respond with another CIS, an active CM's coherent output. Two similar lasers optically coupled directly to each other by virtue of each laser's coherence can have the same noise profiles, but normally one will lead the other, with a delay reflecting the light's propagation time. How can multiple nodes (lasers or apertures) become coupled with zero-lag synchrony, i.e., with coincident noise profiles? As described above, a multinodal network in the CNS cannot have significant time lags among the nodes if the network is to achieve isosynchrony in spite of significant synaptic and conduction delays. The three modes and their laser models are discussed further in Supplemental Materials.

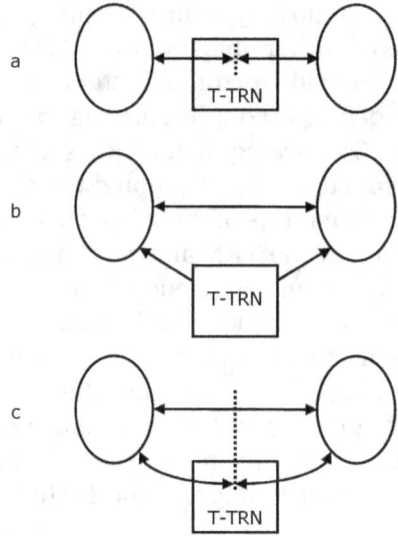

Figure 35. Three synchronizing modes: a) thalamic-thalamic reticular nucleus (T-TRN) as *intermediary* between apertures, b) two apertures synched by a *common* T-TRN source, c) partial reflection of activity between apertures by T-TRN in *peer bidirectional* synchronization.

Intermediated synchronization

In an intermediated mode (Figure 35, a), structured activity passes among VCZ nodes through the thalamus that serves as an intermediating node, with coupling among thalamic nuclei and TRN (T-TRN). Information reflecting the apertures' modulations will project to the thalamus, modifying its relaying properties [696]. The thalamus relays activity between aperture VCZ nodes, achieving ensemble synchrony through the synchrony of long-range thalamocortical oscillations resulting from cortico-thalamo-cortical loops [79, 353, 358]. Uhlhaas et al. [5] noted that Fischer et al. [633] demonstrated an equivalent three-laser system (see Supplemental Materials) in which the third common cohering laser is an intermediary between the other two lasers, which become synchronous. The T-TRN can be modeled as cross-coupled lasers of each CN, functioning essentially as a coupling node. The two outer lasers (apertures) do not communicate directly, passing their outputs through the intermediate laser (T-TRN equivalent) via a beam splitter. The two outer laser outputs interact within the intermediate laser to become synchronous. The path lengths, hence the delays, between lasers must be closely matched for this behavior to occur. The thalamocortical delay is quite consistent across the cortex [709]. The intermediated state may be preliminary to a common drive state.

Common drive synchronization

Intermediated synchronization can progress to a second mode with direct synchronous bidirectional interchange among VCZ nodes, synchrony being maintained by a weak common T-TRN drive. Apertures may become, or remain, synchronous by receiving drives from a common source (Figure 35, b), e.g., the T-TRN system, which is an alternative, but not mutually exclusive, interpretation of thalamocortical synchrony [79, 353, 358], the difference being the apertures' direct communication with each other in a common drive model. The source of the synchronizing drive (thalamus) and the paths of information flows are separate, synchrony being required for information transfer. A common drive can be modeled as the synchrony of two communicating outer lasers driven by a weak common source. Synchronous outer lasers driven by a weak common source can simultaneously cross-communicate different information directly, as demonstrated by Zhou and Roy [722]. Similarly, the model of Santhanam and Arora [723] demonstrated synchronization of two coupled lattices driven by a mutually coupled third map lattice.

Peer bidirectional synchronization

Intermediated or common drive synchronization may progress to a third mode of a strongly cohered ensemble wherein all VCZ node apertures are cohered, with less involvement of the more central structure as an ensemble synchronizer. The latter state may be infrequently encountered, as new activity is constantly flowing into the system, either as sensory inputs or through the recruitment or intrusion of other apertures. Peer-to-peer laser synchronization has been demonstrated. If two [700, 724] or three [725] lasers are directly coupled through a partially silvered mirror, simulating a thick dual phase conjugate mirror model (Ch. 5), they can become synchronous, while transmitting information bidirectionally. These systems are tolerant of differing communication delays. Cortical apertures may communicate directly, peer-to-peer (Figure 35, c), if synchronized. Sadeghi and Valizadeh [726] modeled synchronization among directly coupled neural oscillators in the face of multiple sources of delay inhomogeneity. The combined effect overcomes the local differences, much as slight differences in the natural frequencies of oscillators will be overcome through coupling in a synchronous network [11]. If one considers the entire composite node as a unit, a peer-to-peer model is applicable, as coupling between (and among) coherent (VCZ) composite nodes occurs over the entire CN structure, which is appealing, as it effectively encompasses the intermediate and common drive models. The limitation of a peer-to-peer model is bringing nodes to VCZ status, with subsequent recruitment and rejection. The other models serve as necessary preliminaries.

Synchronization mode progression

It is possible that there are three modes of ensemble synchronization: the first develops the synchronization/coherence through an intermediary; the second maintains coherence with a common synchronizer, permitting the direct exchange of information between (among) coherent aperture nodes; the third operates as a self-synchronized network. The establishment and functions of a cortical network are dynamic, incorporating multiple synchronization modes that align aperture phases of VCZ nodes. This model, particularly its extension to a higher number of nodes, needs further investigation; however a basic framework emerges. The intermediating system (drawn from Fischer et al. [633]) synchronizes the two aperture nodes by exchanging information through a common thalamic-TRN node. The common synchronizer (analogous to the work of Zhou and Roy [722]) maintains that coherence through coherent signals projected to each of the apertures. Sufficiently synchronized VCZ nodes can maintain an ensemble through peer-to-peer coupling [700, 724, 725]. Remarkably, but consistent with the intermediating (Fisher et al.) and common drive (Zhou and Roy) modes, there can be a zero time lag in neuronal synchrony between cortical areas despite conduction delays [700]. Multiple cortical areas, once synchronized, will have a propensity to maintain synchrony in a peer mode. As noted, the composite node interactions blur the differences among these modes, resulting in a fluid system with coexistent modes, presumably incorporating entropy minimization.

If two [724] or three [725] node peer modes with no synchronizing element(s) are sufficient, what is the applicability of the intermediate [633] or common [722] synchronizer modes? As described above, the work of Destexhe et al. [358] and Contreras et al. [79] and colleagues shows that the interaction of the cortex and thalamus is required for the control of synchronizing oscillations. All three modes may operate within the composite node model, so separation of the three may be a distinction that represents the evolution of ensemble synchronization. The composite node model illustrates significant cross-coupling among the composite nodes, so the differences between synchronizer and peer forms are diminished. This is consistent with models

[*80, 88, 727*] incorporating the T-TRN as key in integrating the cortical areas into consciousness. The intermediate and common driver synchronizing models differ from the peer synchronization models in the role of the thalamus in the peer models. Alkire et al. [*481*] found that the thalamus must operate for an animal to maintain consciousness; thus the CN model that combines several of the modes is an appropriate conceptual simplification.

v) Bidirectional Exchanges Regulate the Composition of Ensembles.

Introduction

An ensemble is a network structure with participants. Which apertures participate in an ensemble? To accomplish something, an ensemble must dynamically recruit the appropriate apertures while excluding the rest. Distinct patterns of connections within the cortical graph—the connectome—form for different tasks [*728*]. In the visual system the ventral pathways are associated with object processing. Several paths within the dorsal pathways are associated with visospatial processing, with some sub-specializations for spatial working memory, visually guided action, and navigation [*687*]. Specific areas in the cortex are associated with specific tasks, for example, the fusiform face area (FFA) responds to faces whereas, in contrast to the FFA, the parahippocampal area differentially responds to houses [*479*]. Sensory inputs modify ongoing cortical activity [*76*]. Changes in visual tasks result in changes in the areas in the visual cortex that are recruited, removing those not used [*112*]. Phase shifting of the activity (presumably of the CISs) alone in a cortical area can reflect stimulus feature selection [*486*]. A perception and a recall may have ensembles that share many of the same apertures but are initiated by different apertures.

An ensemble is formed from coherent apertures (VCZ nodes), excluding non-coherent apertures (Figure 30). Ensemble composition is controlled by the ability of nodes to achieve and maintain VCZ node states (Figure 28) while maximizing K (Figure 34) correlations. Two communicating apertures must have moderately correlated communicated Ks (as subsamples of CISs) for both to be in VCZ node states (Figure 31). Uncorrelated Ks would disorder corresponding CMs because, although being but a small subset of the entire aperture's population of elements, they have topological registration between apertures, playing significant roles for all apertures.

Ensemble composition control

Relevant VCZ nodes form ensembles in response to input activity. An ensemble has incompletely connected apertures, although no aperture is an island. Such a network, with a low degree of connectivity, can self-cohere, forming an ensemble (see review by Arenas et al. [*17*]). The limited connectivity within the cortical graph naturally connects only related nodes [*683*], facilitating node recruitment [*324, 617, 688*] into a corresponding subgraph. From the standpoint of early stage sensory processing, initially a CIS (e.g., sensory input or a higher level projection directing attention) will be projected to a single aperture, typically through the thalamus. CISs and Ks will be projected from this initial aperture to other apertures (through one or more corticocortical projections) at about the same time limited corticothalamic projections from the initial apertures are spread by intermediating thalamocortical projections to multiple potential target apertures.

At each level in a hierarchy of cortical areas[21] there will be a spread of projections within the aperture due to the divergence of axonal arborization and the convergence from multiple axons onto dendritic trees (Ch. 5), analogous to a point-spread function [*138, 485, 512, 522*], producing the receptive fields of local functions. Since ensembles have limited connectivity, K relationships will have decreasing alignments with increasing node separation, as divergences and convergences will accumulate, and functions will differ among apertures, although some specific tasks, such as

---

[21] Hierarchy presumes more abstraction the farther from the sensory input, not a unidirectional information flow.

recognition of a face, may have similar Ks among a common subset of preferentially connected apertures (a motif [451]). The functionality or CMs of each aperture in a dynamically assembled ensemble must relate to a common task, either present or developing; otherwise internal experiences might be meaningless—or erroneous. The common task serves as a filter via the Ks.

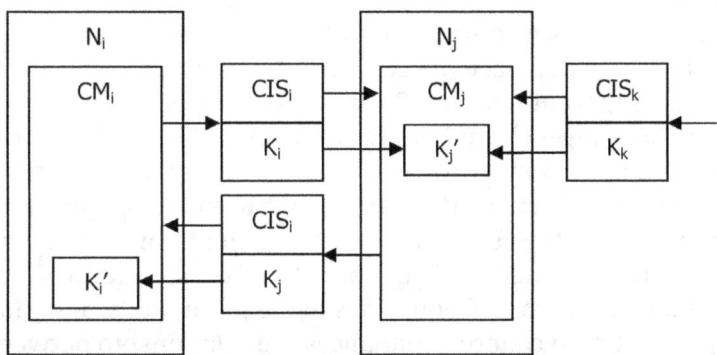

Figure 36. Bidirectional exchange of CISs and Ks simplified. CISs may modulate CMs. Correspondences between Ks tested. See Figure 27.

In the process of forming an ensemble (Figure 31), an aperture attempts to exchange CISs and Ks that arise from active CMs (Figure 36) with other apertures, which may cause other apertures to cohere (Figure 28), their subsequently being recruited into the ensemble. The internal processes of CNs interact at multiple cortico-thalamic levels (Figure 34) exchanging CISs and Ks while synchronizing through the CN synaptic, gap junction, and electrotonic couplings (Figure 33). Cross coupling enhances common features (see below) in the apertures. The enhancement will occur in the apertures—and thalamic nuclei—as the apertures cohere. An aperture that produces CMs in response to the combination of the received CISs and Ks, and from the thalamocortical projection (with T-TRN driving toward synchronization), has a VCZ node status, sending CISs to other apertures in simultaneous bidirectional exchanges, within the graph connection constraints. The originating aperture(s) will continue to project evolving CISs and Ks, modulating VCZ statuses. A VCZ node's coherence is not separable from the ensemble's activity, being coupled to it, so interference within an aperture among incoming projections (constructive or destructive) may play a role in the aperture's coherence. Failure to form an active CM is reflected in the failure of an aperture to cohere, with a concomitant failure to have VCZ node status, with its subsequent exclusion from the ensemble, effectively becoming invisible. A CN with a random pattern of activity will self-inhibit through its corticothalamic cross inhibition, reducing its overall output of structured activity back to the corresponding cortical apertures. The decreased structured activity (in this case, noise: see above) reflected back to a non-cohered aperture will lower its proximity to a critical threshold (critical point) for internal coherence, further removing it from possible participation in an ensemble. It is filtered out. As is the case within the cohering aperture, ensemble coherence creates a greater sensitivity to change in nodal timing (phase) than the effective period of the underlying quasi-oscillator [17, 160, 187]. An increasing phase discrepancy increases the chance that a node will be excluded from the ensemble. The evolution of synchronization produces a refinement of the nodes and an increase in K correlations, resulting in a lower ensemble entropy.

As a CM becomes more complex, the natural key less accurately represents the CM and CIS due to sampling error. As an example of recruitment involving Ks, an invariant will be simpler, and more orderly, than its immediate instance; therefore the CM of the invariant will be less complex with a more generally representative K. A particular face is more complex than a generalized facial structure. A complex CIS, projected to an aperture with an invariant CM, may contain the invariant;

therefore a substantial portion of the invariant CM's K may be included as a subset of the CIS's K, preferentially activating the invariant CM, raising or maintaining the aperture's activity to, or above, the critical point, effectively recruiting it into the ensemble.

The dual roles of noise

Noise has several roles in the CNS. As discussed in Ch. 3, noise can contribute to aperture cohering; yet here, noise is considered to decrease the coherence within an aperture. Different sources and targets may resolve this apparent conflict. Raising the level of spontaneous activity increases coupling in an aperture, bringing it closer to the critical point. With inadequate inhibition, noise can be destructive, allowing a large population of elements, and of apertures, to become recruited without structure, e.g., epilepsy. Increased activity due to brainstem inputs [349] would be diffuse, probably not concentrated on the functions' receptive fields but at other levels or in distal dendrites. These inputs are not coherent. They serve to raise the general synaptic activity, hence coupling, and also increase the natural frequencies of the pyramidals, and thus the LFP pumping, both increasing the propensity to cohere. It enhances the relationship between the thalamus and the cortex, sharpening CMs in the process. The output in one aperture will be projected onto the receptive fields of the target aperture's functions. Noisy (disordered) inputs will suppress coherence in an aperture's functions' responses. Noise perturbations of CISs and Ks can both impede the synchronization process and give it more power as it expands the range of CMs, and hence of apertures, that may participate in an ensemble while increasing the range of synchronization frequencies. Noise supports homeostatic drive by providing small perturbations needed to progressively test for energy, and thus entropy, minimums.

vi) An Ensemble Seeks a Low Entropy State.

Introduction

The CNS will strive to reduce energy expenditures while increasing specific receptive sensitivities by maintaining activity in its components, a dynamic balance. I propose that the operation of the CNS can be fruitfully considered from the standpoint of entropy, which reflects both complexity (information entropy) and energy expenditures (thermodynamic entropy). One must consider uniformity in this description. The process of entropy minimization has similarities with annealing, although annealing seeks a maximally uniform state of elements in equilibrium. Wright [729], using an annealing model, considers the uniformity of the cortical state to be reflected in the presence and strength of the LFP (or EEG) in the gamma band. Entropy is considered relative to elements' activities. I find this description misses the significance of the differences among the apertures as keys to the information instantiations. These differences reflect different functionals within each aperture. I propose that one not consider element activities, but orderliness, as describing entropy. Disordered elements are excluded from entropy consideration as noise, which is somewhat counterintuitive (Ch. 4). A low entropy system has a quasi-stable equilibrium reflected in a quasi-stable orderly subset of elements.

Active information is instantiated within CMs distributed among coherent apertures in an ensemble. An ensemble forms dynamically, recruiting apertures as they cohere in response to projections received. The ensemble will seek to resolve to a minimum joint entropy (hence energy) through the simultaneous mutual reduction of aperture entropies in response to input projections, without regard to the information content, while retaining responsiveness to internal and external changes. The more removed from sensory inputs and motor outputs, the more undecipherable the information instantiations, presumably embodying invariants and abstractions at higher levels. The states of the ensemble's component apertures are resolved into a combined state (a consensus), as apertures comodulate each other, minimizing joint entropy by increasing mutual information [730].

With new information constantly flowing into the system, entropy minimization is a constant process, potentially recruiting apertures into, or excluding them from, the ensemble. There are two processes for entropy minimization: within and among apertures. I shall first address entropy minimization within an aperture.

Entropy minimization within an aperture

The efficiency of a system in incorporating information may be characterized by the minimum information entropy in a structure necessary to convey a specific amount of information [731] relative to the maximum potential information entropy of a system. A maximum entropy system of complete disorganization (disorder) and uncorrelated activity has information indistinguishable from noise. Within an aperture, a decrease in entropy may correspond to an increase in the correlations among elements [732], Weaver's organized complexity. Consistent with Shannon and Weaver's [22, 433, 733] work, in their neural field model Wu et al. [422] found that as the amount of information decreases to the point the entire network is in the same state, synchronous, the system has an entropy minimum that they define as zero (0), indicating an absence of information. It also has a minimum variance.

Not all information in an instantiation may be useful. For example, a photograph does not need to resolve every whisker in a man's face to make the face identifiable. An efficient instantiation will strive to minimize the entropy while enhancing significant information [731]. A high resolution image will have greater information entropy than a lower resolution image that has lower complexity; thus lower entropy is indicative of greater instantiation efficiency. Representation of an image in V1 may have enhanced edges and smoothed surfaces, increasing organization while decreasing complexity, thus decreasing the entropy with the loss of fine-grained structure that is usually of low information value—noise. This can increase the robustness of the remaining information, as can redundancy, which would be reflected in increased organization. Error (noise) is inversely related to the amount of information; therefore, to keep the error small while reducing the total amount of information load, the system must reduce the number of degrees of freedom in the representation, discarding "likely" error, hence increasing the signal-to-noise ratio (S/N). The entropy is reduced by decreasing complexity.

An aperture's entropy has two facets: the amount of activity and the degree of orderliness. Decreasing the f_n, or activity, of the neurons decreases the entropy, hence the energy expenditure. Cohering elements increase orderliness. f_n reflects the state of activity, which may include the effects of unstructured inputs (e.g., brainstem). An increase in the f_ns of the elements can result in an increase in propensity to cohere (Ch. 3), the critical point. Proximity to the critical point increases the likelihood of a coherent response to inputs. Therefore, increased orderliness may be a response to the orderliness in the input projections, forming a CM. The CM divides an aperture's elements into coherent and disordered components, largely through phase quantization. This phase field, by itself, has a strong orderliness, with a subsequent decrease in entropy and energy expenditure. Thus, paradoxically, raising the input noise level can increase <u>or</u> decrease the entropy, depending upon the circumstance.

The coherence, orderliness, and phase structure of an aperture are manifest in its CMs, CISs, and Ks, and subsequently in its information entropies (Ch. 4). A CM instantiates some aspect of information. Many structures may be described using invariants [565], which are stable underlying formations that can form from multiple exposures to similar inputs as a sort of average that contains the essence of similar instances. Facial structure is an invariant with some genetic origins [479], however that might occur. A particular face has information modifying or referring to the invariant. A familiar face or family resemblance may itself become an invariant, with specific expressions or membership modifying it. An invariant in apertures is a well-established persistent CM, or set of CMs. The combined responses of feature-sensitive functions in a functional [734] may produce coherent organizations reflecting the input(s), increasing the S/N. Such organizations may nearly match the CMs or Ks of a latent invariant in an aperture, raising the probability of the inclusion of that CM in the information instantiation. A highly persistent invariant CM, e.g., a phoneme [471], may have a K with a specific, and perhaps localized, topologically stable distribution. An invariant may be distributed as CMs over multiple apertures, each with functions responsive to specific features, linked by Ks through association CMs (Ch. 7). A robust invariant will have lower information entropy, hence thermodynamic entropy, than a specific instance that contains it, as consolidation into persistence will have been responsive to homeostatic processes [605], decreasing metabolic loads by reducing ion flows and modifying synaptic functions. For example, under some circumstances nearly coincident synaptic inputs will produce increased sensitivity in subsequent coincident activations.

The entropy of the subset K is a sensitive indicator of aperture information entropy (providing a particularly attractive metric for computer simulations). An active CM will continue to produce CISs and Ks in response to the synchronous inputs, being persistent and modulated. Changes in an active CM, and its CISs, for example, increasing orderliness, will have corresponding changes in the K, indicating a decrease in the information entropies of the CM and the CISs it designates. By virtue of being a smaller set, K can have a lower complexity among its elements than the entire aperture population; however, the increased separation, and presumably increased independence, of the elements increases the entropy sensitivity. As the number of coherent elements in K decreases, its entropy will start to increase due to noise, as the distinction between noisy coherent elements and uncohered elements is less clear, in spite of phase quantization.

Entropy may be minimized dynamically. Corticothalamic interactions reduce entropy in an aperture through sharpening [550, 696, 705]. Modulation within an aperture was discussed in Chs. 3 and 5, with an emphasis on coupling among elements within the cortical plate. Coupling will affect the local timing relationships among elements, e.g., the local phases. Local phases are subsequently reflected in the properties of the local functions. Synaptic modulations occur due to activity from both within and external to the aperture. Modulations due to changes in the ECS are internal to an aperture [265, 266, 563], reflecting neural activity that may include external influences. Homeostasis can drive coupling modulations [257, 265, 563, 570, 602, 605] in seeking an energy expenditure minimum, potentially influencing spike timing [480]. Homeostatically driven modulation may be rapid [603].

Entropy minimization among apertures

To discuss information entropy in an ensemble, we need a model of information instantiated in multiple cortical areas—an ensemble. Information in CMs and CISs has a level of undecipherable complexity [564, 735] (Figure 37 and Ch. 4) that increases with the level of propagation through cortical areas, returning to some level of decipherability for a concrete output, such as motor output. Although information in higher cortical levels is undecipherable, having no reliable a priori observables across subjects, we may still posit its instantiation, participating in interactions among composite nodes, potentially reflecting relationships rather than features. Perhaps the greater the undecipherability of information instantiations, the greater the plasticity (hence memory formation). Although the functions within an aperture may appear simple, e.g., edge detection [530], their interactions within and among apertures create an unpredictable result [65], particularly as the functions undergo minor modulations. Attractor and similar [669] models have large information capacity potential as dynamic, coordinated, self-forming patterns of activity extending over an area, but are limited to a single aperture. These would also be both complex and undecipherable. Pribram [493, 736] has proposed a distributed representation of information, although the representation's specifics are unclear, perhaps reflecting undecipherability.

Figure 37. Decipherability-undecipherability-decipherability information flow.

Although activity in the CNS may have complexity, the coherence of the elements in the system collectively instantiates information [735]. *Mutual information* is the information in common in two spaces [372], an expression of the level of dependence of variables between two apertures. Two apertures with exactly the same information would have a maximum mutual information. This will never occur because 1) apertures have different functionals and 2) no aperture is connected to only one other aperture. The mutual information will be reflected in the cross-correlation between the Ks exchanged between two apertures (Figure 34) by virtue of their topological correspondence. There is a relationship between mutual information and joint entropy [737]. The *joint entropy* of the system is the entropy of the information in the union of the two apertures. As mutual information increases, the size of the union decreases (i.e., (A + B) > A ∪ B) and the joint entropy decreases.

Ensemble entropy reduction

I propose that the CNS operates to minimize the joint entropy of the current ensemble, while maintaining a "readiness to respond" to new internal and external activity, which requires ongoing energy expenditures. Joint entropy can be minimized by the resolution of the minimum individual aperture entropies with the maximum mutual information, increasing information efficiency. If the mutual information among apertures is low, orderliness is reduced relative to the total activity [731]; consequently entropy has a high minimum. As apertures' CMs instantiate different aspects of the information, often combining inputs from several apertures through differing functionals, alignment of Ks between aperture pairs can never be complete, the Ks being surrogates for CMs. The exclusion of non-coherent apertures (v) increases the orderliness of the remaining ensemble, further decreasing its information entropy. In short, the entropy of an ensemble can be decreased through increased orderliness and coherence in the apertures, and through increased mutual information among apertures.

An invariant is simpler than the immediate instance that contains it. As the processes resolve toward simpler states that include invariants, the ensemble entropy will decrease further. As the ensemble's information entropy decreases, no one aperture can reach its absolute minimum. I suggest the resolution process itself may constitute—or at least contribute to—the internal experience. Voila and Wells [738] demonstrated that two representations, in their case a 3D object model and a 2D image with clutter, can be aligned through entropy manipulations to maximize the mutual information based not on the features of the object but on the physical characteristics of its representations, e.g., intensities. This illustrates that a system can operate on an instantiation of information through purely physical processes that manipulate entropy, without regard to the actual information instantiated.

The concurrent comodulation of apertures through bidirectional causation is proposed as a process for minimizing individual aperture entropies while resolving toward an ensemble minimum joint entropy, a maximum mutual information. Bidirectional causations among apertures are effected through the bidirectional exchanges of CISs (and Ks) described in iv, above, operating within the three modes of synchronization: intermediation, common drive, and peer. A linear solution for maximum mutual information among multiple incompletely interconnected apertures does not appear feasible once the membership in the ensemble is greater than two; therefore a method for a concurrent resolution is required if ensemble entropy is to be minimized. Apertures are modulated by input projections, producing output projections. For simultaneous comodulation, bidirectionally causal interactions between apertures are required, as an ensemble will include several apertures; thus linear and reciprocal causal exchange models cannot apply. Initially there will be unbalanced information flows among apertures as an ensemble forms, as indicated by transfer entropy asymmetries. This fits a general model of liquid computing, as reviewed by Maass [615], who questions use of the concept of "computation" [155] in biological systems, instead suggesting a more filter-oriented model.

Connections between cortical areas are bidirectional [59, 100]; therefore information, instantiated in CISs, is exchanged bidirectionally. During ensemble formation, and with new inputs, the flow of information is from the source apertures through the subgraph connections. Transfer entropy asymmetry indicates directed information flows in systems [739] that Papana et al. [740] used to characterize information flow in EEGs. Wagner et al. [66] used phase synchronization to detect the direction of information flow between two coupled systems, applying the concept to the detection of information flow in EEGs. Transfer entropy asymmetries between apertures may indicate discordant (low cross-correlation) Ks. Isochronally synchronized apertures would not reveal information flows. As activity permeates an ensemble, information flows will become more directionally balanced. As there are more than two apertures within an ensemble, not connected linearly, detection of directed information flows becomes problematic.

It is not information that drives modulation but minimization of entropy in the instantiations, the <u>result</u> being the manipulation of information [738]. Apertures simultaneously comodulate each other toward a low information entropy, described by Georgiev and Georgiev [741] as the *Least Action Principle for an Organized System*. Processes of modulation were discussed in Ch. 5, often being homeostatic processes that decrease the metabolic load by reducing local differentials and increasing the efficiency of response to correlated inputs. Increases in local synchrony minimize ion transport requirements; conversely, increases in the motility of ions within the ECS can reduce local phase differences. A synapse set attuned to coincidence will have an increased sensitivity to correlated activity, and conversely a higher threshold for uncorrelated activity.

Modulations affect timing, determining whether a spike occurs within a critical window (phase quantization), hence whether an element participates in the coherent phase of a CM, contributing to the CISs and Ks. CM comodulation also increases the alignment of Ks, minimizing joint entropy. Reduction in CM complexities of communicating apertures can increase their mutual information, and subsequently reduce the asymmetry of the exchanged Ks. I suggest that discordant Ks produce local differentials that increase modulations in both apertures. Coherent activity provides opportunities for modulation. Interference is a phase effect. Analogous to light wave interference, spatio-temporal interference has been proposed as a model of modulator patterns within an aperture [368, 437, 488, 489, 533, 554, 569].

Undecipherable complexity creates fluidity. For example, the deterministic nature of functions in the CNS may lead to chaotic behavior [65] that empowers attractors [54, 169, 201, 367, 368, 369] that may seek a minimum entropy [372], increasing organization. At an intermediate level of organization, Freeman [417] proposes that vortices modulate phase modulation patterns at a mesoscopic level. Bressloff [56] proposes self-organizing neural fields as bridging the span between local coupling and topographic structure. Such bindings would be reflected in a decrease in the independence of the states of the elements, hence a decrease in the entropy. Processes of modulation that produce entropy minimization need further exploration.

Composite nodes—conceptual entities—constitute the interacting elements in cortical sensory processes (iv, above). Two composite nodes may be cross-coupled[22] through five mechanisms (Figure 33, Figure 34):

---

[22] The model of two coupled composite nodes may be extended to the interactions of more than two composite nodes, not included here for the purpose of clarity. A simplex graph may be used to include more apertures and thalamic nuclei.

1. A cortical aperture projects to another cortical aperture (12, Figure 33).
2. A thalamic nucleus projects to more than one aperture (9, Figure 33) [*691*].
3. A cortical aperture projects to a limited set of other thalamic nuclei (10, Figure 33) [*673*, *719*].
4. Inhibitory neurons from within one TRN sector may project to non-identical composite node thalamic nuclei (11, Figure 33) [*697*, *721*], although the rather diffuse nature of those projections would not carry substantial information, but serve to modulate activity over some finite spread.
5. Electrical gap junctions among inhibitory neurons in the TRN provide weak positive coupling that supports cohering (8, Figure 33) [*248*, *706*].

A_i and A_j will project bidirectionally. The excitatory cross-coupling input to thalamic nucleus T_j from T_i (Figure 33) will result in a change in activity projected back to both the composite node aperture A_j and to TRN, R_j. The cross inhibition within T_j and the enhancement in N_j will further increase the local contrast. This increase in contrast will be limited as the composite node cross coupling is sparse. The cross-correspondences of A_i and A_j, reflected in T_j and T_i respectively, establish a feedback system that will respond to the coherences of the activity patterns in the two apertures. This is essentially a dynamically emerging filter for the two composite nodes. As the thalamic and cortical activity patterns increase in local contrast, they decrease in local complexity.

Causation has two effects: an input contributes to an aperture's response, and an input modulates the response characteristics of an aperture. Several causal interactions among apertures in an ensemble may be considered: Granger causality, circular cause, and bidirectional causation (Figure 38). In Granger causality, one system causes an effect in another with no reciprocal effect [*742*] ($A_i \rightarrow A_j$, Figure 38a). As all intra-aperture connections are bidirectional, one-way cause does not occur. Circular causation [*743*] is a loop of one or more links of Granger causality among one or more items closing back on the original source. Reciprocal causation is a circular causation with only two items ($A_i \rightarrow A_j \rightarrow A_i$, Figure 38b). Conceptually two apertures may be able to reciprocally project in alternation, exchanging information; if more than two apertures are involved in a network, this would be problematic. Granger and circular causation are implausible. Many apertures are involved in an ensemble, for example, the visual system has 32 centers, 25 of which are devoted solely to vision [*52*]. During bidirectional causation, apertures concurrently comodulate each other through simultaneous bidirectional exchanges ($A_i \rightleftarrows A_j$, Figure 38c). A_i modulates A_j while A_j modulates A_i. By the time the effects on A_j from the projection A_i result in a new projection from A_j, A_i has already changed due to the effect of the simultaneous projection from A_j. In the CNS, bidirectional causation is enabled by synchrony between two (or more) apertures [*744*, *745*] in an ensemble.

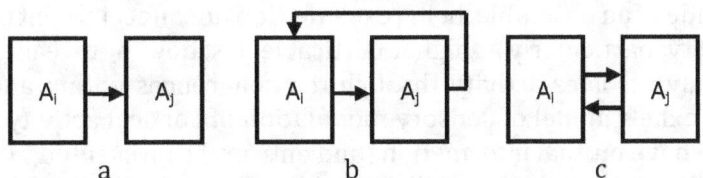

Figure 38. Three causal models. a) Granger causality ($A_i \rightarrow A_j$); b) circular cause ($A_i \rightarrow A_j \rightarrow A_i$); c) bidirectional causation ($A_i \rightleftarrows A_j$).

The requirement for synchrony (as aperture phase synchrony [*746*]) among apertures is a result of the exchange of temporally structured blocks (CISs) (Figure 29) that must occur with equal AFP frequencies in both directions without collision or reciprocal or circular causality. These bidirectional exchanges tend to decrease entropy within each aperture, with a decrease in energy (virtually by definition). Continuous bidirectional activity results in the comodulation of the apertures, sharpening [77] the coherence maps, lowering the entropy in both apertures and of the system. The joint entropy, as the common information among apertures, has organization through correlation, resolving toward a solution. The non-participating apertures have decreased activity, further lowering system entropy. Information, in its instantiations, becomes distributed among apertures as it is allocated for different operations, i.e., through divergent transformations, presumably parsing out invariants. Image processing and storage by phase conjugate mirrors [497, 553, 555, 557, 651] and dual phase conjugate mirrors [747, 748, 749] are useful analogies for the development of information through bidirectionally causal interactions between (and ultimately among) apertures, which can manipulate, store, and retrieve multiple images [497,] as they are coupled in a non-linear medium sensitive to the coherence of the radiated (or projected) energy.

For illustration of interaperture causality in Figure 33, the operation within A_j is presumably modulated by A_i, other apertures, and internal factors. As information flows continuously, A_i and A_j will simultaneously project information between themselves in bidirectional causality [664], although an initial causal source is required in the relationship [740]. Destexhe [76] considers sensory input to modulate an ongoing activity; hence it would modulate bidirectional causality. Such modulation will introduce new instantiated information that must propagate in the system, resulting in potential modification of the ensemble's participants. This new initial input (e.g., sensory CISs) produces essentially Granger causality,[23] with a transient reciprocal causality within the composite node, with a subsequent transition to bidirectionality as coherence develops in the subgraph of apertures to form an ensemble. Initially aperture A_i has meaningful information it projects to another node, A_j, or nodes (from layer VI), through the white matter, perhaps involving the full composite node. If this projection provides a pattern of activity into A_j that is relevant to the operations of both apertures, A_i creates Granger causation [640] in A_j: the activity in A_j is influenced by activity in A_i. A_j "reflects" transformed information back to A_i, the thick mirror model (Figure 26), analogous to a phase conjugate mirror [456, 494, 555, 556]. A_i's behavior is influenced (modulated) by this back projection as it continues to project CISs. Modulation will occur within the aperture A_i. A pair of apertures in a bidirectionally causal relationship has an analog in dual phase conjugate mirrors [19, 497, 749].

Only those components that can be meaningfully modulated participate in an information instantiation, although an instantiation may include invariants, which are not modulatable on a short term, serving as resources. An invariant provides a framework but does not instantiate a particular instance. If coherence is the context for the instantiation of information, then disordered elements, and their modulation, do not affect the instantiation of information unless such modulation causes them to become coherent. As information entropy expresses the ability to instantiate information, the entropy of the cohered elements reflects information instantiated; consequently the cohered elements in a CM are the source of the aperture's entropy. By similar logic, apertures outside of an ensemble, being excluded, do not affect its joint entropy. The inability to decipher the activity, particularly at higher cortical levels, does not weaken the argument that the system continually resolves activity through the coherences within and among apertures, consistent with Destexhe's model of sensory modulation of cortical activity [76]. Bidirectionally causal apertures will have mutual information (and entropy) represented in the Ks and CISs. The initial information flow directionality will be evident in entropy transfer asymmetries, the asymmetries being reduced as the ensemble forms, seeking a joint entropy minimum.

An entropy minimum is an expression of the instantiation's efficiency. There are competing factors in forming and in minimizing ensemble entropies. Entropy decreases must occur in spite of structural hindrances. Input projections diverge in each aperture, and functions have inputs

---

[23] Ignoring the small amount of reciprocal efferent activity to the primary sensory organ.

converged onto their receptive fields, with a resulting degradation of strict topological mapping between apertures, analogous to a point spread function, potentially increasing entropy. Bidirectional and intra-aperture modulations minimize the total information entropy by reinforcing areas of coherence and gating out less coherent activity. It is highly unlikely that an aperture that participates in an ensemble of two or more apertures with one or more CIS inputs and differing functionals can achieve its lowest possible entropy state. There can be an ensemble entropy minimum in which all participating elements have simultaneously negotiated low, non-minimal, entropy states. Bearing in mind the topological correspondences among apertures (mappings) with differing information content instantiated, the negotiation may be modeled as incorporating the Ks and subsequently their entropies. The negotiation is the process of optimizing the cross-correlation of Ks, thus the mutual information among the apertures, through bidirectional causation. When an ensemble has resolved to the lowest joint entropy state, it has optimized a solution to the task, for example, recognizing a face. The ongoing nature of experience does not normally allow a final resolution; thus the internal experience is continually unfolding in an imperfect state.

vii) The Internal Experience Emerges.

Introduction

Experientially we <u>are</u> what our brains are doing in certain conditions. We are not aware of the development of an experience, only the end result [*38, 626, 643, 735*]. Models of consciousness and its exploration can be complex. Crick and Koch [*2*] have developed a framework of consciousness composed of 10 neural correlates, e.g., the (unconscious?) homunculus. Consciousness and the internal experience differ. Usually the internal experience occurs during consciousness, although dreaming may be considered an internal experience. I am not considering unexperienced unconscious processes. Although I am only addressing perceptual issues here, the concepts may be extrapolated to abstract constructs (e.g., verbal, mathematical, and conceptual thought). During the internal experience the instantiation flow is from a disordered graph to an ensemble, to ensemble entropy minimization, as the various parts of the brain operate on information instantiations in different manners to create a unified percept or construct that is emergent as a whole [*5, 614, 750*], just as the murmuration of a flock of starlings becomes unified, emerging from the individuals' properties [*751, 752, 753, 754, 755*]. Often the flow will be continuous, as new information creates and updates the instantiations; thus the internal experience is continuous.

Models of brain function fall into three categories: processes that are outside of consciousness [*23, 47*]; an alternative to consciousness based on processing modes [*87*]; or processing that is centered on consciousness [*26, 51, 756*]. The coherent apertures model is in the latter class if one interprets that to mean that some processes are concurrent with consciousness. In a review particularly relevant to the model here, Turk-Browne [*566*] describes four aspects of working memory: 1) representations are distributed across brain regions; 2) the regions interact dynamically; 3) the interactions vary according to cognitive state; and 4) the space of possible interactions has high dimensionality. Treisman and Gelade [*713*] propose that the process of attention integrates features into a whole, hence would be part of the internal experience.

Anatomical correlates of consciousness have been widely explored. For example, Tang et al. [*757*] describe brain states of resting, alertness, and meditation, with the anterior cingulate cortex and striatum maintaining brain states, and the insula active in switching between states. Långsjö et al. [*750*] envision consciousness as the activation of a subcortical and limbic core network functionally coupling the frontal and parietal cortices with motor intentions. The thalamus is intimately involved in attention [*377, 696, 702, 758*] and experience [*759*], participating in information transmission between cortical areas [*699, 721*]. In fact, disabling the thalamus in rats by injection of anesthesia (a GABA blocker) results in immediate loss of consciousness, regained by injection of minute amounts of nicotine (a GABA agonist) into the thalamus [*481*], although intra-aperture EEG synchrony increased during loss of consciousness from anesthesia [*640*], suggesting synchrony alone may not be synonymous with information transfer. The TRN in particular has been implicated in thalamocortical synchrony [*621*] and the internal experience [*80, 697, 702*], although separation of thalamic and TRN functions would be difficult. One must be careful not to presume that the activity of such a core is considered as "consciousness" itself. The light is not in the flashlight's battery.

Perception

Perception is a significant contributor to the internal experience. Crick's [*702*] searchlight hypothesis of attention is consistent with Destexhe's [*76*] characterization of cortical activity not being driven by the senses, but modulated by them. Koch and Crick [*626*] proposed visual awareness as the binding of distributed areas. Dhamala et al. [*760*] find that distributed binding occurs within and across perceptual modalities. Integration of image elements into a perceived whole is not instantaneous, as it requires propagation into multiple apertures and recurrent processing [*78*] with specific areas recruited by attention [*479*]. The processes of perception are not constant, perception fading with constant retinally stabilized images [*761*]. Perhaps this is due to long-term entropy minimization without perturbation.

Specific areas responding during visual perception are often considered correlates of the internal experience. There is evidence of a relationship between visual perception and memory, and the medial temporal lobe [*762, 763*]. Kravitz et al. [*687*] describe three linked pathways emerging from the dorsal pathway projecting to the prefrontal, premotor, and medial temporal cortices as a framework for visual processing, akin to a network. The same cortical areas may be used for different perceptual states, e.g., wakefulness and sleep [*472*]. It is tempting to conclude that the same processes and structures are involved in perception (wakefulness) and reconstruction (sleep, recall). An afterimage, of constant size on the retina, will have different perceived sizes according to the distance of the surface on which the afterimage is perceived. The size of the image represented on V1 will change according to the perceived size [*667*]. Fox et al. [*112*] found that small cortical areas in V1 may be recruited into a task, although the small sizes may reflect image structural components in the retinal projection into the aperture that may have been within a single CM. At a more extreme level, single [*764*] or small sets of neurons, which Quiroga [*439, 765*] refers to as "concepts cells" or "Jennifer Aniston cells," akin to "grandmother cells" [*766*], have been proposed as foci of specific perceptual information. I consider such reductions unlikely as they would make the CNS vulnerable to small localized losses.

Synchronous subgraphs

The coherent apertures model describes a potential process for the synchronization of a limited specific aperture set into an ensemble. The ensemble emerges as a significant descriptor of awareness, hence of the internal experience, which appears broadly based [*767*], incorporating a network [*62*] of cortical apertures and other structures [*768*]. Models of variable subgraph networks that focus on the connectivity of areas are associated with a global workspace of attention and cognition [*32, 39, 85*]. Emergent coherence, including synchrony, is a major concept in models of human brain activity, supported by considerable experimental evidence, as reviewed by Uhlhaas et al. [*5*]. Chialvo [*157*] supports the brain as being "naturally poised near criticality" to become synchronized. Varela et al. [*365*] describe the synchronization of multiple brain regions as a "brainweb." Consistent with this model, Tononi et al. [*735*] and Baldauf [*479*] consider the internal experience to occur when multiple regions cohere on some common task. Desimone [*622*] correlates the synchrony among multiple cortical areas with selective attention, supporting the searchlight model of attention of Crick [*627, 702*]. The synchrony of a coherent subgraph as an ensemble has been discussed above (iv).

Experience and EEG oscillations appear to be concomitant. EEG oscillations during the conscious experience [*5, 155, 418*], usually considered correlates of awareness and cognition [*38, 326*], reflect the binding synchronization among cortical apertures [*86, 365*], higher frequencies differentiating wakefulness from deep sleep [*355*]. Coherent oscillatory activity among separated areas correlates with attention [*623*] and sensory processing [*460, 677*]. EEG activity during vision has been explored extensively because of the large amount of dedicated cortex and relative ease of stimulus presentation, revealing phase-locked oscillations during visual perception [*156, 485, 627*]. Hipp et al. [*167*] found that EEG synchronization in two frequency bands (beta and gamma) in two large-scale networks predicted subjects' percepts of ambiguous audiovisual stimuli and the integration of auditory and visual information. Two vertical bars approached each other; a click sounded as they passed through the centerline. Subjects reported either "pass-through" or "bounce" perceptions. The differences were correlated with EEG pattern differences, some of which occurred before the bars met; in other words, they predicted the perception, consistent with the results of Yamamoto et al. [*624*] that explicit awareness of working memory content is evident in a transient appearance of high gamma synchrony in the EEG. Ehm et al. [*639*] found EEG changes preceding shifts in the perception of ambiguous figures. Melloni et al. [*643*] found shifts in neural activity only when test words presented visually were consciously perceived.

Conclusion

The concept of the internal experience has been widely explored and debated. Ensembles emerge during perception, cognition, and behavior [*365*], the AFP, and hence EEG, reflecting their size and strength. A strong EEG will be a transient state, as a stable coherence without further entropy minimization fades [*761*] or transitions into another subgraph. The process is ongoing. The cortical ensemble contributes to the internal experience, providing "content." The coherent apertures model proposes that the internal experience is concomitant with the formation and entropy minimization of ensembles in response to dynamic percepts, constructs, and cognition. Consciousness, unconsciousness, preconsciousness, attention, preattentiveness, priming, awareness wakefulness, dreaming, working memory, and the global neural workspace [*39*] are not considered as separate processes here [*85, 86*]. The model framework encompasses these differing concepts, as all involve structured information instantiation and coherence. Outside the current scope, they must be considered separately in future explorations.

6.4. Results

Principles and Corollaries

Principles:
- Cortical apertures form a network graph.
- Communications in the cortical graph are bidirectional (with a few exceptions).
- A cohered aperture is a node in a subgraph.
- An aperture that is synchronous or about to synchronize will filter to accept CISs that are in the appropriate phase.
- Synchronized apertures seek a low net information entropy.

Corollaries:
- There is no coherent communication if both apertures in a pair are not synchronized.
- A disordered aperture does not synchronize with other apertures.

- A subgraph of synchronized apertures will synchronize into an ensemble.
- Cohered apertures that are synchronized in an ensemble will comodulate through bidirectional causation.
- The internal experience is defined as the current ensemble.

Summary

The processes of ensemble formation and its entropy minimization resolution comprise the internal experience. Coherent apertures are considered as nodes in a graph of apertures through application of the van Cittert-Zernike theorem (VCZ). A composite node, which is an abstraction composed of corresponding structures of a cortical area (aperture), a thalamic reticular nucleus sector, and a thalamic nucleus, is a useful construct in describing how ensembles are formed. A coherent aperture is grossly synchronous. Synchronous apertures, VCZ nodes, in subgraphs produce synchronous oscillations when active. Cortical ensembles, as synchronous subgraphs, are dynamically formed from coherent apertures in bidirectional communication seeking an ensemble entropy minimization through simultaneous aperture entropy minimization. I model subgraph synchronization through three mechanisms with coherent optics analogs: intermediated, common drive, and peer. A subset of the axon output projections from an aperture acts as a topological algorithm for producing unique natural key identifiers—abstractions—for each CM. These natural keys create linkages among CMs in multiple apertures specific for an ensemble in both active and latent states.

7. Operations

7.1. Introduction

How does the coherent apertures model describe operations that produce an internal experience? In the model proposed here, the cohering of apertures creates an ensemble of a subset of all of the cortical areas. Their collective activity resolves to an internal experience, formed from perception or memory. The model is limited here to declarative sensory experience, principally of visual objects, with some notes on auditory perception. I am not discussing cognition and how it occurs in working memory.

Perception is a complex process. Seeing a new object starts a flow of activity from the retina through the brain, from perception to the formation of a long term memory of the object. The retina, specialized cortex, responds to specific facets of the image—in spatial frequency bands—projecting the results to the lateral geniculate nucleus (LGN) in the thalamus, which sharpens the temporal coherence of the activity, projecting it to the primary visual cortex, V1 (Brodmann 17 [102]). V1 responds to higher level facets, such as oriented edges, projecting into a complex of bidirectionally interconnected cortical areas, apertures, that for object perception, largely flow forward over the ventral cortex to progressively higher apertures. As the activity spreads, specifically responsive cortical areas become coherent, becoming dynamically recruited into the growing synchronous subgraph, an *ensemble*, which constitutes perception in working memory—the internal experience. The earlier cortical areas have largely intrinsic dedicated functions with resultant rapid responses supporting rapid perception. As activity persists, the perception moves into a stabilized memory, freeing working memory in a continuous flow. The stabilized memory may be accessed freely for recognition and recall back into working memory. Stabilized memory is rather fragile and easily disrupted. During sleep the stabilized memory becomes more robust as consolidated memory, instantiated in maps within multiple cortical areas. Memory, particularly consolidated memory instantiated among multiple cortical areas, can be reconstituted through inter-area associations.

The important aspects of the coherent apertures model are discussed in the preceding chapters: aperture coherence (Ch. 3. The Coherent Cortical Aperture), information instantiation (Ch. 4. Information Instantiations in Apertures), functionals composed of response functions (Ch. 5. The Aperture Operator), and cortical aperture integration (Ch. 6. Integration of Coherent Cortical Apertures).

A cortical area is considered an *aperture*. A coherent aperture is grossly synchronous over its area, with instantiated information bound together within a *coherence map* (CM), composed of phases of synchronous and disordered minicolumns. A CM may be active or latent, transient or persistent (Ch. 4, Table 1). Active CMs within apertures project *coherent information structures* (CISs) that contain *natural keys* (Ks) that uniquely identify CMs (Ch. 6, i), characterizing their *information entropy* (Ch. 4). Apertures project and receive CISs from multiple apertures. The net affect of incoming Ks may activate latent CMs that will subsequently project CISs if the match is sufficient.

The formation of an ensemble involves multiple components related to each aperture. An aperture and its corresponding structures in the thalamic reticular nucleus (TRN) and thalamus comprise a *composite node* (Ch. 6, ii). Through application of the van Cittert-Zernike theorem (VCZ), a coherent aperture, and by extension its composite node, is considered a *VCZ node* in an ensemble.

The composite node provides underlying mechanisms for its ensemble participation (Ch. 6). An ensemble is a dynamically formed synchronous subgraph of VCZ nodes within the cortical graph (Ch. 6, iv), performing a unified operation. Sensory input results in a synchronous ensemble of coherent apertures that forms as a coherent subgraph out of the total cortical graph of the connectome. This ensemble resolves to a solution—a percept in working memory. The ensemble solution may be biased by potential outcomes. You quickly recognize faces you know, you may see a new face as resembling someone you know—"He looks like Joe!"—or you may see family resemblances. As the ensemble forms and progresses toward an entropy minimum, the internal experience emerges.

7.2. Operations Model

Introduction

I shall focus here on the roles of apertures within ensembles in performing meaningful *operations*, creating the *internal experience* in working memory. During the formation and entropy minimization of an ensemble no aperture operates in isolation. Each communicates bidirectionally with a limited number of other apertures, some of which become and/or maintain coherence and synchrony in an ensemble (Ch. 6, Figure 31), with a subsequent emergent internal experience. The functions within each aperture are limited (deterministic) and relatively simple, the basis for a state machine. The transitions between disorder and coherence—most particularly synchrony— are abrupt in ensemble and aperture contexts, a characteristic of state machines [40, 616] composed of deterministic components.

A subgraph of bidirectionally interacting coherent apertures dynamically assembles into a synchronous ensemble that performs a sensory process. The constituents, formation, and operation of an ensemble are inextricably interwoven (Ch. 6). The cortical graph in humans has evolved to have over two hundred functional areas [114, 769] from which subgraphs of ensembles and preferential motifs form to solve specific problems, "solve" and "problem" being broad terms expressing the CNS operation through entropy minimization. The ensemble resolves to a "solution" among its apertures; even perception is a constructive process, integrating many brain regions to form a percept (see review by Tononi et al. [735]).

General Network Structure and Aperture Classes

Figure 39 illustrates a generalized model of a network of nodes (a graph) for early stage sensory processing in the CNS. Each node represents a class of apertures:

S_d: Primary sensory cortex such as V1 or A1, although the retina may be considered an aperture in its own right;

F_j: Feature subapertures (e.g., F_1, F_2, F_3) that may be separate or included in an aperture, responding to one or more specific features;

D_i: Domain apertures that contain domain-specific coherence maps (CMs), which may include invariants, of varying persistences and activity levels;

K_q: Key registry, a structure in the medial temporal lobe (MTL), that includes the hippocampus, parahippocampal cortex, perirhinal cortex, and entorhinal cortex [770], that incorporates natural keys (Ks) with varying degrees of persistence;

R_p: Association aperture that contains CMs of various persistences and activity levels that incorporate relationships among CMs of other apertures;

C_e: Outcome aperture dedicated to a domain output such as the premotor cortex.

CMs within some of these apertures typically have high persistence and may comprise a library of components, including invariants, from which a complex multi-aperture de facto CM is produced.

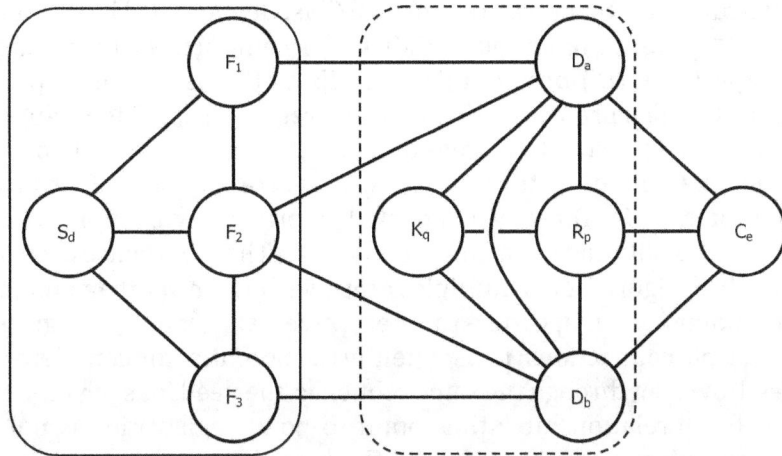

Figure 39. General functional subgraph (ensemble). Apertures represent classes of apertures (see text). Communications are bidirectional. Solid box is an aperture with functional subapertures. Apertures within dashed box are plastic to varying degrees.

Perception and memory use many of the same apertures. For clarity the types of apertures have been isolated, although such a discrete classification might not always exist. The class in which an aperture belongs, which may change, will reflect its immediate operation in the ensemble. A coherent sensory input creates coherent activity (Ch. 6, Figure 28) within a sensory aperture, S_d, in vision the retina or V1. Feature response subapertures, most of which are stable over the time scales of this discussion, may exist within almost any aperture type. The (primary) sensory aperture (S_d) and the first level of feature responses may be conflated into a single aperture that contains multiple feature subapertures (F_j) that communicate with CISs to domain apertures (D_i), creating coherence in those apertures for which the input projections are appropriate. The coherence within an aperture and the synchrony among apertures in a subgraph provide justification for this dichotomy of separate and conflated feature apertures. S_d and F_j have been separated for clarity. At relatively short time scales only the complex within the dashed box—the domain D_js, the K_q registry, and R_p association apertures—has plasticity that is bidirectionally modulated (see simplified model in Ch. 6, Figure 27). D_is may include more complex feature response apertures. There may be functional overlap of the D_js, the K_q and R_p apertures. The K_q registry and R_p association apertures may actually exist in a piecewise fashion within other apertures. A hierarchy of small world R_ps is speculation beyond this discussion.

Process

The operation of the CNS in the coherent apertures model is realized in a process with a flow, biased by potential outcomes, that resolves through entropy minimization toward a solution, the process creating the internal experience. Apertures with active CMs are coherent (grossly synchronous) and hence participate in the ensemble. Two factors determine if an aperture will become coherent and participate in the ensemble: 1) its level of activity, and 2) its reception of one or more projections with CISs to which it is responsive. If its level of activity is high enough and its response to CISs is strong enough the aperture will form a CM that projects CIS(s). The activity level of an aperture is reflected in the f_ns of its elements due to input projections from brainstem and limbic structures and other apertures. A sensory input, such as from the retina, projects to a primary sensory aperture (S_d) "through" a nucleus in the thalamus (such as the LGN), which enhances coherence and select features. Subaperture feature responders (F_j) in S_d project CISs to domain-specific apertures (D_i) that create or contain CMs. Deeper level latent persistent CMs, e.g., invariant hand, face [479], mouth, or phoneme CMs, will become active if the inputs are appropriate.

The domain apertures (D_i) project CISs with Ks to the association aperture(s) (R_p) and outcome aperture(s) (C_e), also projecting Ks to the key registry (K_q). The R_p aperture responds to the coherence of the CISs projected to it from domain CMs, establishing its own CM that largely reflects the common Ks. Common natural keys are those components of several Ks that are the same: the overlaps of the Ks. Directly connected apertures will approximately preserve a topological registration, facilitating the commonality of Ks. The divergence of an input from a point within an aperture (Ch. 5, Figure 22) to multiple receptive fields in another, and the convergence of inputs from several points on an aperture to a receptive field on another, gradually diffuse the registration over multiple connections; consequently the potential for complete registration of Ks naturally decreases. However, through the same sequence, the need for such registration decreases, as the Ks may have fuzzy relationships; thus point-to-point registration is not required. Noise, external factors, and internal factors constantly modulate an aperture's local functions; consequently the patterns in an aperture will not be exactly the same for the same stimulus presented twice.

The transforms performed by apertures may not preserve the registration of the original pattern as the functionals operate on their inputs, for example, edge orientation and movement responses, or facial detection dividing the general facial invariant from the specifics. Information instantiations, and their processing, are increasingly abstract in the deeper level intervening D_i apertures, including the instantiations of invariants. The association aperture, R_p, maintaining some of the topological aspects of the Ks and input CISs, reflects the specific relationships of the domain CMs and their input features and the outcomes; thus they put the right mouth in the right place in the face, so to speak. In addition to participating in subgraph synchronization of the apertures (Figure 34), the K_q registry serves to transfer working memory to a stabilized state, as the Ks have a relationship with CMs. The Ks are incorporated in CMs (Ch. 6, Figure 27); thus the Ks become persistent with the persistence of the CMs. As stabilized memory is consolidated during sleep, Ks in K_q will gradually diminish, the time course dependent upon strength and repetition of experience. Well-consolidated memory does not need the K_q registry Ks. The remaining apertures suffice for recognition and recall of consolidated memory, which is distributed over the ensemble, when presented with appropriate Ks for CMs from other sources, although some apertures may have concentrations of Ks.

Flow

The flow of processing is reflected in the flow of the creation and modification of the ensemble. Each aperture interacts with multiple apertures in the dynamically assembled ensemble (e.g., Deco et al. [693] and Dhamala et al. [760]) (Ch. 6, Figure 31). Prediction of specific aperture associations, as reflected in CISs and Ks, is at best problematic, although nearest and next nearest neighbors have the most immediate connections [691]. Gong and van Leeuwen [568] have modeled information processing in the cortex as arising from the interaction of coherent patterns, which can be interpreted as the interaction of active CMs. The combination of undecipherably complex CMs across apertures (Ch. 4), sparse K structures reflecting entropy, and bidirectional causality creates a process from which it is not possible to understand (i.e., decode) activity beyond the most basic levels early and late in an information flow (Figure 39, Figure 37), progressively less so for higher or more abstract cortical apertures (D_i). The malleability of an aperture, particularly in the association class (R_p), will be reflected in the undecipherable complexity [*65, 433*] of activity projected from its CMs in CIS-Ks. The structures within an individual's apertures, although having considerable regularity, are still unique for that individual, further increasing the undecipherability. The binocular interdigitation patterns in the visual cortex are as unique to an individual as his or her fingerprints. Only post hoc analysis of the activities resulting from persistent CMs, particularly Ks, might be possible, as one would be essentially looking at an index in a hash table [*656*] as it is accessed, having no inherent meaning, only a possible correlation.

As persistent CMs consolidate, their entropies will reduce (Ch. 6), with a subsequent reduction in the complexity of the Ks. As described by Ayzenshtat et al. [*657*], only small samples from the aperture, in this model Ks, will suffice to differentiate among CMs. A smaller K will have a more limited, but specific, topological representation; thus the location of a K may indicate its CM's function. As memory consolidates, the function of the MTL (K_q) decreases as smaller Ks distributed among apertures will comprise memory access (a de facto distributed hash table [*771*]), a robust noise tolerant system particularly suited to peer-to-peer architectures.

Outcomes bias

Through bidirectional causation, aperture states <u>anywhere</u> in the ensemble are reflected <u>everywhere</u> in the ensemble (Ch. 6, v). The role of sets of potential outcomes as creating predispositions is consistent with Desimone's [*622*] model of attentional feedback biasing competitive cortical interactions in favor of behaviorally relevant stimuli. Sensory input has some form of topological representation in early stages. Working in reverse from the preceding description, the outcome aperture(s) has (have) within it (themselves[24]) a number of latent persistent CMs such as motor patterns, or cognitive states such as recognition, for example, the propensity to see faces in clouds and tree branches or to readily accept the four digit hands of cartoon characters such as Homer Simpson and Mickey Mouse. Transient CMs may become more persistent, some achieving the status of feature responders, such as phoneme [*471*] responses. An ensemble may be composed of subgraphs or motifs such that an outcome is a de facto source for other aperture(s); for example, a phoneme is an output of a complex of features which interact with a deeper level semantic context. Thus a subgraph may be a compound aperture. An input, biased by a finite set of existing outcome "matches," will resolve the ensemble to a low entropy solution that has a high probability of matching an existing outcome. Strong persistent outcomes will drive rapid resolutions. We recognize a very familiar face without noticing any delay. Potential outcomes are as much a part of the operation as the sensory inputs, biasing the "resolution" of the ensemble toward preexisting outcomes.

Extending the declarative CM model to activity, the role of outcomes is evident in motor behaviors. Just as input functions become shaped by experience, so does output as motor memory, familiar to musicians and athletes. Motor output has a topological mapping. Activity may increase or decrease for specific ensembles in anticipation of behavior for specific tasks [*375, 672*]. Outcomes that are preparatory, such as preparatory activity in the dorsal premotor cortex, can be detected before a motor outcome, movement, occurs [*772*]. Wong et al. [*672*] report rapid formation of ensembles among cortical areas that are predictive of the movement outcome. Outcome CMs create predispositions for activations of specific domain CMs through the Ks, consistent with the findings by Parameshwaran et al. [*150*] of coherence potentials in specific somatosensory areas in anticipation of actual movements.

The research of Guo et al. [*773*] would seem to contradict the idea that outcomes bias the operation of the entire ensemble. Mice were trained to respond, after a delay, to one of two whisker

---

[24] Apertures referred to in the singular may also refer to a plurality.

stimulations with a decision to move their heads in one of two positions to receive a reward. During trials the activity in the sensory system was separated from the motor cortex, demonstrating the separation of activity into two cortical areas. As the animals were operating in a well-trained binary task with a delayed forced choice response, it is not surprising that the two areas had separate patterns of activity. The sensory system would be trained to discriminate only two possibilities. The motor system had a forced delay. There was a significant number of actions that occurred during the delay period. This would seem to indicate that the separation of the gross ensemble into two separate subgraphs is a bit tenuous. In a related experiment Li et al. [774] found that the motor cortex does have a planning phase that can occur before the action. Thus, at best one can say that an ensemble can be broken down into subgraphs, but there appears to be a persistent connection. The patterns of activity in the cortices of freely moving animals experiencing a continual flow of sensory input and able to make a wide range of decisions are probably much more unified. The research on cortical activity during training by Masamizu et al. [375] is more consistent with the continual flow situation. The differences between delayed forced choice and continual flow point out important differences in the formation and composition of ensembles in different contexts.

Resolution

The emerging solution in an ensemble is the basis of the internal experience, a resolution involving the whole, as seen in activity flows. Siegel et al. [775] found that, in monkeys performing a choice task, following an initial upward sweep of cortical activity resulting from a visual stimulus, sensorimotor choices emerged in a network from an integration of bidirectional activity flows of sensory and task information. I propose that the bidirectionality of causation among apertures results in energy—and hence information entropy—reduction. The resolution of the ensemble to a solution is a reduction in its entropy. One may consider the resolution of an ensemble as a decision. Simple situations are straightforward and rapid. Complex situations that involve the resolution of ambiguities are slower decisions: cognition. Ensembles are complex, with more than two apertures; hence, the process is beyond conventional computation. On top of this, no two individuals have the same experiences. There are random events that occur early in development, some of them prenatally, which create unique structures that are subsequently elaborated on. These factors contribute to the undecipherability of an individual's CNS activity beyond the most primitive levels. We must accept this limitation on knowability, as uncomfortable as I am with it. There will be no universally readable code from the CNS.

7.3. Perception

The operations of visual perception, memory, recognition, and recall are tightly bound. Due to the extensive research in the area, perception may be considered separately, although the separation is artificial, as perception occurs in working memory. Visual perception is an immediate internal experience that involves a substantial subgraph. 20–30% of the human cortex is devoted to vision, 50% in the macaque [776]. The visual system has a complex graph of at least 32 apertures, 25 exclusively for vision. An integrated vision subgraph operation differs from parallel processing [777], as the interactions among apertures are bidirectional [78] with multiple processing streams [687, 778]. Visual processing is distributed among cortical areas [52, 116, 507, 779] that are richly interconnected [59, 61, 780], although only 31% of possible inter-aperture connections exist [52], almost all of which are bidirectional. With respect to the ventral visual pathway, which is dominantly associated with the processing of object (declarative) inputs, V1 (the primary visual cortex) is a vertex in an early stage cortical graph [781], connected to V2, V3d, V4d, V3v, V4v, PO, PIP, and the MTL, which is focal for these areas. Within this graph V2, V4d, and V3d are connected, as are V2, V3v, and V4v, such rich interconnections elaborating up through higher areas [691, 782].

Within my current model there is no persistent memory within the early stage apertures, ignoring transient effects such as contingent aftereffects. The early stages contribute functional responses as components of higher level associations; responding rapidly, their functionals produce transient CMs that project CISs with little delay, a model proposed by Kravitz et al. [*781*] and others.

Although initial projections in the visual system have significant topological correspondence with the input, i.e., the retinal image [*513, 783*], functions, and CMs (Figure 39) are increasingly abstract with greater distance from the primary sensory organ, e.g., the eye [*784*], with increases in receptive field sizes [*512*] of functions [*116*]. Apertures do maintain some topological correspondence; thus as instantiations change, the surfaces upon which they are mapped have correspondence. The original image on the retina has a feature vector (a form of configuration space, which allows cross-coupling of dimensions) of some 300 million dimensions defining its feature space, each of the 100 million photoreceptors having a value and a location, x and y, ignoring color. These dimensions are reduced to some 5 million through the ON-center and OFF-center responses of the retinal ganglion cells (1 million RGCs each with dimensions of x, y, type, spatial frequency, and value). RGC receptive field sizes occur in four octaves [*515, 517,*], responding to local differences in intensity [*513, 518, 701*], each effectively performing a local first derivative on the image. RGCs project to the lateral geniculate nucleus (LGN) in the thalamus [*507*], where stimulus coherence and select features are enhanced before projection to the primary visual cortex (V1) or aperture (Figure 40, S_d). Subsequent feature operators (i.e., F_j) further reduce the dimensionality, dividing and projecting the results to other apertures (D_i) that perform further feature dimension reduction (Figure 21, Ch. 5).

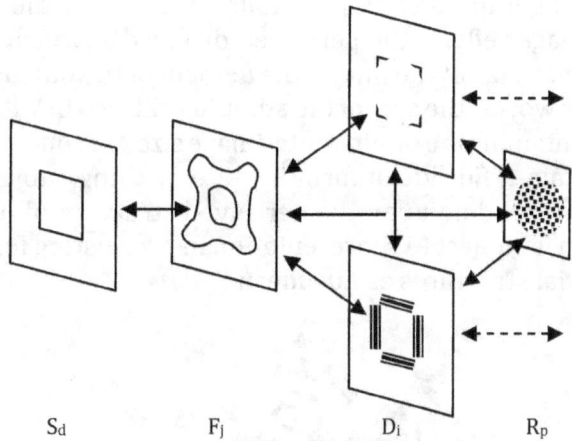

S_d F_j D_i R_p

Figure 40. A simple ensemble. S_d = Primary sensory aperture, F_j = Feature apertures, D_i = Domain apertures, R_p = Association aperture.

The retinal ganglion cells have a generally consistent distribution in the projection to the cortex [*514, 545, 785, 786, 787*], ignoring foveal magnification [*511, 788, 789*], which is a conformal mapping. The V1 input projections fall within various overlapping feature response fields (F_i) of pyramidal cells in minicolumns. The degree to which an input at a location in V1 coincides with the minicolumn's feature response characteristic, e.g., a moving oriented edge, will increase the activity level of the minicolumn's layers V–VI pyramidals through synaptic activity. If the collective response across the aperture, in strength or number, brings it past its critical point to coherence (Ch. 3), an active transient CM will result (Ch. 4). The only changes that might occur in V1 are elastic in the time scales considered here. These changes would most probably happen within macro-columns (Ch. 3, Figure 5) as transient increased coupling among minicolumns which become synchronous through increased f_ns, LFP pumping, and pyramidal-chandelier circuit (PCC) gating. The synchrony occurs at a gamma frequency, effectively modulated by the lower frequency (sensory) input. Entropy is minimized as a result of sharpening of the CMs through both inter-aperture interactions and PCC phase quantization (Ch. 3, Figure 13, Figure 16).

Perception reflects not only the image features but also the dynamics of the processes producing coherence [790], as stationary images result in a fading of perception [761]. The retina produces correlated [272] time-structured activity [391] that results in coherent activity in V1 [425] that extends across the graph of cortical areas [677] during visual processing, with sustained feature-selective activity along the dorsal visual pathway [339]. The stereoscopic outputs from the two retinas are bound together into a single, cyclopean, percept [791] when the two images fall within Panum's area of fusion [792]; otherwise there is binocular rivalry, switching perception from one eye to the other. As retinas are specialized cortex, hence apertures, this provides evidence that apertures bistably are (fused)—or are not (rivalrous)—participants in the same ensemble.

V1 maintains a retinotopic organization with overlaid functional maps of orientation and spatial frequencies [454, 793]. Perception involves excitation and inhibition in realizing the functions in V1 [540]. Ignoring binocularity [791] for the sake of simplicity, V1 functions, which may be dominantly innate [612], respond to edges with orientation [327, 526, 530, 533] and motion sensitivities [116, 119, 519, 520, 549, 783] with coherence across spatial frequencies [525] and spatial separation [13]. These response patterns, as CISs from CMs, are projected to other apertures with more complex functions [521, 523, 782, 794, 795, 796], with deeper level apertures instantiating abstractions and invariances [430, 565]. Invariants, as persistent latent CMs, constitute de facto intermediate outcomes that influence the process of perception. The perceptual process reflects the stimulus and the context. Sperandio et al. [667] found the size of the pattern of activity in V1 associated with an afterimage reflects the perceived distance at which the image is "projected," consistent with the perceived size of the image, the domain of the outcome influencing the entire subgraph process. In other words, the size of the stimulus pattern in V1 <u>changed</u> according to the perception of size, not maintaining the original afterimage size. Wason [15] found that the perceived structure of space can retain a Euclidean form in spite of being subjected to significant affine (linear) transformations, depending upon the perceived structure of objects within that space, illustrating that perception is subject to representation in preexisting forms. We live in Euclidean space, so we perceive spatial structures as Euclidean.

Figure 41. Hidden cowboy on a horse.
Original source of image undetermined.

It is not enough that the bits and pieces be coherently bound but that their specific spatial relationships are bound, although one must presume in a transformed manner. Perception of a form takes time for association CMs in R_ps to bind together [6] coherent features, and higher level line and area interpolations, in multiple apertures, into an organized whole through bidirectional negotiations (Ch. 6). A set of persistent CMs, as a potential outcome, is an integral part of the perceptual process, or negotiated result, providing a potential framework that is presumably elicited through Ks. The cowboy and horse in Figure 41 are not apparent on an initial viewing, but once perceived, the perception returns rapidly with any future viewing, reflecting CMs of the form of some persistence, presumably a subjective contour [797] (Figure 42). Perception resolves to a single percept. The ambiguous image in Figure 43, the Boring figure, is perceived in only one state at a time, like a state machine [40, 616].

Figure 42. A subjective contour.

Figure 43. The Boring figure.

Illustration in E. G. Boring, "A new ambiguous figure,"
The American Journal of Psychology 42(3), 444-445
(1930); adapted by Boring from W. E. Hill, "My wife and
my mother-in-law," illustration in *Puck*, 6 November
1915, 11, adapted by Hill from earlier images.

Figure 44 illustrates the strength of a facial CM, an invariant. In spite of the strong shadowing of the left side of the image, a complete face is perceived. Baldauf and Desimone [479] found specific areas in the medial cortex to be differentially responsive to faces versus places. Each became selectively synchronous with a frontal cerebral cortical area, the inferior frontal junction, appropriate to the focus of attention, indicating specific structural motifs [451] that can be incorporated into ensembles, consistent with dynamically assembled functional structures [566].

Figure 44. Shadowed face.
Thomas Friedman
https://c2.staticflickr.com/6/5288/5257787441_c7a29578e3.jpg.
Permission requested.

The glint from the glasses provides a stimulus for the symmetry of a face as does the lighting over the forehead. We do not normally consciously perceive facial asymmetry unless we specifically attend to it. Interestingly there is a propensity to perceive faces as symmetrical. The degree to which a face matches this ideal is reflected in perceived attractiveness [*798, 799*]. The Boring figure and shadowed face illustrate the presence of a facial invariant operating with ambiguity [*639*] that is resolved through cognitive contexts [*564*]. Thus deeper level cognitive CMs can influence outcomes. As discussed previously, potential outcome CMs may be an integral part of the perceptual process. We may each know many peoples' faces. Some are very familiar. These faces are favored potential outcomes. We may recognize a family resemblance. When meeting a new person it is not unusual to see him or her as looking like someone familiar if the faces are roughly the same. Over time and exposure this conflation disappears and the new face assumes its own identity.

7.4. Memory

Introduction

Memory is an integral part of the operation of the CNS: without it, unaided survival is not possible. One cannot easily separate memory from its creation and exercise, as its existence is probed through access: recognition and recall. Memory—and its failures—has four principle foci: formation, persistence, access, and reconstruction. In a review, Chaudhuri and Fiete [*800*] propose requirements and questions that theoretical models of memory should address, with emphasis on states in neural systems confronted with noise rather than on encoding. This systems states approach is consistent with the coherent apertures model. Memory implies change, some persistent modification of structures—physical or molecular—instantiating information that can be recovered. A rough canonical microcircuit (Ch. 3, Figure 6) illustrates opportunities for such modifications resulting in the malleability of the behavior of structures. Malleability may be *elastic*, able to return to an original state, or *plastic*, a permanent change. Perception and working memory normally entail elastic modulations; they may result in plastic changes. Both plastic and elastic modulations of biophysical processes in synapses, electrical conduction, and the ECS within an aperture are manifest in synaptic efficacy and time (phase).

I shall adopt the convention of using the term "memory" preceded by an article, e.g., "the memory," as an instantiation of a particular memory. "Memory" refers to the category, of which there may be a specific form, e.g., long-term memory. The former, "the memory," allows discussion of that which was, and can again become, an ensemble or synchronous subgraph. The expression "potential subgraph" is difficult to support from the literature, as a memory becomes fractionated in CMs (Ch. 4) in multiple apertures when not active. It is a logical consequence of the coherent apertures model.

Representative models

Many models consider the role of the MTL in memory. Generally the concept of consciousness is not applied in these models [87]. MTL structures, usually the hippocampus, are considered as circuits with variable connection strengths [801, 802] or dynamic, e.g., harmonic, responses [438]. Buzsáki and Chrobak [803] propose that the short-term dynamic behavior of the hippocampus enables consolidation of memory traces in the cortex. Kosko [744] has modeled associative memories as energy minimums in bidirectionally driven reverberations of neural networks, with the form of the memory providing a content address that includes temporal patterns. The coherent apertures model's entropy minimum is consistent with such an energy minimum and, to a less concrete degree, an addressing method similar to natural keys. Object recognition is a more specific form of memory. Serre et al. [804] and Lowe et al. [805] proposed feature discriminators with some invariant properties, inspired by models of the visual cortex, as underlying object recognition. Neven et al. [806] proposed combining temporal coherence with feature detection as a model for both representing and detecting objects. This is consistent with the concept that perception and recognition are both constructive processes that use many of the same apertures, also proposed by Yli-Krekola [668]. Horikawa et al. [472] demonstrated that activity patterns in dreaming sleep shared those patterns during specific stimulus perception, concluding that dreaming uses areas in a manner comparable to perception. Grossberg and Pearson [373] propose that the same cortical circuit design that serves vision, in other parts of the cortex supports memory with laminar differences between spatial and non-spatial working memories.

Chaudhuri and Fiete [800] reviewed issues in computational models of memory, with particular emphasis on states of complex systems, not inconsistent with the CM model with its phase states (Ch. 4). Psaltis et al. [554] propose an optical analog to memory as a hologram in a phase conjugate mirror [555, 557, 651, 807], which is a nonlinear optical mirror in which an outcome is not the simple linear sum of intersecting activities [553, 651]. For example, the index of refraction of a medium may change according to the intensity of light, so one laser beam might "bend" another in the medium where they intersect. Yariv and Kwong [497] describe associative memories in optical phase-conjugate mirrors. Dual phase conjugate mirrors (DPCMs) [648, 649, 650] more appropriately model the bidirectional causality among apertures [19, 749] (Ch. 6) with a noise-induced threshold [747] consistent with the noise-induced critical point in a network such as a cortical aperture (Ch. 3).

Memory types

Retained memory implies a change in a "recording medium"; thus the instantiation of object memory is a transition from perception to memory. Memory will be considered to be of three types (Figure 45): *working* memory [808] or short-term memory of up to 5 seconds; *stabilized* memory that can be maintained until sleep, even if not currently active; and *consolidated* [809] or long-term memory that, obviously, persists for a long time. Dehaene and colleagues [32, 86] use the term global neural workspace (GNW) to describe working memory. Dehaene et al. [85] propose a taxonomy of different GNW processing states that I shall not address here. Perception occurs within working memory. Crick and Koch [627] associate awareness itself with working memory. I agree. Working memory can transition into stabilized memory, relying on the MTL. Stabilized memory is consolidated into long-term memory during sleep, as Henke [87] suggests. Consolidated memory may be enhanced with repeated exposures. Very long-term changes are outside the time scale described here, although they may be an outgrowth of the more intermediate processes. Henke cautions that memory does not necessarily require consciousness, thus should be understood from the standpoint of processing modes. Where memory resides has been a topic of considerable exploration. I shall attempt to address this issue.

Memory in Coherent Apertures

Memory is complex, involving many aspects of the functioning of the CNS. I have described these aspects in the foregoing chapters. I shall discuss working, stabilized, and consolidated memory in subgraphs of multiple apertures, and the corresponding roles of memory control in the MTL and cortex. I shall consider here the simple case of object or declarative memory with the operations of formation (instantiation), transitions among types, persistence, and access, as recognition and recall. Perception, recognition, and recall implicate the internal experience. Several themes emerge: memory is distributed within and among apertures, retrieval equals reconstruction, modulation can be elastic and plastic, and information instantiation may be undecipherably complex. I shall not address errors during memory formation, persistence, access, and reconstruction.

What is a memory?

In the coherent apertures model, a memory is an association of CMs, some of which are memory-specific (i.e., malleable), some of which are intrinsic, and some of which form as the result of feature response functions. Memory occurs in the context of aperture coherence, either actual or potential. Memory resides both within cortical apertures as CMs and among apertures in the relationships among CMs in (an) appropriate subgraph(s), an ensemble when active (Ch. 6). A memory-specific CM, resulting from malleable structures, has two functions: instantiating some aspect of information, and instantiating some relationship/association with other apertures, often CMs in other apertures. Although a separate association CM aperture is indicated in Figure 39, such a CM may be incorporated in, and distributed among, multiple apertures, some of which may also carry persistent CMs. A memory may include motifs—a set of intrinsically associated apertures.

Memory types in the coherent apertures model

One can consider memory as similar to energy, having kinetic and potential states. Working memory, as the evanescent internal experience, has an active (i.e., kinetic)—and changing—ensemble with interacting CMs in multiple apertures. Paradoxically, the interaction of the apertures causes the ensemble to emerge, as discussed in Ch. 6. A latent memory is quiescent (i.e., potential), it has no ensemble; it is instantiated among apertures in latent CMs. At best one can think of such a state as a "latent ensemble." It will, under the correct conditions, form an ensemble with the appropriate CMs active; hence their appropriate apertures will be coherent. The CMs empower the

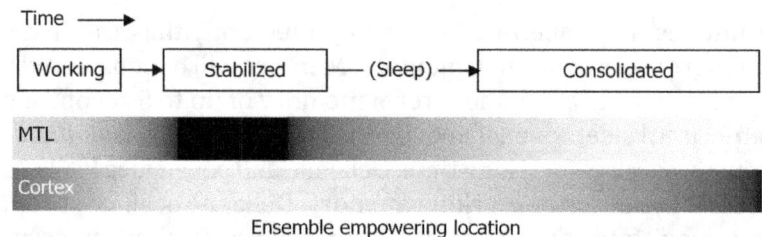

Figure 45. Ensemble empowering locations (dark is stronger).

existence and content of the ensemble, but must actively interrelate to form a working memory (Ch. 6). Memory instantiations correspond to the three memory types. For clarity, I shall consider a *memory* as latent and an *ensemble* as active; thus a persistent memory can be dynamically *resurrected* into an ensemble, and therefore into working memory. Memories are distributed widely rather than being in localized areas of the cortex [*768*]. The location of control and the form of memory within the subgraph differ for working (ensemble), stabilized, and consolidated types (Figure 39), with locations performing changing roles in maintenance and access (Figure 45). By "location" I am referring to the structures necessary for the persistence and recovery of a memory, not necessarily the actual location(s) of the memory.

Declarative object memory types incorporate many of the same components, with notable differences, as shown in Table 2. Memory, as defined here, is instantiated in malleable components. These malleable components may recruit intrinsic (fixed) components into an ensemble in working memory. The variability of participation of the MTL and associations in working memory reflects the ensemble's source. If working memory results from perception, then, initially, the MTL and associations do not participate strongly. The involvement of the MTL may be transitional for declarative memory. Stabilized memory requires the involvement of an MTL structure, essentially as a hub. Consolidated memory is distributed over multiple apertures in a network and does not require the MTL. A few areas, such as the pre-frontal association cortex, may act as hubs, or key registries, within the network in consolidated memory, maintaining relationships among structures (CMs) within multiple apertures.

	Working	Stabilized	Consolidated
Intrinsic			
Features	X		
Invariants	X		
CMs	X		
Motifs	X		
Malleable			
CMs	X	X	X
MTL	V	X	
Key registries		X	X
Ks	X	X	X
Associations (R_p)	V	X?	X
Transient			
CIS	X		
K	X		
Ensemble	X		

Table 2. Components in memory types; X = included, V = variable, X? = Uncertain.

Malleable components may become intrinsic with repeated exposure and/or a strongly aroused state, becoming potential resources for other ensembles. I am considering intrinsic CMs not to be part of a <u>specific</u> memory, but as members of a memory resource pool. Given that intrinsic and transient components are not part of stabilized or consolidated memory, but are necessary for a functioning ensemble in working memory (Table 2), one can say that a memory does not "exist," per se, but is instantiated in quiescent CMs and associations.

Memory in the CM

Introduction

The CM performs two functions. It instantiates some facet(s) of information in an aperture (Ch. 4), and it associates with other CMs in other apertures (Ch. 6). The CM is a result of the responses of the functions in an aperture; it may be undecipherably complex. A CM may result from locally modulated functions, varying along a spectrum from elastic to plastic changes. Memory—and perception—have two apparently disconnected mechanisms: synaptic strengths and phase modulation. They work together. The CM incorporates both. It is convenient to think of the dynamics of a pyramidal neuron in terms of its natural frequency (f_n) and phase (Θ) (Ch. 3). The natural frequency is an expression of how often, on average, a neuron will spike. It is an expression of the level of excitation of a neuron, here typically a pyramidal. The phase is spike timing relative to some reference, such as the AFP, LFP, or a local group average (e.g., the macrocolumn). The suggestive cortical circuit in Figure 6 in Ch. 3 illustrates many opportunities for small changes in synaptic and electrotonic coupling to affect the response of an element (i.e., minicolumn) to input projections. Synaptic plasticity and other modulation mechanisms, summarized below, are discussed in greater detail in Ch. 5.

Synaptic efficacy

Synaptic efficacy is the response, typically in net post-synaptic current, of a neuron to a specific synapse or a group of excitatory and inhibitory synapses and a temporal structure of their activations (neurotransmitter releases). The efficacy of synapses may be modified. Chaudhuri and Fiete [*800*] have produced a useful review of the current state of research and the issues on memory, particularly on long-term synaptic plasticity. Outside of lesser effects of ECS neurotransmitter levels, synaptic activity modulates the short-term level of a neuron's excitation, as its natural frequency (f_n). The natural frequency of a neuron (principally pyramidals here) will be reflected in the overall frequency of spiking. Synapse creation, modification, maintenance, and loss are influenced by astrocytes. As the synaptic efficacy is changed by long-term potentiation (LTP) and long-term depression (LTD), memory can be instantiated in those changes. LTP may occur when multiple inputs occur at nearly the same time, either requiring smaller inputs or providing a stronger response for the same inputs in the future. The converse is so for the LTD. The efficacy of a synapse may be modulated by the presence or absence of a neural spike within the target neuron's soma or dendrites, hence back-propagated from the axon initial segment (AIS), concurrent with an excitatory or inhibitory post-synaptic potential (EPSP or IPSP). Thus the activity level of a pyramidal in the future may be modulated by past experience—memory. Increase of the f_n of a population of neurons raises the effective coupling (as power due to spike frequency and number), increasing the potential for coherence across an aperture. An increase in f_n may increase the participation of a neuron in the cohered phase of a CM.

Timing

Spike timing will be influenced by the LFP, which is reciprocally pumped by the pyramidal's spike (Ch. 3) and other factors. Phase will be grossly modulated by the net synaptic input: the stronger the excitation the sooner a neuron will spike. In addition to LFP threshold shift effects, spike timing will be modulated by the electrical activities of local neurons coupled through the ECS and astrocytes [*239*]. Astrocytes may also modulate timing through gap junctions and buffering of the local

ECS—even at the synaptic level of granularity. Thus astrocytes may play a role in fine phase modulation. The pyramidal-chandelier circuit (PCC, Ch. 3) will preferentially gate a neuron's phase-modulated spike as within or outside a coherence window in a macrocolumn, and thus in or out of the cohered phase of a CM.

The CM

I will consider the CM in memory separately from the ensemble it occurs in. As described in Table 1 in Ch. 4, a CM may be described as latent or active, persistent or transient, extending the concept of a map of coherent elements to a map of potentially coherent elements in the latent states. A CM has an associated topological natural key, K, as a unique identifier (Ch. 6, i). Unless noted, a CM is active, projecting CISs and Ks. A CM is a phase field, a map of the dispositions of elements in the cohered and disordered portions of a coherent aperture (as described by Kuramoto and Battogtokh [442]) as reflected in elements' neural spikes' phases relative to the AFP. The formation and operation of CMs is described in Ch. 4. The AFP is synchronized across the coherent aperture in the gamma band (30–80Hz); the relative phase of the neurons' spikes within the gating window within the AFP of 3–8 ms constitutes coherence. Those neurons that are coherent comprise the cohered potion of the CM. Information is instantiated in the pattern and phase of the elements in the cohered portion. Small changes in phase may move a spike into or out of the gating window due to PCC binary phase quantization, discussed in Ch. 3 (see also Ch. 6, Figure 29. In addition to shifts in the internal cell potential affecting f_n, largely attributable to synaptic activity, the phase of spike timing is strongly influenced by the LFP; pyramidal spiking and the LFP are linked (LFP pumping, Ch. 3, Figure 18). As discussed in Ch. 3, a neuron's spike train may be considered a pulsatile discharge of the cumulative depolarization, probably linked to the LFP [347].

Active CMs occur in working memory during the internal experience. In response to CISs and/or Ks projected into an aperture, the PCC phase gating (Ch. 3, Figure 13) of the neurons' spiking produces an active CM, an analog to a binary phase hologram [470, 810], a distributed instantiation of information with an "output" of CISs and Ks. The active CM's gating window will coincide with the appropriate (negative, rising) AFP phase, as the two are interdependent (Ch. 3). During transitions to stabilized or consolidated memory, the coincidence of projections input from different apertures (Figure 39D_i, S_d, K_q, and R_p) will reinforce or disrupt synaptic coincidence, with a subsequent modification of synaptic efficacy (Ch. 5). The pattern of cohered neurons which comprise a CM will initially be stored in stabilized memory as LTP and LTD synaptic modulations that will affect f_n—the probability of spiking—and subsequently the phase in the context of PCC gating, when stimulated with appropriate input CISs and/or Ks. The result will be an active CM.

A CM may form from the pattern of responses of the aperture's functions (Ch. 5). There are specific cortical apertures that respond to specific sensory inputs [747, 471, 479, 811] with CMs that produce feature-specific CISs (e.g., oriented edges). Apertures with relatively fixed response properties have low malleability and hence low memory potential, although they may contain relatively fixed CMs (e.g., instantiating invariants). Some apertures may undergo relatively short-term modulations, e.g., contingent aftereffects, shaping the response characteristics of its CMs.

CMs in an aperture, e.g., F_j (Figure 40), produce CISs and Ks that are projected to multiple apertures, e.g., D_i, that have functions that respond to different features (Ch. 5) in the CISs. If the response, e.g., in D_i, produces coherence, an active CM, then a CIS (with K) is projected back to its source, e.g., F_j, and to other domain apertures; otherwise it projects only noise. Each domain aperture responds to different features, potentially receiving CISs from more than one domain aperture. The relationships among aperture CMs need to be maintained. One or more association apertures, R_p, map the relationship of CMs among the disparate domain apertures in the ensemble. The R_p's aperture CMs would be quite complex, approaching complete undecipherable complexity. D_is project to outcomes apertures (C_e), incorporating them into the ensemble, biasing the trajectory of the ensemble's resolution, as discussed above. The apertures that participate in a memory's ensemble may participate whole or in subsets in many memories with different CMs. A fixed, or intrinsic, CM responsive to a specific input, e.g., to a face invariant, could participate in multiple ensembles.

I speculate that consolidated memory incorporates both synaptic efficacy and phase modulation to instantiate information in latent persistent CMs, shifting progressively from synaptic to phase mechanisms over time. A stabilized latent persistent CM can be activated by an appropriate CIS or K, synaptically generating higher f_ns of the CM's neurons. Once activated, the stabilized CM can be converted to a robust consolidated form by incorporating more passive phase modulation, which results from homeostatic processes that reduce local phase differences. CM spikes are temporally converged toward a common PCC gating window in the process; those spikes which fall within the window are coherent. The resulting local phase cancelling is an inverse local phase modulation when the CM is no longer active, the phase modulations being retained in the consolidated latent state. Latent Ks are robustly retained in this form, obviating the need for the MTL.

Spiking neurons exchange ions and molecules with the ECS. Local spike phase differences in an active CM cause local ion and molecular concentration differentials that often require active energetic transport to restore normal balanced local CSF concentrations. Processes that reduce local phase differences move toward homeostasis by altering coupling and diffusion, subsequently providing phase modulation that counters the active local phase differences. Coupling through gap junctions with other neurons and astrocytes can directly influence spiking occurrence and timing, making astrocytes among the candidates as phase modulators. Astrocytes may buffer ionic and molecular ECS concentrations through selective channels in their end feet. As a neuron approaches spiking, local ion concentrations will change as voltage-activated ion channels open. An influx of Na^+ ions will decrease the external potential, increasing spiking probability. An outflow of K^+ will decrease the probability. Changes in Ca^{++} will affect the dynamics of the ion channels. Neurotransmitters, most particularly glutamate, will be scavenged or released by astrocytes. Thus modification of the terminal and internal properties of astrocytes may modify local phase relationships. The actions are complex, well beyond the scope of this paper. The result will be a modulation of local phase. The synaptically activated stabilized CM effectively "turns itself off" through local phase cancelling by synaptic modification, and diffusion and transport modifications, probably through astrocytes, moving to a consolidated state; hence, the CM requires lower energy expenditures to be activated.

The potential role of astrocytes in phase modulation and phase retention in memory is not well understood, but is a promising area of research [239]. I envision a model of two pyramidals connected by multiple separable unidirectional (e.g., diode-like) and modulatable conductance paths through astrocytes (by virtue of end foot properties and morphology). External ion concentrations affect spiking threshold, hence spike timing. If two nearby pyramidals spike at the same time, there will be little ion flow between their respective ECSs. If the input spikes have a time (e.g., phase) difference, the ion conductances between separated, but local, points will become asymmetric, tending to converge the spiking of nearby pyramidals into coincidence, reflecting a cancelling of the phase difference between synaptic inputs to the pyramidals. This aligns them such that phase quantization can bring them into the gating window. The modification of conductances provides phase modulation for consolidation of a memory instantiation. Subsequent input spike train pairs, with the appropriate time delays, will tend to align due to the pre-existing ion conductance differentials. The phase difference is "inverted" in the latent CM.[25] A pattern of coherence and phases of an income projection that matches the passive phase cancelling of part of a CM will activate a CM, as the aperture will cohere, activating the remainder of the pyramidals in the CM.

---

[25] Analogous to a phase conjugate mirror.

A CM is generated or activated by one or more input CISs and/or Ks that activate a pattern of neurons that cohere. If the activity is sufficient, the aperture will cohere, as evidenced by an LFP that is synchronous across the aperture, consistent with working memory. A stabilized CM may be reactivated by input CISs and/or Ks that partially match its synaptic efficacies, similar to a neural network model. A neuron's excitation level, as f_n, that promotes spiking within the local LFP gating window, decreases the spiking threshold, thereby increasing the probability of inclusion in the cohered phase of the CM. This strengthens the CM as the aperture coheres. Bidirectional causation among apertures will promote this development. A consolidated CM is activated when a specific CIS or K input preferentially evokes neural activity with local phase relationships that may not be strong, but with phase modulations that tend to move a neuron's spike in to or out of the PCC LFP gating window, which does not have sharp edges. The PCC gating promotes or inhibits spiking of multiple local elements with a subsequent strong phase differentiation through binary quantization. If the relative timing (local phase difference) between two spike thresholds is greater than the PCC temporal gating window, only the leading element will spike. For a preferentially phased modulated neuron, a weakly increased f_n will translate into participation in the cohered phase of a CM, requiring less energy for the activation of an instantiation.

Phase modulation that nulls the local phase differences in the latent state of consolidated memory allows the subsequent resurrection of the CM. The symmetry in the lead-lag relationship of spike order in PCC gating produces the same pattern of local activity in an active CM (Ch. 4) irrespective of phase lag or lead. The phase modulation in the latent CM will have reversed the phase relationship among coupled neurons. Local phase reduction through homeostasis provides an inverse local phase CM that can be responsive to (an) appropriate input CIS(s), analogous to the equivalence of a phase hologram and its inverse, each producing two symmetrical results, one of which can be masked [812]. A holographic medium may support multiple images; thus small phase modulations, the result of the superposition of multiple holograms, can still yield distinctly different results according to the characteristics of the illumination. Phase inversion is a property of phase conjugate mirrors [553, 651]. Dual phase conjugate mirrors [648, 649] can store [19, 497] and process [749] multiple images. The relationship between phase-reversed holograms and phase conjugate mirrors in this context bears further investigation. All memories occur in the context of ensembles that serve to drive construction or reconstruction through bidirectional causation. Phase modulated consolidated memory makes that particularly apparent with its dual phase conjugate analog. The bidirectional relationship between any two apertures in the ensemble makes comparisons to dual phase conjugate mirrors as interacting components in memory even more attractive. The topological alignment of Ks provides stable registrations among CMs. As noted, such a model is speculative.

Binary phase quantization can enhance recovery of a specific CM. Local phase relationships that define a persistent CM will arise from the modulations of responses to different inputs across an aperture. Different input projection patterns will result in slightly different phase structures. The slightly different phase structures will result in different CMs, as the local gating in the PCC will perform a local binary phase quantization. This allows multiple latent CMs to be superposed. The superposition of multiple phase CMs results in a blurring of all of them. Ahonen et al. [318] demonstrated that blurring of an image (face) can be reduced in the phase domain of the Fourier transform of the image by quantizing the local phase. The face image itself was considered the invariant to be recovered. There is a parallel here. The local phase difference was reduced to two states, 1 (π) or zero, analogous to the PCC phase gating. This process must occur on a transformed version of the image, not on the image itself. We can expect some similar process to operate in D_is in the deeper invariant forms of instantiations. This phase quantization should increase the separation among latent CMs during memory access as small local differences, elicited by an incoming CIS or K, will cause locally specific phase differences based on the synaptic and electrotonic instantiations of the CM.

The Memory System

Introduction

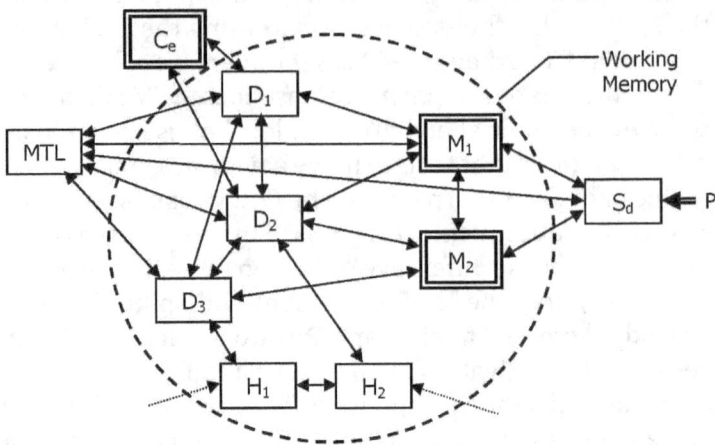

Figure 46. Suggestive relationships of apertures. See text. MTL = medial temporal lobe, S_d = primary sensory aperture, M_1, M_2 = motifs, D_1, D_2, D_3 = domain apertures, H_1, H_2 = high level apertures, C_e = outcomes motif, P = perceptual input.

What happens in an aperture, the CM, is only part of the story of memory. Having considered the underlying mechanisms that form CMs, let us turn to the operations of memory. Memory is a system. Working, stabilized, and consolidated memories have somewhat different forms. Memory is best understood relative to the ensemble, which will stabilize itself, although it is dynamically changing. A working memory ensemble—from any origin—has relationships among its apertures. During perception they are inherent in the perception itself. Recovering any memory to working memory results in an ensemble with its relationships. It is inappropriate to speak of a memory as if it were an entity. It is a set of relationships among CMs in a subgraph of apertures. Chapter 6, describing the integration of cortical apertures, provides much of the conceptual underpinnings for this section, particularly working memory. A working memory is composed of an active synchronous subnetwork, or subgraph, of fixed and malleable apertures (Figure 46). Fixed apertures (F_j) respond to specific features or facets of inputs with resulting transient CMs producing CISs specific to the particular input (s) (Ch. 5). Apertures instantiating invariants, some D_is, are similarly fixed. More malleable apertures (D_i) can retain CMs in either stabilized or consolidated CMs.

Memory type models

Memory has a divide-and-recombine strategy: memory is distributed among apertures. Oscillations among apertures are associated with learning [813]: memory formation.[26] In a broad review Sporns [43] addressed the usefulness of structural network models in providing an understanding of the underlying organization of CNS activity. Graph theory methods of vertices (apertures) and edges (connections or tracts) have been used to characterize connectivity models [679]. The Connectome project, in which Sporns is deeply involved [100], is mapping out connections in the brain, supporting network connectivity models, also supported by experimental evidence [208, 683, 685,

---

[26] Admittedly a simplification.

814]. CNS network models have two major structural forms, hierarchical [*81*] and hub [*324, 682, 683, 684, 686*], the two not being necessarily separate [*815*], as embodied in the TrueNorth neomorphic chip [*25, 606*] designed and built by IBM as part of DARPA's SyNAPSE project to model the human brain. *Motifs* [*647*] are functional subgraphs. I am interpreting "motif" to define a permanent or preferentially assembled group of functionally interconnected cortical areas that respond in some task domain, e.g., vision. Motifs may be parts (i.e., "building blocks," Milo et al. [*450*]) of larger structures [*449*] (Figure 46, M_1, M_2, and C_e), forming [*645*] or activating as pre-existing substructures through synchronization [*440*]. Intrinsic motifs responsive to specific features may provide rapid responses to inputs [*816*]. Such intrinsic motifs have been located in the cortex [*112, 207, 451, 644, 769*]. Dynamically assembled functional motifs rely on underlying existing connections [*117, 693, 700, 718, 760, 781, 817*].

Working memory is evanescent. Working memory occurs when an active subgraph (ensemble) has formed, with entropy minimization underway. During working memory all of the apertures necessary to "build" the internal experience are active; the sequence in which they are incorporated in an ensemble may differ between perception and memory access, the result being the same: an ensemble comprising the internal experience. For descriptive purposes, an aperture, in the context of an ensemble, is part of a composite node (CN). In addition to the cortex, corresponding parts of the thalamus and thalamic reticular nucleus, as core components of the CN (Ch. 6, ii), are deeply involved in the formation and maintenance of ensembles, providing necessary couplings.

A stabilized memory is latent, bound together in the MTL, which acts as a hub. Unique identifiers, Ks, in the MTL link to CMs in apertures. Those apertures that <u>can</u> respond to the Ks can form a latent ensemble. Not all of those apertures are malleable; thus not all are "true" memory. Hippocampal networks (in the MTL) may facilitate the coordination of multiple cortical areas through a more scale-free network, or some other network structure that is not isomorphic with a particular cortical aperture, decreasing the requirements for topological correspondences. Bonifazi et al. [*154*] found that such scale-free graphs in the hippocampus contain functional hubs. Combinations of such functional hubs may support the multiple associated natural keys necessary to bind together the specific CMs within multiple apertures. The ratio of K to CIS sizes will vary by aperture class, being high for MTL projections as the Ks projected to various apertures must be sufficient to address specific CMs (latent or active) (Figure 27), binding them together as stabilized memory, providing later access. Stabilized memory does not survive as such over a sleep cycle.

In consolidated memory the relationships among CMs in multiple apertures are embodied in Ks and CISs that have formed in negotiations through bidirectional causation (Ch. 6, iv). The Ks will be distributed among apertures. Some cortical areas, such as the pre-frontal association cortex, may instantiate more K relationships. Consolidated memory is persistent as long-term memory.

Memory formation: relationships among memory types

In Figure 47 I suggest the relationships among memory types and their formation. The relationships among memory types have anatomical correlates. All cortical areas are connected directly or indirectly to the thalamus. The thalamus, as part of the CN, is involved in the ensemble, linking cortical areas. The MTL is a cortical system [*770*] containing the hippocampus and entorhinal, perirhinal, and parahippocampal cortices, underlying working and stabilized memory. The importance of the MTL increases as memory progresses from working to a stabilized form, decreasing in a consolidated form (Figure 45). It is usually essential for the transition from a stabilized to a consolidated memory unless the strength (e.g., emotional impact) of a working memory is sufficient to consolidate it directly. The perirhinal cortex appears to play a role in perception itself [*763*]. Perception of declarative information (often referred to as "what") relies on bidirectional communication between other cortical areas and the perirhinal and lateral entorhinal areas; contextual information (often referred to as "where" or "how") flows through the parahippocampal cortex and medial entorhinal area. Both systems ("what" and "how") communicate bidirectionally with the hippocampus [818], with ablative tests of any of the subcomponents producing differential recall and recognition deficits: not a simple relationship. For example, damage to pathways may also induce changes in memory performance that are difficult to separate from the MTL proper [819]. In addition to other cortical areas, the MTL interacts with the thalamus [820]. Tsivilis et al. [819] found that communication from the hippocampus via the fornix and mammillary bodies to the thalamus was necessary for the formation of long-term (consolidated) memory, but not for short-term (stabilized) memory. One might posit that it is this latter communication that underlies transition of stabilized memory to consolidated. A loss of mammillary body volume was accompanied by loss of recall but not recognition performance. Explorations of memory must consider the time between learning and testing because stabilized and consolidated memory differ in form. Structures explored and the relative timing of acquisition and testing make studies of memory sensitive to experimental methods [821, 822] and memory type (object versus episodic)[823]. Spatial and episodic memories also rely on the hippocampus [676, 824] but are outside of the scope of this paper.

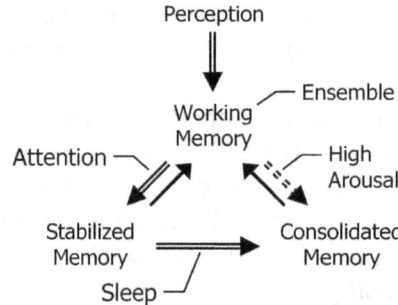

Figure 47. Relationships of memory formation
(double line) and access (single line).

Perception occurs during working memory [763]. Perception incorporates the associations among existing persistent structures (e.g., invariants) and features (e.g., feature responders) in apertures into an ensemble, exchanging CISs and Ks within a subgraph (Figure 27, Figure 39). Working memory entails the coordinated synchronization of multiple cortical apertures [625, 825] recruited into, and excluded from, an ensemble, as described in the flowchart in Figure 31, Ch. 6. As described above, during perception of declarative information, in working memory there is no instantiated memory beyond the transient activity of CMs in the current transient ensemble, essentially a resonance phenomenon that can persist for a brief period. Mitchell et al. [820] found that the magnocellular mediodorsal thalamic nucleus (MDmc), which has sensory and pre-frontal association cortical connections in rhesus monkeys, responds to fast, low spatial frequency visual inputs with high frequencies. This high temporal frequency characteristic contributes to the establishment of a coherent subgraph (ensemble) during perception, essential to the establishment of a working memory of the image—working memory necessary for the establishment of a stabilized memory and subsequent memory consolidation, even though the MDmc is no longer necessary for later access. Mitchell et al. did not test on the day of acquisition, presumably while the animals recovered from the lesioning surgery, so it is not certain what the role of the MDmc is in stabilized memory (i.e., same day). I presume that the MDmc will have a transient role during the transfer of associations from the MTL to other cortical areas. Initially the medial temporal lobe (MTL) is not involved in working memory, but is required for stabilized memory; thus there will be a transition period of MTL involvement.

As working memory occurs in the cortex, cortex must be continually released from its current active ensemble for working memory to progress to incorporate new information. Working memory must be either lost or transferred, most commonly to stabilized memory. An ensemble is able to instantiate only small amounts of information for short time periods [*808*]. Stopping the ensemble, for example, by stopping thalamic function (Ch. 6), stops consciousness [*481*], as working memory, and prevents its stabilization. Optogenetic disruption of working memory (in the mouse) immediately after a learning period impairs information retention, with a subsequent loss of the ability to perform cognitive tasks [*826*]. Memory stabilization requires persistence of the relationships in an ensemble's subgraph structure that are formed during the internal experience, presumably via the MTL. fMRI scans demonstrate the significance of the MTL for working and stabilized memory in humans in recognition tests on the same day as learning [*827*]. Attention is required to graduate working memory into stabilized memory. Attention can arise from several sources. Emotional (limbic) activity will raise the level of activity, fostering persistence of the CM; such an increase in activity itself raises an aperture toward coherence by raising the f_ns of its pyramidals (Ch. 3). Attention may arise from higher level apertures embodying cognition joining the ensemble through matching of CMs, or Ks, thereby reinforcing them, and increasing the persistence of the ensemble and of the CMs within the coherent apertures comprising the ensemble.

Ks are links among CMs either directly or through the MTL. MTL is differentiated here from other cortex for functionally descriptive purposes. The MTL (e.g., hippocampus) provides stabilization of, at a minimum, object memory, and thus has a role in the transition from working to (latent) stabilized memory. During the transition, as the duration of the ensemble extends past the period of working memory, the apertures are projecting Ks to the MTL, achieving stability by virtue of the linking of Ks, which acts as a hub. In establishing stabilized memory, CMs in different apertures generate Ks (subsets of CISs) that are registered and cross-mapped within the MTL through synchrony. The MTL and domain apertures bidirectionally exchange projections, some contents of which are coherent as CISs and Ks. These Ks are cross-mapped within the MTL during the ensemble's brief life, if attention is adequate. Every K does not need to map to every other K, the mapping probably reflecting the network connectivity of the apertures. It is all that is needed. As the Ks in each aperture have similar topological registrations, upon convergence in the MTL, and in other apertures, local synapses firing coherently will reinforce each other, potentially increasing synaptic efficacy through LTP, decreasing non-coincident synaptic activity through LTD (Ch. 5). As projections among apertures spread, exact registration of all points comprising a K are not required, allowing less distinct, "fuzzy," Ks. The number of projecting elements is high; consequently statistical and proximal matching is sufficient. Inhibitory interneurons will serve to sharpen the K responses in the MTL through cross-inhibition. The net result is a structure of synaptic efficacy that is well-defined, a limited K association CM. This is the start of stabilized memory. As the structure linking Ks stabilizes in the MTL, the reliance on the primary sensory cortical activity to maintain the ensemble decreases. Clearly, synchrony of coherent structures converging in the MTL and in domain apertures is necessary to achieve coincidence-based synaptic enhancement. The ensemble provides this synchrony (Ch. 6).

The MTL provides a transition from working to stabilized memory, providing access to stabilized memory [*828*], but it is not required for consolidated memory, as is evident in memory tests at varying times after learning. There is a delay-dependent decline in the effects of loss of the hippocampus in the formation of long-term contextual memory, i.e., mouse freezing shock situation, with losses decreasing with ablation delays over a day [*822*]. Once a stabilized memory has formed it may be transformed into consolidated memory during sleep [*809, 829*], probably entailing reactivation of some of, or the entire, ensemble, perhaps in dreams.

Dreaming is consistent with the notion that the structures and mechanisms underlying a memory are malleable only when they are components of an ensemble, an internal experience, with little resulting persistent awareness. The consolidation of a stabilized memory into long-term memory requires change someplace. Boyce et al. [830] found that optogenetic disruption of GABA neurons in mice during REM sleep disrupted the consolidation of stabilized memory into long-term memory. This is consistent with a model of partial or full ensemble resurrection during REM as being necessary for re-establishing the associations among apertures' CMs to transfer those relationships from the MTL to CMs, diminishing the role of the MTL. The process is largely unknown, but one can speculate that CMs' neurons' spike timings (phases) are modulated toward a coherence gating window (Ch. 3), lowering the effective threshold. As astrocytes are involved in synaptic modulation, electrical coupling, and maintenance of the ECS, one can presume they play a role in establishing consolidated memory.

Carr et al. [828] suggest that memory consolidation occurs in both awake and sleeping states as replays involving the hippocampus, with interactions with the cortical apertures [831]. During sleep the hippocampus participates in the replay of subgraph sequences of visual experiences while awake [472, 832]. Sleep may not be required for the consolidation of strong emotion-laden memories [833, 834], raising the issue that arousal during dreams may be related to memory consolidation during sleep [835], although this does not necessarily mean that all dreaming is involved in memory consolidation. We frequently dream of things long past or with no apparent current reference. As thalamic involvement is required for the internal experience [80, 481, 727] (Ch. 6), its role in the transfer from stabilized memory into consolidation may be through a reactivation of the ensemble or some of its fragments into a brief working memory (e.g., dreaming), forming associations in the cortex. This is supported by the work of Mitchell et al. [836] (see above), who report a large thalamic nucleus (MDmc) with sensory and pre-frontal association cortical connections. The MDmc is active during some visual experiences, e.g., working memory. Consistent with the natural key model, Behrens et al. [837] propose that complex response patterns in the hippocampus in some way represent information that is transferred to the cortex to promote memory consolidation. The formation of persistent memory has a hippocampal-cortical interaction that Lesburguères et al. [476] term "tagging," supporting early associations among apertures that are required for subsequent long-term (consolidated) memory, not inconsistent with natural keys. The Ks in the MTL assume less importance as memory consolidates, gradually extinguishing. As a memory becomes more consolidated during access to the stabilized memory, its distributed CMs have more sparse Ks, their power residing in their increased alignment. The probability of less distinct (fuzzy) Ks in multiple apertures being simultaneously exercised provides specificity while retaining robustness, akin to signal reconstruction from sparse sampling as described by Candès et al. [838].

Memory access

A memory is worthless if it cannot be accessed due to failures to record, retain, or access it properly. I shall not deal with such failures here. I'll not discuss the lability of consolidated memories during access and potential reconsolidation. This description of memory access is speculative as it is. I shall briefly consider recall and recognition. Working memory, stabilized memory, and consolidated memory incorporate controls that enable ensemble operation through the synchronous bidirectional exchanges of CISs and Ks among coherent apertures (Ch. 6, iv, v). Working memory is an ensemble; the other two memory types must enable the re-formation of an ensemble for the recall of a perception. During the re-establishment of an ensemble, in addition to cortical apertures, the thalamus and thalamic reticular nucleus (TRN) are involved as components of composite nodes (Ch. 6, ii).

Memory access reinstates the ensemble's internal processes, differing for recognition and recall. A memory may incorporate non-malleable apertures as reusable components, for example, feature responders and invariants. The order in which feature responders are incorporated into an ensemble is one of the factors which separates a perception from an accessed memory: leading in perception, lagging in recall. During working memory, an ensemble resonantly reinforces aperture CMs through bidirectional causation, largely embodied in the transient components, with neural excitation, as f_n, resulting from synaptic inputs and the apertures' functionals' responses (Ch. 5), seeking an entropy minimum (Ch. 6, vi). We can consider two paths for resurrecting a working memory for recognition or recall, one for stabilized memory, and another for consolidated memory. Stabilized memory access will recruit the component apertures' synaptically stabilized CMs by linking them through projections from an MTL K association CM map, as the inter-aperture associations are weak, if at all. If the ensemble is derived from stabilized memory, MTL participates in the ensemble by maintaining its binding K relationships among the CMs in malleable apertures and intrinsic components. The role of association apertures other than in the MTL in stabilized memory is uncertain, as the K structure in the MTL may be sufficient. Motifs would incorporate some internal links. The MTL does not participate significantly in consolidated memory. Accessing consolidated memory occurs directly through associations in some or all of an instantiation's apertures. The Ks in consolidated memory may—or may not—differ from those in stabilized memory. Presumably a working memory could influence the instantiation(s) from which it was resurrected, following the general progression through stabilized and consolidated stages. Such an idea is unverified.

Not all elements in a CM are necessarily connected. How can a fragmented latent CM be resurrected? What is fragmented in one aperture can be effectively bound together in others. Although there may be islands, minicolumnar elements will have some binding within their macrocolumns, providing a spread of coupling. We can presume that there are divergences from a strict topological mapping between apertures due to axonal arborizations and extended receptive fields. As evidenced by colinearity perception, some of these can be extensive. Using the natural key abstraction, the natural key in one aperture which is part of the maintenance of the associations that binds CMs together in multiple apertures will span the unconnected spaces in other apertures. Thus activation in one aperture can, through the multiple Ks in other apertures and itself, bind together the fragmented CMs, building bridges among the islands.

I'll first discuss recognition, roughly modeled as a matching process at an abstract level. This will provide concepts useful in discussing recall, modeled here as a response to an internal or external prompt. In both cases access will differ according to the type of memory: stabilized or consolidated. I shall not deal with access within working memory. I suggest that it is a direction of attention within an operating ensemble; Crick and Koch have addressed this in a searchlight hypothesis [626, 627, 702]. In doing so I am also sidestepping a differentiation between the internal experience and attention.

Recognition and recall differ substantially. Recognition is a matching of a sensory input with a stored memory, not necessarily recalling it, while recall is a reconstruction of a perception. I suggest that recognition without recall entails a partial overlap of an ensemble (working memory) with some of the latent CMs that would comprise an instantiated memory of the target, reinforcing the ensemble sufficiently through those CMs associations to cause the experience of familiarity. "Familiarity" is a rather vague term I'll not explore further. My intention is to provide a plausible concept which, although it might not be the right one, contains appropriate components.

Recognition, the perceived registration of a new stimulus with an existing memory, illuminates the interaction between perception and memory, as the implication is that a new stimulus matches the memory. Such a 1:1 correspondence is unlikely at the level of the primary sensory cortex, as scale and orientation differences alone would interfere [10]. DiCarlo and Cox [430], and Afraz et al. [839] argue that the solution to object recognition arises in a population that reflects invariant properties of the object rather than in the responses of a single neuron. I agree. Perception and memory use largely the same apertures and therefore the same CMs within those apertures; consequently recognition and recall use the same apertures and CMs as perception. The association CMs capture the relationships among features and higher level constructs (i.e., CMs) that comprise an invariant object representation. This is consistent with the argument of Neven and Aertsen [806] that coherence embodies invariant relationships in object recognition. Lowe et al. [805] modeled object recognition as based on image features largely invariant to transformations. Invariance is preferential for an object memory versus its figural representation, as invariance corresponds to shape constancy in spite of representation transformations. Kourtzi and Kanwisher [784] demonstrated this, finding that shape representation was invariant in the lateral occipital lobe for different contours, e.g., partial occlusion versus unoccluded contours of the same object, but not when the shapes differed in 3D when the contours were the same but the objects differed through a stereoscopic figure-ground reversal.

Recognition must involve instantiations including higher, and more abstracted, levels. As higher levels involve essentially distributed feature responses, one must presume that recognition—and hence perception and recall—does not occur within a single aperture. As such an operation occurs within an ensemble, it is likely that primary sensory cortices are involved in much the same way that outcomes bias ensemble resolutions (which is beautiful, if it is true). I have not attempted such a functional decomposition. It may not even be possible beyond general descriptions of aperture functionals: such is undecipherable complexity. I may be accused of dodging an essential issue here. I do not think this to be the case, as neither perception nor memory appear to be "encodings." As noted earlier, I have adopted the term "instantiation" to make clear the idea that embodiments of information in the central nervous system are physical, not symbolic, distributed within and among apertures. As described in Chapter 4, information is instantiated in CMs in the infragranular cortical layers, consistent with the finding by Lopez-Aranda et al. [840] that infragranular layers in V2 are implicated in object recognition memory. The experiments involved manipulations of proteins specific to the infragranular layers, supporting an infragranular model of instantiation. This may confuse the responding elements with the sites of instantiations, however, as instantiations are modulations of functions, which may involve supragranular layers as well.

Recognition, matching a stimulus with an internal representation, involves the formation of a subgraph of coherent apertures with active CMs that collectively largely match the subgraph of latent CMs, or their Ks, that comprise a memory, demonstrating a relationship between a memory and perception. A person may be aware of the match, perhaps to the point of resurrecting the matching instantiation to reconstruct the original percept, a directly prompted recall. As perception and object memory occur in the same place, recognition can occur within a few hundred milliseconds [841]. Sehatpour et al. [677] found that during object recognition, the coherence of networks that included the hippocampus was significantly higher during the recognition of recognizable sparsely fragmented images than scrambled images of familiar objects. This illustrates a convergence of perception and memory, although it calls into question the considerable body of evidence that the hippocampus is not required for recognition. As the stimulus exposure and recognition tasks for each subject were performed in a single day, one must presume stabilized, hence MTL-based, memory. One might presume that during recognition in this paradigm the organized fragmented images could rapidly form a stabilized instantiation based on the associations of the original CMs. The transition of working memory into stabilized memory entails an increasing MTL involvement (Figure 45), as the original percept was not resurrected.

Recalling a memory resurrects its ensemble in response to some external or internal prompt. Recall can be initiated by an internally generated "probe" of a partial ensemble or an ensemble serving some other cognitive function, incorporating coherent apertures that contain sufficient active CMs producing CISs and/or Ks that activate latent CMs in target apertures either directly or through the MTL. If a probe's projections are able to elicit a CIS back from a connected aperture, that aperture will have become coherent, available to be recruited into an ensemble (Ch. 6, v), enhancing the CMs' orderliness through bidirectional causation (Ch. 6, iii). Disordered apertures will be excluded. In the process the ensemble forms or changes, working memory progressing. As a perception forming a memory may be built on invariants, recall does not need to recreate the actual initial perception, but a comparable internal experience; thus one might envision a 3-dimensional object seen from a new direction. Recognition can serve as a prompt for a recall. The ability to rapidly re-perceive the hidden cowboy on a horse (Figure 41) is apparently a combination of recognition and recall. The image probably produces subjective contours in forming the perception. This reflects an ensemble with distributed CMs and Ks, the apertures necessary for the perception itself being numerous and with different functionals, as the perception of subjective contours (Figure 42) also has a moderately high level of abstraction. The active CMs in the probe ensemble will project to the MTL or other apertures, as apertures project to all connected apertures, although the coherence of the projections will change as the ensemble emerges. For example, by thinking about the target, enough of its subgraph is instantiated to initiate its ensemble as a recall. As Ks are imprecise (i.e., "fuzzy"), recall can have intrusions from similar internal experiences, potentially contaminating the reconstruction. Stabilized memory requires MTL interactions for recall, as the MTL acts as a K association hub with other cortical apertures in a coherent subgraph, evident in frequency and pattern characteristics [676]. This is not true for the recall of consolidated memory, for which association hubs are distributed elsewhere in the cortex.

7.5. Results

Principle and Corollaries

Principle:

- An aperture may contain a pattern that, when active, is appropriate for activating one or more CMs, including those in other apertures.

Corollaries:

- An active or latent CM may provide associations between or among other active or latent CMs.

- A CM may be latent, able to contribute to the self-assembly of an ensemble when receiving a critical level of appropriate excitation.

- Outcomes with higher probabilities of occurrence influence the resolution of the ensemble.

- Not all apertures in an ensemble may have instantiating CMs when the memory is latent.

Summary

The CNS does not perform "information processing": it performs biological processes on biological structures. The coherent aperture model presented here provides a framework for basic operations of perception and memory of declarative objects within the cortico-thalamic network. The model proposes that working, stabilized, and consolidated memory are related subgraphs of apertures, not a completely new idea; however the dynamic of the subgraph builds on coherent apertures, specific to this model. A coherent aperture is a functionally distinct cortical area. When coherent, an aperture is grossly synchronous, able to communicate bidirectionally with other coherent apertures, potentially creating coherence in them. Synchronization provides a de facto filtering of aperture communications. This synchronous communication underlies the formation of the ensemble, a resonant subgraph of coherent apertures, which are features of the coherent apertures model. The ensemble model is drawn from concepts demonstrated by researchers using multi-laser systems. Perception of declarative information (i.e., visual perception of objects) occurs as an ensemble, which is often referred to as "working memory." An ensemble, through entropy reduction, resolves to a solution of its components' activities. As the ensemble is an emergent integrated whole, its solution is shaped by both the input apertures (e.g., perception) and outcomes apertures (e.g., decisions). Other apertures provide ensemble building blocks of various forms: motifs, invariants, fixed feature responses, and other domain apertures.

An ensemble comprises the internal experience (working memory) through which intermediate (stabilized) and long-term (consolidated) memories can be produced and accessed. Information is instantiated in apertures in coherence maps (CMs), patterns of cohered (or potentially cohered) elements in a field of disordered elements. Some apertures will contain instantiations that are undecipherably complex. An active CM may be formed within a coherent aperture as its functions' responses to input projections from other apertures and the thalamus, producing projected coherent information structures (CISs), comprising the synchronous communication among apertures. A CM may assume a latent persistent state derived from an active state, embodied in synaptic and phase modulating structures. Instantiated latent memory can be modeled as latent CMs in multiple apertures associated by natural key (K) links either directly among CMs in consolidated memory, or indirectly through the MTL in stabilized memory. Recall and recognition memory access occur through partial matches of latent Ks and/or CMs of the originating ensemble as responses to external and internal prompts that provide partial subgraph probes.

Epilogue

Introduction

The coherent apertures model grew out of my work in visual-spatial and auditory perception. Consolidating a body of existing experimental and theoretical research by others in a theoretical model was my objective: I am pleased by what has been accomplished. I am aware of issues yet to be addressed, and of some incomplete explorations. I welcome the prospect of collaboration in many areas.

The coherent apertures model proposes an explanation of how biological structures and processes generate the internal experience. The cortex is composed of many functionally distinct areas—apertures—containing many elements—minicolumns. An aperture may become synchronous over its area, as indicated by a local field potential (LFP) that is synchronous over the aperture, an aperture field potential (AFP). Information can be instantiated in an aperture through the complexity of individual elements' activities in the context of the synchrony; therefore the activity is coherent with respect to the synchronous AFP. A virtually infinite number of different instantiations may occur in the context of an aperture's synchrony. Different aspects of information will be distributed among multiple apertures. Apertures will have links specific to the information jointly instantiated. Linked synchronous apertures form a synchronous instantiation network—an ensemble—which is the internal experience. The aperture instantiations, with the links, can be stored in memory. A memory can be revived through the links among instantiations, which will cause the original ensemble to be resurrected, bringing the memory back into the internal experience. I have considered the internal experience to encompass consciousness, awareness, working memory, the global neural workspace, attention, and perception, without differentiating among them. The brain may be considered a hyperspace of aperture subspaces. A subset of these subspaces, an ensemble, forms the internal experience as an intermediate hyperspace. Connected subspaces. which may contain subspaces, may have some common subspaces through bi-directionally causal interactions.

I shall review some principles and their corollaries, discuss some issues, speculate on the implications of the current model, and suggest some future explorations.

Principles and Corollaries

I propose ten principles that emerged in the process of developing the coherent apertures model. Chapters 3–7 each embody one or more principles, with some resulting corollaries.

Ch. 3. The Coherent Cortical Aperture
Principles:
- The cortex is composed of functionally distinct areas—apertures (parcellation).
- A cortical area may synchronize.

Corollaries:
- Not all elements need to synchronize for the cortical area, or aperture, to synchronize.

- Synchronized elements may have local phase differences, i.e., (cohere).
- Aperture coherence may occur in response to coherent inputs.
- Aperture coherence may emerge.
- Concepts from radiant energy systems may be applicable.

Ch. 4. Information Instantiations in Apertures
Principle:
- A synchronous aperture can produce a coherence map (CM), a phase field of a pattern of synchronized and disordered elements.

Corollary:
- A CM may be characterized by its information entropy.

Ch. 5. The Aperture Operator
Principle:
- The responses of the cohered functions within the CM produce one or more coherent information structures (CISs) that are embedded in the projections of activity from an aperture.

Corollaries:
- A functional of an aperture is composed of elements performing some essentially common functions.
- The CIS instantiates some aspect of information in the synchronized portion of the aperture's projection.

Ch. 6. Integration of Coherent Cortical Apertures
Principles:
- Cortical apertures form a network graph.
- Communications in the cortical graph are bidirectional (with a few exceptions).
- A cohered aperture is a node in a subgraph.
- An aperture that is synchronized or about to synchronize will filter to accept CISs that are in the appropriate phase.
- Synchronized apertures seek a low net information entropy.

Corollaries:
- There is no coherent communication if both apertures in a pair are not cohered.
- An disordered aperture does not synchronize with other apertures.
- A subgraph of synchronized apertures will synchronize into an ensemble.
- Cohered apertures that are synchronized in an ensemble will comodulate through bidirectional causation.
- The internal experience is defined as the current ensemble.

Ch. 7. Operations
Principle:
- An aperture may contain a pattern that, when active, is appropriate for activating one or more CMs, including those in other apertures.

Corollaries:
- An active or latent CM may provide associations between or among other active or latent CMs.

146

- A CM may be latent, able to contribute to the self-assembly of an ensemble when receiving a critical level of appropriate excitation.

- Outcomes with higher probabilities of occurrence influence the resolution of the ensemble.

- Not all apertures in an ensemble may have instantiating CMs when latent.

Issues

The concepts I have proposed offer a variety of issues for future exploration. The CNS has biological and physical processes with information physically instantiated. A computer model is appealing, as the +1's and -1's of synapses appear similar to a binary system, a digital logic system. A computer may be used for information processing. One puts in data, runs a program, and out comes some product: text, numbers or images. From a programmer's standpoint, data is structured and the processes are realized as operations. His or her objective is to define the proper set and sequence of operations and run the data structures through them—a gross simplification, of course. At a deeper level, the data structures are simpler. The high level processes are composed of a set of low level instructions, at the lowest level, processor op codes. The low level stuff is what allows the computer to operate. I propose a model that is not consistent with a computer model. What is the validity of this approach?

Biophysics of Neurons and Local Circuits

In the context of the coherent apertures model, the production and modulation of neural spikes are core subjects. The generation of spikes is modeled as the result of an integrate-and-fire process: when the potential of a neuron has been raised to a threshold, it produces a spike. I offer a modified version of this model. Consistent with the aperture synchronization processes, a neuron may be considered an oscillator with a natural frequency, f_n, which is a function of the activity level of a neuron. Clearly the f_n and integration will be associated: the higher the f_n, the more frequently the neuron will reach a spiking potential. A perceived limitation of the integration component of the coherent apertures model is the relationship between electrical potential and spiking, namely, why do pyramidals often produce spike trains? One can posit that synaptic inputs producing excitatory (EPSP) and inhibitory (IPSP) post-synaptic potentials change the potential in a continuous fashion. A spike is a unitary discharge of that potential: a pulse. A specific amount of charge is transferred when a neuron spikes. A strong input may produce such a large net influx of current (charge) that a single spike may not be adequate to discharge it, thus initiating a train. What are the effects of limited ion motility within and immediately exterior to a neuron? How might this affect the timing of a spike train? How does this interact with the pyramidal axon initial segment-chandelier cell gating circuit (PCC)? Neurons preferentially spike during a particular rising phase of the LFP, presumably reflecting some change in the threshold. A discrete pulsatile spike will produce an external decrease in potential, presumably the basis of the proposed LFP pumping and a subsequent pumped AFP. A pulsatile discharge model is particularly effective in LFP pumping. How valid is a monopole spike model, given its apparent violation of Kirchhoff's circuit laws?

Synchronization in the Aperture

Synchronization of a cortical area, an aperture, is well-documented, coherence less so. What is the temporal relationship between synchronization and coherence? In the conventional Kuramoto model, array synchronization rests on a critical level of coupling strength being required to achieve self-synchronization. In my expansion of the Kuramoto model, a relationship between neuron-to-neuron coupling strength and neuron activity, as f_n, is proposed. How valid is it? What are the causes of f_n? The pumped AFP model of synchronization I propose needs rigorous validation. The distributed, non-oscillatory, quasi-random, phasic pumping has an analog in lasers; that is not a proof. In the coherent aperture model I propose that the expanded array and pumped AFP models operate in concert to create synchronization. Either might suffice. The laminar cohering model is speculative. It represents an attempt to resolve the different frequency profiles in the SG and IG layers. The possibility that the supragranular layer could sustain its own synchrony as a propensity for the infragranular layer to synchronize as a laminar cohering mechanism is attractive, but needs more experimental and theoretical support.

Information Instantiation

Although the brain performs physical processes on physical structures, there must be some relationship between information and its instantiation. I am proposing that the coherence of activity—or potential for coherence in a latent form—is a significant aspect of information instantiation. This is modeled as a coherence map (CM), a phase field of synchronous and disordered elements. A CM may instantiate information not only in its phase field, but also in the temporal relationships, or phases, among its synchronous members. While synchronization with two phases has been modeled and demonstrated by others, the relationship of two-phase synchronization to information instantiation is less clear. The coherent information structure (CIS) projected from an aperture, emanating as it does from a CM, is proposed to contain that instantiation—or part of it. Further, as instantiations are distributed among apertures, a linking identification system of natural keys, subsets of CISs, is proposed. Is this simply a convenient concept or does it have any reality? How would one know? As noted, the brain performs processes on structures. These processes are modeled as transforms of an aperture resulting from its functionals, which are composed of individual functions. The results are CMs and CISs. These have information entropies, putatively related to the energies necessary to create and maintain them. It is proposed that within and among apertures, processes operate to minimize energies, and hence entropies. How can this be tested?

Multi-Aperture Processes

Multiple synchronous apertures form a synchronous subnetwork within the larger cortical network. Which comes first: aperture synchronization or subnetwork synchronization? Conceptually they could co-evolve. A synchronized aperture is modeled as a node through application of the van Cittert-Zernike theorem. This does not seem to be a stretch, but needs study. The suggested composite node (CN) is inferred by combining various reports in the literature. It simplifies explanations, but is by no means proven. For example, the contribution to synchronization through electrotonic gap junctions in the thalamic reticular nucleus is unknown. Lumping structures together in a CN dodges the need to be explicit in the role of each structure in synchronization. As almost all cortical apertures are bidirectionally connected and synchronous in an ensemble, bidirectional causation with subsequent comodulations and ensemble entropy minimization is reasonable, although unproven. The operating processes will result in existing potential outcomes biasing the ensemble's resolution, or consensus. Further exploration is needed.

A Test of the Aperture Coherence Model

Introduction

The coherent apertures model rests on a cortical area being able to instantiate information within a coherent structure comprised of a subpopulation of elements and their relative timings. The coherence occurs within a context of synchrony. Control of synchrony should affect the cohering capabilities, and thus information instantiation. The degree of synchrony should subsequently affect the internal experience. Within the current model, the primary source of the internal experience—either directly or recalled—is perception. An objective of a test of the coherent apertures model is to modulate aperture synchrony, subsequently affecting perception. An aperture that is in, or close to, synchrony should result in a more perceivable stimulus. The visual domain is an appropriate mode, as the visual input is mapped fairly well, through retinal ganglion cells and the LGN, to the primary visual cortex (V1). The assumption that a retinal stimulus maps precisely to a V1 stimulus is not true, but is intended here as a first approximation. The two cortical aperture states, synchronized versus disordered, should correspondingly create improved versus degraded stimulus perception. Visual attention to a display screen will probably result in a grossly synchronous V1. An experimental objective is to create two states, one which fosters synchronization, the other impeding it. Additionally, the probe stimulus itself should have minimal interaction with the synchronizing field.

The Experiment

The visual system is sensitive to overlapping spatial frequency bands. The overlaps support the integration of the visual scene components. Figure 24 in Ch. 5 illustrates the breakdown in stable image perception in the absence of spatial frequency overlaps. The spatial frequency bands (which do not have sharp frequency limits) result from the receptive field characteristics of the retinal ganglion cells (RGCs), generally modeled as circular, having difference-of-Gaussians (DOG) characteristic sensitivities [516]. The RGCs also reflect the distribution of cones in the retina, being denser in the fovea. The RGCs project to the lateral geniculate nucleus (LGN) in the thalamus where, with modulations in time and contrast, the activity is passed on to V1 through magnocellular (M) and parvocellular (P) neurons. M neurons have greater sensitivities to higher temporal, lower spatial, and higher luminance contrast than P neurons, which have the converse. Only the P neurons have color sensitivity.

Two stimuli will be used: aperture stimuli (AS) to modulate apertures synchrony and a probe stimulus (PS). Aperture synchrony is in the gamma band of 30–80Hz, thus the AS should use lower spatial frequencies, higher temporal frequencies, and low contrast. The probe will have high spatial frequencies, low temporal frequencies and high contrast. As described in Ch. 3, not all elements within an aperture need to be firing synchronously for an aperture to be synchronous; in fact there is little benefit above 30–40% participation. This finding is incorporated in the pumped aperture field potential (AFP) model, which speaks of a "scintillating" field of activity. Thus, it is not necessary that any given neuron fire continuously to produce a synchronized field, only that some threshold percentage fires at approximately the same time when they fire, a "volley" model. This description of some characteristics of the visual system provides a basis for testing the hypothesis that a synchronized or nearly synchronized aperture is more sensitive to a probe stimulus than a disordered aperture.

Aperture stimuli (AS) can be created that are large (i.e., low spatial frequency) by mirroring the DOGs in the retina (Figure 48). The DOG structure has a net luminance of zero, to which some mean luminance intensity, I, can be added. Thus any activity of such stimuli will not affect the overall luminance of the display. A monochrome display is assumed. As the conditioning background stimulates P channels, the ASs have low contrast. The intent is to have timing and location of AS activity mirror that desired in V1. AS sizes can vary over approximately two octaves, being larger than the probe stimulus (PS). The effective receptive fields of ASs of the same size will not overlap, although the fields of AS of different sizes may overlap. The distribution of ASs and sizes can result in a uniform coverage of the display. It will not be possible to match the projection of the AS with the RGC fields (RFs). The RGCs have essentially a 1:1 relationship with the cones, which have a hexagonal arrangement. Although ideally one could have the distribution of AS sites around a visual fixation point reflect the cone density, such a refinement should only be considered in light of the results of this first experiment. Each AS will be briefly illuminated for about 8 ms (presuming a driven synchronization frequency, f_g, of 30 Hz). Briefer would be better, but this already pushes the limits of many displays.

Figure 48. Difference of Gaussians (DOG) aperture stimulus (AS).

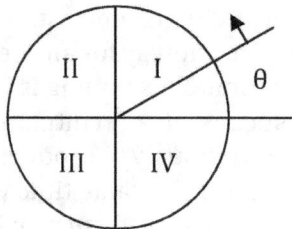

Figure 49. Phase angle θ and quadrants.

A gamma cycle may be considered ideally as a sinusoid with a frequency of f_g. Such a waveform can be represented as a vector rotating through four quadrants, Q_j (Figure 49), with j = I,...,IV. In a synchronized aperture, elements fire preferentially in Q_{IV} (Figure 17). The aperture synchronization/disorder state should then be affected by a field of ASs that have controlled quadrant probabilities ($P_I,...,P_{IV}$). In a disordering state, firing in all quadrants is equally probable. In the synchronizing state the P_{IV} firing is greater than the others. Each AS will have an underlying natural frequency (f_n) of firing. These will be distributed about some mean (f_m) with some standard distribution, σ. When the phase of the AS is such that it may fire, the phase of f_g and the probability P_j will determine the firing of the AS. P_j has two groups, j = IV and j ≠ IV. We can set the latter equally probable, thus P_js of the latter group are (1/3) x (1 – P_{IV}). As noted, f_n is a function of f_m and σ. Thus the behavior of an AS can be characterized by its f_n and P_j (j = I,...,IV) and the f_g. The activity of the population of ASs can be characterized by f_g, f_m, σ, and P_{IV}. A strongly synchronizing condition will have an f_m that approaches f_g, a high P_{IV} and a small σ. A disordered condition may have a difference in only a lower P_{IV}, approaching 0.25. Thus f_g, f_m, and P_{IV} must be carefully adjusted such that the differences between the synchronizing and disordering conditions are clearly perceived, but not overwhelming. A low f_m and large σ may be desired, producing a background of randomly located flickering sites. Two aperture conditioning states will be established: synchronizing (S) and disordering (D), as subjectively reported. This may—or may not—be established for each subject (forced choice discrimination). As attending a visual display can be expected to result in a synchronized V1, it is the ability disorder V1 that is sought. A PS will be introduced against this aperture conditioning background. It will have a high luminance contrast relative to the grey background and a high spatial frequency. A small well-defined square (Figure 50) is suggested. It will be significantly smaller than the AS DOGs to ensure non-overlapping spatial frequencies. It will be presented as superposed on the aperture background, i.e., the ASs will show through it. It will have a slow onset (2 cycles, > 66 ms).

Figure 50. High spatial frequency square probe stimulus (PS).

The aperture condition will be randomly varied between S and D for each trial. For each trial, following a random delay, the probe will occur at some random, but essentially central, position. The subject will press a button when the probe is detected. The delay between onset of presentation and subject response is the metric. It is difficult to estimate the number of trials and the S versus D differentiation necessary for an adequate test. Pilot studies will be required. The expected result, detection delay x aperture condition, is that the probe will be detected later in the D condition. An exploration of the variables will be desirable.

The auditory system can be presumed to offer similar opportunities, as it should have the form of a map of log(freq) x time. The cochlea is effectively comprised of linked active bandpass filters, analogous to the retinal ganglion cells. A time domain may be produced through "delay lines" constructed of neural circuits, including chandelier cells.

Speculations

Decisions

Ensembles resolve toward a low entropy state. A decision can be a resolution of an ambiguity. One can speculate that meaningful cognition, involving decision-making, will entail resolving ambiguities. The more complex the situation, the more ambiguous. The greater the ambiguity, the closer the two (or more) states are to each other. This does not explain all decision-making, but may apply to some circumstances. Recognition of a familiar face has little ambiguity. The outcome bias influences the state of the ensemble, which will rapidly resolve to recognition or a fast reaction. A less familiar face will not resolve as fast. Experiments in which animals face forced choices in well-learned tasks must be viewed with caution in interpreting the results. One would expect the resolution of an ensemble to be reflected in the EEG [624]. The end state EEGs of rapid and slow decisions might appear similar—the intervening time of the slower process having more EEG complexity. The perception of more complex scenes may also reflect such temporal differences. The development of CMs in cortices may add complexity to the LFPs, with subsequent complexity in the EEG, which will incorporate a net effect. The relationship between ensemble resolution and decisions will require significant exploration. Perhaps tolerance for greater ambiguity and more time can result in more rational decisions. How might one slow such resolutions without distorting them?

Seizures

The relationship between inhibitory neuron types and seizures would indicate that it is the larger basket cells that would dampen the pyramidals' activity. The chandeliers' activity would not be expected to have as large an effect. The chandeliers synapse on the axon initial segment. They do not significantly alter the level of excitation of the pyramidal, only the timing of the spike. The inhibitory basket cells, however, affect the level of activity of the neuron itself.

Uncoherent Instantiations

I speculate that uncoherent instantiations, e.g., color, always occur in the context of temporarily structured information instantiations (e.g., form). CISs are exchanged in an ensemble. Apertures not within an ensemble are suppressed. Thus, a non-suppressed aperture would be coherent, able to project uncohered information with cohered instantiations. Note that I consider spatially structured instantiations as, of necessity, temporally structured; else the structure will lack coherence. Vision has to have the coherence of the perception to tie the pieces of an image together.

Phase Conjugate Mirrors

Is information instantiated in phase structures that are analogous to phase holograms? This requires a medium that can sustain such modulations, which it appears the cortex can do. Perhaps a model similar to a double phase conjugate mirror [19, 495] is appropriate. The relationship between phased and phased-reversed implementations also needs study.

Evolution and the Composite Node

The composite node (Ch. 6, CN) is an abstraction that reflects the anatomical relationships among a cortical area and its corresponding areas in the thalamus and thalamic reticular nucleus. The composite node presents an interesting hypothesis: might one trace out the evolution, and perhaps development, of the brain through the repeated divisions of composite nodes? If one considers a cortical area as the top of the composite node cone, then one can model the elaboration of the cortex as occurring when an aperture reaches some critical size such that either it can no longer reliably maintain coherence over its area, or the size is such that the functions differ significantly over the aperture, losing the continuity of the functional. In either case, the separation of the aperture occurs through a functional pinching of the logical aperture (there being few aperture morphological demarcations in the cortex restraining the process), creating the same separation in the thalamus and thalamic reticular nucleus. As cortical apertures separate, one might presume that the axons previously within the original aperture now span the new apertures. These eventually form intercortical tracts and subcortical connections. It might be possible to map out the evolutionary parcellation of the cortex by tracing the successive parcellation of the cortex within an individual, within a species, or through evolutionary relationships among species. The connectome [100] or CoCoMac [780] may singly and together provide the resources for tracing these developments.

Future Explorations

Introduction

Many stones have been left unturned; who knows what beasties may lie beneath them? Some key concepts are parts of the model and not presented as topics of future investigations per se, as the model as a whole is presented for future investigations. Core issues should be addressed, as the model requires a firm ground. The coherent apertures model can be developed further as a framework to explore operations in the CNS beyond the perception, and memory and its access, of visual declarative objects. There are several potential, interlocking, scientific domains for future coherent apertures research. Each domain contributes specific resources of knowledge, technologies, and talent. It is not expected that all domains will exist in future research efforts, or that they will have equal strength. Theoretical neurobiology is the core domain. The others overlap, reflecting the topics addressed. Technological implementation concepts are outside the scope of this paper, implying the generation of an abstract model. It will be interesting to follow the work of Merolla et al. [25] with their development of the very large-scale integrated circuit (VLSI) implementing a million spiking neurons. Their architecture supports the interconnection of large numbers of these programmable tiles.

A general outline of research is proposed. The specific topics addressed, and how they are addressed, will reflect the people available, their interests, and resources. Although many topics can be undertaken separately, almost all future research will be collaborative, including an ongoing seminar. The first task is to evaluate the model. What are its strong and weak points?

Domains

There are five domains that may be included in future research. Each domain can productively interact with the others. The specific topics undertaken will determine which domains are included, and the people and resources involved.

Theory

All research rides on a core of ongoing theoretical work.

Modeling

Modeling builds on theory, experiments, and analytics. It is a major driver of technological implementations.

Analytics

Analytics will use both new data and data available from other sources, frequently referred to as "big data," being repositories of data collected with new tools. Big data begs for interpretation. Analytics rests on theory, modeling, and experiments.

Experiments

Biological experimental work tests theory and modeling. It attempts to fill in holes exposed by analytics.

Technology

Technological implementations provide two major thrusts: verification of theoretically based modeling, and explorations of new technologies based on the CNS. Although technological modeling may initially use existing systems, such as graphics processing units (GPUs) and high capacity computers, it is assumed that the theoretical models will eventually outstrip these technologies.

Research Topics

Aperture cohering

Association aperture

Bidirectional causation

Big data analysis

Cognition

Coherence map (CM)

Coherent information structure (CIS)

Composite node

Decision-making

Ensemble

Ensemble phase conjugate mirrors

Entropy minimization

Episodic experience

Extended Kuramoto model

Gating and the LFP

Heuristics

Information entropy in the aperture

Information instantiation

Internal experience

LFP pumped aperture

Laminar cohering

Memory in phase

Multi-frequency complex in the AFP

Noise and the natural frequency (f_n)

Perception

Phase CM

Phase quantization, minicolumns and macrocolumns

Prediction

Undecipherable complexity

van Cittert-Zernike (VCZ) node

The research topics are avenues of inquiry that ask different, and overlapping, questions about the role of coherent apertures in the internal experience.

Supplemental Materials

Introduction

Chs. 3 and 6 are complex. For readability, I moved some materials to this Supplemental Materials section. The core concepts expressed in the chapters carry the weights necessary to understand the overall models. Array synchronization, as a network phenomena, is based on the work of Kuramoto. Exact solutions of his equations for large arrays are not possible. These expanded equations incorporate broader concepts including the pumped LFP model. It begs for future expansion. The monopole model is expanded here. The ensemble synchronization models are supported by solid state laser models and experiments that are included here as more complete explanations of concepts briefly presented in Ch 6.

Ch. 3. The Coherent Cortical Aperture

Array Synchronization

The coherence of a cortical aperture underlies much of the coherent apertures model. Synchronization underlies aperture coherence. As noted in the text, it is well established that an aperture may synchronize. The question addressed here is, how does it synchronize? Although it relies on the underlying neural circuits and tracts, this model is not focused on such essential structures. There are two major models of cortical aperture synchronization: network and field. The network model here is an expansion of work by others. The field model, as the pumped AFP, is described in the text, and incorporated in the network model here. The laminar coherence model is a combination of the network and LFP models, and includes coherence. The potential conditions for the emergence of an aperture-wide synchronous AFP are described here.

Kuramoto [11, 174] developed a model of a large scale-free network (SFN) in which all N elements are oscillators with similar, but not necessarily identical, natural frequencies, and which are reciprocally coupled to all other $N-1$ elements, with a resulting phase modulation:

$$\dot{\theta}_i = \omega_i + \frac{\sigma}{N}\sum_{j=1}^{N}\sin\left(\theta_j - \theta_i\right)$$

Equation 1

where N is the number of elements in the network, i is the element index, and j is the coupled element index for $i = 1,...,N$, $j = 1,...,N$, $i \neq j$. σ is the coupling constant, θ_i is the phase of the ith element, and ω_i is the natural frequency of the ith element. $(\theta_j - \theta_i)$ expresses the phase difference between the ith and jth elements. The coupling works to minimize that phase difference. The solution for the system is the σ for which all θ_is are the same, i.e., all elements are synchronous, the value for which this occurs in the critical point. For all but small networks, this system of deterministic equations cannot be solved. Computer modeling and statistical models have been used to show that at very small values of σ/N the system will synchronize [17]. This value is usually expressed as λ for large arrays. Kuramoto's model (KM) was refined, as described by Arenas et al. [17], to a more general form:

$$\dot{\theta}_i = \omega_i + \sum_{j}\sigma_{ij}\alpha_{ij}\sin\left(\theta_j - \theta_i\right)$$

Equation 2

for which α_{ij} is a connection and σ_{ij} is a coupling coefficient.

A cortical aperture may be modeled as an Ising-like array of elements on triaxial coordinates, a hexagonal network. In such a configuration, the coupling matrix will limit the connectivity to being local. There are multiple coupling mechanisms (k) involved in the array model: excitatory and inhibitory synapses, and electrotonic (local) coupling. Each of these has (potentially) a different coupling strength (σ_{ijk}), coupling connections (α_{ijk}), and delays (τ_k). The total phase modulating influence on θ_i is thus the sum of all of these.

For notation simplicity, for the next evaluation:

$$\omega_i' = \dot{\theta}_i.$$

Equation 3

Triplett et al. [185] demonstrated that arrays with equal coupling delays, τ, can synchronize. The value of ω_i at a time τ in the future is:

$$\omega_i'(t+\tau) = \omega_i(t) + \sum_j \sigma_{ij}\alpha_{ij}\sin(\theta_j(t-\tau) - \theta_i(t)).$$

Equation 4

The phase adjustment is a function of the three coupling types. Delays for the k coupling types can be estimated:

$k = 1$: excitatory, $0.5\text{ms} < \tau_1 < 2.0\text{ms}$

$k = 2$: inhibitory, $0.5\text{ms} < \tau_2 < 2.0\text{ms}$

$k = 3$: electrotonic, $0.0\text{ms} < \tau_3 < 1.0\text{ms}$.

Combining these coupling types requires accommodation of different τ_ks. Using Equation 4 would involve projecting to different times in the future. Equation 4 may be rewritten to reflect current time, t, for a particular coupling type, k:

$$\omega_{ik}'(t) = \omega_{ik}(t-\tau_k) + \sum_j \sigma_{ijk}(t-\tau_k)\alpha_{ijk}\sin(\theta_j(t-2\tau_k) - \theta_i(t-\tau_k)).$$

Equation 5

The coupling coefficient, $\sigma_{ijk}(t-\tau_k)$, indicates that coupling may change. This is consistent with modulation. Coupling symmetry is not expected, as $\sigma_{ijk} \neq \sigma_{jik}$ and $\alpha_{ijk} \neq \alpha_{jik}$.

Summing all of the k instances together must accommodate their differences in $\omega_i(t-\tau_k)$. We can consider the summed portion of Equation 5 to be an adjustment term:

$$\Delta\omega_{ik}(t-\tau_k) = \sum_j \sigma_{ijk}(t-\tau_k)\alpha_{ijk}\sin(\theta_j(t-2\tau_k) - \theta_i(t-\tau_k)).$$

Equation 6

The shortest delay will be the basis for the ith element's frequency, ω_{ik}, it being the most recently established. The shortest τ_k is 0. One can presume the effects of frequency adjustments decrease in time as they are overridden by more recent effects. In summing over the ks, they can be weighted linearly in time against the most recent τ_k, which is 0:

$$W_k = \frac{\tau_{max} - C\tau_k}{\tau_{max}}$$

Equation 7

with C a constant,

$$0 < C < 1 \qquad\qquad \text{Equation 8}$$

resulting in the weighted summation:

$$\Delta\omega_i(t) = \sum_k W_k \Delta\omega_{ik}(t - \tau_k). \qquad\qquad \text{Equation 9}$$

Rosenblum et al. [374] have approached the synchronization of oscillators as coupled through a field: a global coupling model. Much of the mathematics is beyond my expertise; however the basic equations are effectively the Kuramoto model. Under a field model, LFP coupling is global, which is more consistent with the pumped AFP model, approaching a scale-free network (SFN) model.

An aperture may be synchronous, but that does not mean that information instantiation activity transfer across it is instantaneous, presumably involving changes in the phase structure. Synchronization and changes occur across an aperture in far less time than synaptic transmission would support for rapidly changing phase structures involved in coherence. Thiagarajan et al. [151] and others [150] find that the synchronization of well-separated neurons within an aperture are synchronous within a few milliseconds, despite lack of direct synaptic connections. This is consistent with the Kuramoto model for steady state conditions, possibly providing a reference phase, but limited in the speed with which changes may be communicated across the aperture. Nobili [161] makes a good case that the EEG (and by extension, the LFP) is more than simply an epiphenomenon of neural activity: it is part of the aperture synchronization system.

An additional coupling mechanism, incorporating the AFP, may be added:

$$k = 4: \text{LFP, } \tau_4 \approx 0.0\text{ms,}$$

although RC considerations may give some frequency dependence, which I shall not consider here. The AFP may be considered a single node, a, that is coupled to <u>all</u> the elements in an aperture: a simplification converged on during synchronization. In a sense, the aperture <u>itself</u> is an element. Ignoring the complex frequencies, the frequency of the AFP node is the same as the synchronous frequency. Normally the term "global" refers to all elements in a network or population. As I am using global to refer to the ensemble's frequency (Ch. 6), I shall adopt the term "aperture" when referring to the aperture's characteristics, including frequency (ω_a) and phase. "Global" in the <u>current</u> context refers to an aperture-wide network characteristic, consistent with network terminology. This creates some unfortunate confusion.

An aperture has a population of P_a elements. A subset of that population, P_s, will synchronize. The aperture population has a mean frequency, ω_a, of its elements, with a distribution of $g(\omega_a)$, which is broad, such that there is no sharp frequency peak, being "pink" noise-like (i.e., disordered). Similarly the synchronizing subpopulation, P_s, has a frequency of ω_s, with a distribution of $g(\omega_s)$. Presume that ω_s is greater than ω_a; $g(\omega_s)$ is narrower than $g(\omega_a)$, as P_s is a subset of P_a. P_s may change, hence $P_s(t)$.

The orderliness of the activity of elements in an aperture can be described by an order parameter in the manner of Kuramoto and Battogtokh [442], dividing the population into synchronized (P_s) and disordered elements—the equivalent of a CM (Ch. 4). I shall use only the modulus, R, ignoring the phase portion (Θ). The order parameter, R, can be defined as:

$$R(t) = \frac{P_s(t)}{P_a}. \qquad\qquad \text{Equation 10}$$

As described in the LFP pumping model (Ch. 3), all elements can potentially contribute to the maintenance of the AFP, even if they do not have a ω_a mean frequency as high as ω_s. They contribute as a function of their frequencies. R may also be considered as:

$$R(t) = \frac{\omega_a(t)}{\omega_s}$$

Equation 11

as ω_a may change. ω_s is presumed to be stable. This simplification is adequate for this analysis, as I intend to show the logic of the relationship of the AFP to synchronization. ω_s is determined by the synchronized population, P_s, the remaining population being noise. The global coupling and connection are confounded, as the AFP node, a, is globally coupled to all the elements: a is a virtual node. The amount of drive (AFP coupling) is related to the proximity of ω_a to ω_s.

R is an indicator of the strength of $\sigma_s\alpha_s$, the global coupling of the synchronous AFP drive, G:

$$G(t) = BR(t).$$

Equation 12

B is a constant. This leads to an equation for the contribution of the synchronous AFP to synchronization:

$$\Delta\omega_{ia}(t) = G(t)\sin(\theta_s(t) - \theta_i(t)).$$

Equation 13

This may be combined with Equation 9:

$$\Delta\omega_i(t) = \Delta\omega_{ia}(t) + \sum_{k=1}^{3} W_k \Delta\omega_{ik}(t - \tau_k).$$

Equation 14

Combining Equation 9 and Equation 14 to sum for the phase-based modulations of ω_i:

$$\omega_i'(t) = \omega_i(t) + \Delta\omega_i(t).$$

Equation 15

The larger the N, the greater the synchronous AFP coupling relative to the narrowly scoped individual elements; consequently, a large aperture favors the AFP if an adequate synchronized population, P_s, exists.

A planar triaxial Ising-like array may be presumed, it being evident in retinal and other areas; thus each element has six nearest neighbors. Presume that each element reaches out to its 12 next-nearest neighbors, resulting in a degree (connections) of 18, α_{ijk} being defined locally. An expansion of Equation 15 results in $N \times k$ simultaneous equations. An exact solution is not expected for any reasonably representative N; even a small aperture will have 5,000 elements. Strogatz et al. [842] provided an approximate solution to this class of problem by using the law of large numbers to replace summations with integrals, and assuming uniform coupling among all elements.

Under most Kuramoto models (Equation 1), synchronization is not frequency (ω) sensitive, being a phase-modulation model of a network of elements, with natural frequencies normally distributed around a stable mean frequency. By definition the mean frequency of all N elements is the aperture mean frequency:

$$\omega_a = \frac{\sum_{i=1}^{N} \omega_i}{N}.$$

Equation 16

The network ω_a does not change significantly under the model thus far (Equation 15). The synchronization of an aperture can be influenced by the mean aperture frequency, ω_a. If ω_s is unchanged, an increase in ω_a results in more elements' frequencies, ω_i, falling within $g(\omega_s)$.

An increase in ω_a will increase the order parameter, R, (Equation 11) which will increase the AFP coupling, as P_s will also increase (Equation 10). The effect of an increase in ω_a on R (Equation 11) is to increase G (Equation 12), and subsequently to increase the strength of phase locking through Equation 13; thus inputs to an aperture that raise its activity level, as ω_a, increase the phase locking. Such inputs may be cortical, thalamic, or other subcortical structures (e.g., locus coeruleus), contributing either structured (e.g., CIS) or general excitatory (or inhibitory) inputs. Clearly, raising the natural frequencies of an aperture's elements increases its synchronization, if there is a stable AFP: a familiar circularity in emergent phenomena. This global AFP model is consistent with Nobili [161], who proposes intrinsic subthreshold oscillations to promote synchronization, which is also consistent with the laminar coherence model (Ch. 3).

In addition to some baseline natural frequency, the aperture frequency will reflect external inputs from several sources—other cortical areas, the brainstem, and the thalamus. Each of these inputs, I_p, can translate into some change in the aperture frequency:

$$\delta\omega_p(t) = f_p(I_p).$$

Equation 17

Under a linear assumption these can be combined:

$$\Delta\omega_a(t) = \sum_p f_p(I_p).$$

Equation 18

One would expect hysteresis in such a system. As the strength of the AFP increases, it will recruit elements from the otherwise disordered population. This will recursively increase the AFP and separate the aperture's population into synchronous and disordered subpopulations, consistent with the CM and the results of Kuramoto and Battogtokh [442]. Decreasing this coupling will require a reduction in ω_a.

The aperture-wide synchronous pumped AFP creates a globally synchronized network, consistent with the results of Strogatz et al.; their global frequency is equivalent to the aperture phase's frequency discussed in Ch. 4. Not all pyramidals need be active for the aperture to be synchronous. The response of a particular pyramidal cell is a function of its current activity level and the phase of the AFP. Thus a coherent CM can exist. An I_p may be negative. Inhibitory inputs, perhaps acting through large inhibitory basket cells [286], will prevent runaway oscillations through damping. Perhaps inadequate inhibitory basket cell activity relates to epilepsy. The effects of activation of a latent coherence map are not included here. This model describes convergence to a single global frequency. Such convergence is not the case. The frequency of ω_s is not resolved here. As discussed by Buzsáki and Draguhn [384], different states and tasks are associated with different frequency bands, which may result from sources not addressed here, including the specific members of the ensemble. As Buzsáki and Draguhn discuss, the CNS evidences frequencies spanning over two orders of magnitude, with multiple frequency components present simultaneously, although there appear to be some underlying constraints, e.g., the competition between adjacent frequency bands. The particulars of ω_s test the validity of the model, which, at a minimum, needs expansion.

Monopole Model

If a neuron (pyramidal cell) limits the motility of ions within its soma, the neuron may be considered as a source that redistributes charge (ions) through pulsed discharges, a monopole [335,

334]. A monopole is effectively injecting charge rather than exercising a closed circuit of a di—or higher—pole. A monopole may contribute to a lower frequency LFP. Charge difference accumulates across the neural membrane more slowly than does the charge release, almost as if charge were dumped at a point. The natural outgrowth of such a model is a negative point charge impulse as Na^+ ions move rapidly into a pyramidal, briefly hyperpolarizing the surrounding area, hence the LFP by a very small amount, producing a local electrical "dimple," pumping it. The probability of a neuron's spiking increases as the LFP becomes more negative (lower). Thus as the LFP decreases, spikes increase, further decreasing the LFP in a brief positive feedback. This model is consistent with the work by Zanos et al. [336] on the relationship between the LFP and spiking, although they prefer to digitally "remove" the spike as an artifact. Zanos et al. and Waldert et al. [333] consider the spike to "contaminate" the LFP—an interesting position. It is not unusual in the literature to encounter 1–300 Hz band-pass filtering [333]. Zanos et al. demonstrate a significantly sharper recorded LFP when the filter has a band-pass of 100–5,000 Hz rather than with a 1–1,000 band-pass, used as a basis for removing the spike as an artifact. The higher frequency filter reveals spikes with widths of 1–2 ms, demonstrating that the limits of measurement tools are reflected in the results. Waldert et al., using a 300 Hz <u>high-pass</u> filter, report spikes with peak-to-trough times as short as 0.24 ms.[27] It is this spike "contaminant" that justifies its consideration as a monopole, and as a source of the transient potential dimple. The chattering cells [153], with spike bursts of 400–800 Hz, demonstrate a slow repolarization, exhibiting a repetitive pulsed discharge behavior consistent with a monopole, hence able to "inject" polarization into the neuropile, contributing to the LFP. The chattering pyramidals' burst behavior is phasically linked to the negative phase of the LFP such as when the aperture receives a coherent stimulus, e.g., a drifting grating. The chattering cells are principally in layers II–III, perhaps contributing to the supra- infra-granular laminar differences in Fourier components of the AFP [132]. A pumped AFP model would not require all neurons to fire nor that all the same neurons fire at the same time. A sprinkling across the aperture, if the natural frequencies were high enough, could support synchronization. The driver would be subpopulations of synchronous neurons. This model supports the ability to instantiate information in subpopulations of neurons coupled through the pumped AFP.

---

[27] Capturing this waveform would require, at a minimum, an 8 KHz upper limit on a filter (Nyquist criterion). The captured waveform would be distorted with such a low cut-off frequency as a spike has a waveform that is far from sinusoidal.

Ch. 6. Integration of Coherent Cortical Apertures

Laser Models

Solid state laser systems provide useful models for understanding the relationships among structures in the CNS during formation of an ensemble of coherent apertures, and the exchange of information instantiations. Solid state lasers emit coherent light with a narrow frequency bandwidth. The amplitude is not constant, fluctuating randomly by small amounts: chaotic behavior. Laser noise profiles may be considered as equivalent to meaningful signals. The interaction of lasers in various configurations may cause several of them to have the same noise profile. These provide useful models for understanding the dynamics of ensemble synchronization with nodes with irregular temporal structures, and the relationships among the various forms as the system progresses through synchronization modes.

Intermediate Synchronizer

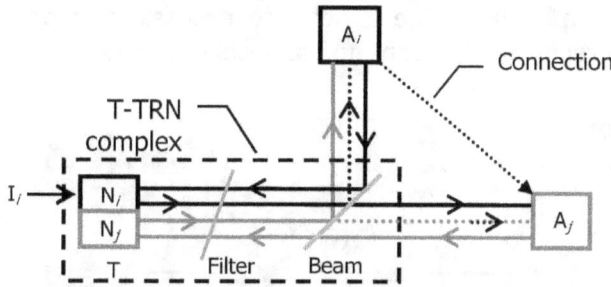

Figure 51. Fischer model [intermediated]. Apertures A_i (black) and A_j (gray) become synchronized with the introduction of a third unit comprised of thalamic (T) nuclei N_i, N_j, and a TRN filter. The filter function is realized within the T-TRN. Adapted from Fischer et al. (2006) [633], who depict two outer lasers coupled through a common laser with a beamsplitter. N_i and N_j comprise a single (common) laser in Fischer et al. Paths from a thalamic nucleus to its associated aperture, e.g., (N_i–A_i), are dotted. I_i is a peripheral input to a thalamic nucleus N_i. Optical paths have been separated for clarity to illustrate neural activity paths. A unidirectional projection from A_i to A_j activates A_j.

Two coherent nodes may be brought into synchronization through an intermediate node. In the CNS an intermediate configuration develops synchronization of two aperture nodes through interaction through the TRN-thalamic system in a manner analogous to the coupling of two outer lasers through a common laser (Figure 51) as illustrated by Fischer et al. [633], an idea they suggest as having potential in the CNS without mentioning specific mechanisms. Their apparatus monitors the noise profiles of three solid state lasers. In the intermediating model of Fischer et al., "a central laser diode (T) is bidirectionally coupled to two outer lasers (A_i) and (A_j) by mutual injection. The central laser, which does not need to be carefully matched to the other two, mediates their dynamics." The noise profile of the central laser (T) lags the outer two. A filter in the pathway ensures that the signal features, e.g., polarization, are common to both outer lasers. The two outer lasers in Figure 51 correspond to the two cortical apertures (A_i, A_j). The combination of the inner laser, beam splitter, and filter correspond to the combined TRN and thalamus, and the corticothalamic crossovers (e.g., A_i–N_j) as might be modeled with the composite node (Ch. 6). The filter is the equivalent of the alignment and coherence of the natural keys necessary for exchange.

Intermediation operates with cohering apertures. Inputs to the thalamic nucleus (I_i) increase the activity of elements in an aperture (A_i) such that it becomes coherent (Ch. 3). That aperture projects to another aperture (A_j) through a corticocortical connection. If that aperture is responsive to the input, it will start to cohere. Both apertures will be projecting through the TRN to the thalamus (T), as each aperture has extensive projections to its respective thalamic nucleus. The activity through a thalamic nucleus is cohered by corticothalamic projections [79, 358, 353]. Lien & Massimo [710], Li et al. [550, 711], and others find significant operations occurring in the thalamus alone, such as auditory frequency sweep and visual orientation sensitivity. The cortex amplifies this behavior, increasing the S/N, enhancing the coherent components. Activity within a TRN sector is driven toward coherence by the coherence of its input projections from the thalamus and cortex. Electrical gap junctions create weak excitatory coupling among the TRN neurons [706] across aperture sectors (6, iv), potentially cohering activity as a network within the thalamus [17, 535, 635].

Zero lag synchrony among apertures can be achieved despite conduction corticothalamo-cortical delays if all delays are the same [700]. As the delays are small (<2ms) and constant, synchronization can occur irrespective of distance between the thalamus and cortical areas [709]. One would not expect cortical apertures without reciprocal thalamic crossovers to synchronize. This limits the structure of possible dynamically assembled networks. Future work will explore frameworks beyond two apertures. The coherence necessary to produce and accept a CIS constitutes the filter. The reciprocating common aspects of the Ks provide mutual CM activation.

Common Synchronizer

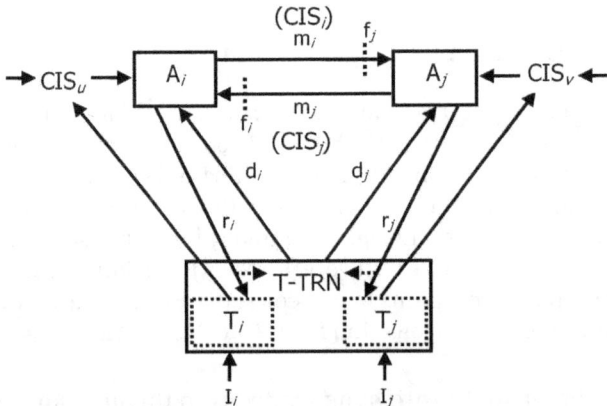

Figure 52. Zhou and Roy model. Chaotic oscillators A_i and A_j become bidirectionally coupled, with common synchronizing drives d_i, d_j. T-TRN coheres A_i and A_j, allowing bidirectional exchange of messages m_1 (CIS_i), m_2 (CIS_j), which are derived from inputs CIS_u and CIS_v. CIS_u and CIS_v include input from the thalamic nuclei T_i and T_j. r_i, r_j are recurrent pathways, f_i, f_j are filters. Adapted from Zhou and Roy (2007) [722].

A common synchronizer can, with a weak signal, sustain the synchronization of two coherent nodes. Bidirectional white matter (cortico-cortical tracts) coupling between apertures builds on the initial intermediate synchronizer to form a common synchronizer. Zhou and Roy [722] provide a model for direct exchange of information instantiations between nodes while maintaining coherence. Information (noise profile) is simultaneously exchanged between two random oscillators (lasers), A_i, A_j, with a weak common driving signal (Figure 52) d_i, d_j, from the common node (laser), T-TRN. The synchronization of common drive signals is reinforced through recurrent coupling (r_i, r_j). These loops $(d_i-r_i$ and $d_j-r_j)$ may support the underlying oscillation. This system maintains coherence, supporting simultaneous bidirectional information transfer, m_i (CIS_i), m_j (CIS_j), in the CNS through the white matter. No significant information is transferred from A_i to A_j through the common node. Each outer node may receive outside information instantiations CIS_u, CIS_v, some of which may be of thalamic origin. Peripheral information, I_i, I_j, may be introduced into the thalamus, and subsequently to the cortical aperture(s). Once the apertures are synchronized they can directly exchange information, presumably through the K and CIS exchanges, if coherence between the apertures can be maintained. The activities in each of the apertures will differ, reflecting multiple inputs. Thus, as the interaction evolves, one might expect the activity patterns to differ more; therefore coherence becomes the more appropriate term. These results are similar to the synchronization of two coupled map lattices driven by a third coupled map lattice, as described by Santhanam and Arora [723]. Gollo et al. [647] modeled synchronicity between reciprocal "resonance pairs of elements"—up to aperture scale—in spite of element parameter and delay differences. They propose that such pairs may synchronize motifs. Such a model is similar to a common synchronizer model.

Peer Synchronization

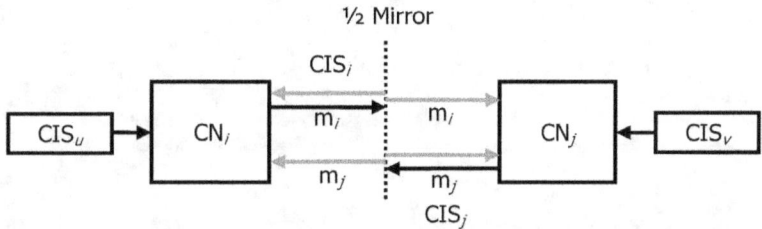

Figure 53. Synchronous simultaneous bidirectional exchange. Lasers CN_i and CN_j directly coupled through a partially reflecting mirror become synchronous. Messages m_i and m_j derived from CIS_u and CIS_v are exchanged simultaneously. Modified from Vicente, R., Mirasso, C. R., & Fischer, I. (2007) [724].

Peer synchronization is the self-synchronization of two coherent nodes. Two chaotic solid state lasers may become synchronized with the introduction of a partially silvered mirror between them, bidirectionally exchanging information (Figure 53). Vicente et al. [724] report that this is a robust phenomenon, relatively resistant to differing time delays due to spacing. A message (CIS_u) introduced into one laser (e.g., CN_i) produces a message (m_i) that will be transmitted through a partially silvered mirror to the other laser (CN_j). The mirror returns a diminished message back to the source. The converse occurs simultaneously. Vicente et al. found such behavior even with delays of tens of milliseconds, consistent with axonal delays among cortical areas, their target model. This is suggestive of a model of an interaction between two composite nodes. The neurophysiological analog to the half-silvered mirror is unclear. The limited commonality of Ks may suffice, keeping in mind that all cortical connections are bidirectional.

Zhou and Roy's work may be considered a special case of multipath synchronization. Englert et al. [646] demonstrated that a system of two coupled chaotic units could become synchronous if there are two coupling paths of different lengths or delays, although the delays must fit specific, although multiple, ratios. Chawla et al. [725] preceded Zhou & Roy with a hierarchical triplet model of reciprocally connected nodes to demonstrate cortical zero-lag synchrony, again a multipath form. This can be considered an alternative form of the Fisher et al. [633] model, with equality of coupling among the three elements. The results of Englert et al. and Chawla et al. support the modeling of Sadeghi and Valizadeh [726], demonstrating that synchronization can occur among coupled neural oscillators in the face of multiple sources of delay inhomogeneity. The multiplicity of couplings compensates for the inhomogenieties. Signals exchanged in delay coupled models may not be required to be essentially identical. Shrii et al. [843] demonstrated that nodes in an arbitrary network with couplings that use dissimilar (conjugate) variables can achieve synchrony in the face of delays, supporting bidirectional coupling among CNs that have different, but related, CISs and Ks.

Glossary

Affine transformation: A linear transformation or function between two spaces that preserves parallel lines.

AIS: Axon initial segment of the pyramidal neuron.

Aperture: A bounded area through which radiant energy, including neural activity, flows.

Aperture entropy: An expression of the information in the coherent aperture array resulting from the independence and organization of the elements across the aperture. Based on Shannon and Weaver's definitions of entropy, but stated relative to an aperture to simplify the incorporation of spatial patterns, e.g., coherence maps.

Aperture field potential (AFP): The mean LFP over an aperture. When the aperture is synchronized the LFP is the same as the AFP at all locations, ω_s.

Aperture frequency: The AFP frequency over the aperture (ω_s).

Aperture phase: The phase relationship of the AFP to the aperture frequency, ω_s.

Bidirectional causality: A influences B at the same time that B influences A.

Bijective: A function that can map from A to B and back from B to A without loss of information.

CIS: I use Lizier et al.'s [165] term coherent information structure (CIS) from elementary cellular automata [166, 506] for the coherent projection from a coherent map.

CNS: Central nervous system.

Coherence: A defined or stable relationship or correlation in time and/or space.

Coherence field, coherence phase field: A phase field demarcating those elements that are coherent from those that are disordered in behavior. Although this is typically described as a matrix of 0s, and 1s, it can contain a continuum from 0 to 1.

Coherence map (CM): A topographic map of coherent (hence synchronous) and disordered elements in a cortical area. The product of a dichotomous phase field of coherence and a phase map. Coherent elements do not need to be contiguous. Through the functional it generates CISs for projection.

Coherence path length: The distance over which elements retain a correlation.

Coherence propensity: A readiness to cohere, typically resulting from a supragranularity state near the critical point.

Coherent: Correlated elements or points.

Cohesive: An aperture that can become coherent, bound together, e.g., in a network.

Column: An arrangement of neurons normal to the surface of the cortex. Unless specified, it refers to a minicolumn.

Composite node (CN): An abstraction composed of a cortical aperture, its corresponding sector of the thalamic reticular nuclear (TRN), and its thalamic nucleus. In the composite node information schematic (extended from the model of Ferrarall & Tononi [697]), spatio-temporal information may be considered as information units which may project, or flow, over pathways among structures.

Critical point: The point at which an array may become synchronous across its extent (or area) as the coherence path length exceeds the largest dimension of the array.

Degree: The number of other nodes linked or coupled to a node in a network.

ECM: Extracellular matrix.

ECS: Extracellular space.

Element: A logical constituent.

Elementary cellular automata: Elements with two possible states and a rule to determine the state.

Ensemble: A coherent, usually nearly synchronous, subgraph of coherent apertures.

Field: A physical quality with values for each point in a space.

Fisher information: The mutual information between the input and the output of a system.

Function: A function is a relationship between an input and an output. In the aperture model, the input to a function at a point (node) is the node's field.

Functional: A composite of functions; a relationship between a vector space and a scalar field. A functional's input vector space is usually composed of functions. A functional, in this context, is a mapping from a vector space of a set of functions into a field or set of scalars.

GABA: Gama amino butyric acid, an inhibitory synaptic neurotransmitter.

Global phase: The phase of a set or subpopulation of apertures in a synchronous ensemble.

Gradient: The vector sum of the partial differentials which may vary from point to point over the space (e.g., surface). It is a type of vector field.

Granger cause: A influences B.

Graph: A set of nodes (vertices) and their connections (edges).

HCNN: Hierarchical convolutional neural networks. A layered model.

Hologram: A stored coherent structure from which a holograph (image) can be (re) constructed. Analogous to a photographic negative.

Holograph: An image or other representation reconstructed from a hologram. Analogous to a photograph.

Hub: Concentration of connections in the CNS graph. It is the hub of a community or small world.

Image: A picture on the retina (in this context).

Instantiate: To make an abstraction concrete.

Interstimulus interval (ISI): The interval between two stimuli.

Invariant: Unaltered by certain transformations, e.g., a topological homeomorphic mapping of a donut into a coffee mug has an invariant of a volume with a hole through it.

Ising-like array: A flat array of connected nodes. In the cortical aperture model the next-to-next constraint is relaxed to local connections in a triaxial network.

Isosynchronous: Phase locked and of identical phase (timing).

Laminar coherence model: Supragranular (SG) and infragranular (IG) cortical layers with linked coherence. SG will form a coherent gamma network against which the infragranular coherent map may form. The SG is a de facto irregular strobe.

LFP: Local field potential (electrical), electrical potential at a point in or on the cortex.

LFP pumping: Reinforcement of a low LFP by a pyramidal cell's spiking, creating a small local potential "dimple."

Liquid computing: A system in which new information for a process arrives all the time, before processing of the previous information has completed. This is an overlapping process.

Local phase: The time deviation of a point (element) relative to the local average.

Local phase structure: A pattern in a small locale composed of relative phase relationships among coherent elements within that locale.

LTD: Long term depression. Decrease in synaptic efficacy.

LTP: Long term potentiation. Increase in synaptic efficacy.

Macrocolumn: A grouping of 40-80 minicolumns normal to, and traversing, the cortical plate.

Memory, working: Short term memory, t<5 sec.

Memory, stabilized: Pre-sleep, can be lost in seizures, accessible

Memory, consolidated: Long term memory. Post-sleep, will survive seizures, accessible

Microphase: The phase relationship of an element relative to the macrophase (aperture LFP phase).

Minicolumn: An arrangement of a column of approximately 80 neurons normal to, and spanning the cortical plate. The existence of this structure is not universally accepted. It forms a useful modeling concept.

Motif: A frequently encountered subgraph of apertures (Sporns).

MTL: Medial temporal lobe. The medial temporal lobe (MTL) is a cortical system (Squire 1991) containing the hippocampus and entorhinal, perirhinal, and parahippocampal cortices underlying working and stabilized memory, and essential for the formation of long term memory.

Natural key: A key that is formed of attributes that already exist in the real world. It is analogous to a hash code, a code for an information structure that does not necessarily have any inherent information.

Neuron: The basic electrically active cell in the CNS.

Node: A connection point in a network, a vertex in a graph.

Operation: In its simplest meaning in mathematics and logic, an operation is an action or procedure which produces a new value from one or more input values, called "operands".

Operator: A mapping from one vector space to another. A more general case of a function or functional. In physics, an operator is a function acting over the space of physical states. As a result of its application on a physical state, another physical state is obtained, very often along with some extra relevant information.

PCC: Pyramidal-chandelier circuit that produces a small gating window for activity among pyramidal cells.

Phase: Relative timing between events or processes. In physics and engineering it is often expressed as a phase angle, presuming it is defined relative to a frequency. In the CNS it is normally refers to time.

Phase structures: A set or group of points that are correlated in time.

Phase field: A description of the points in a field relative to possible states or phases, for example, an Arctic sea of ice floes has liquid and solid phases. A map of two (or more) phases.

Phonon: In physics, a phonon is a collective excitation in a periodic, elastic arrangement of atoms or molecules in condensed matter, such as solids and some liquids.

Piecewise function: In mathematics, a piecewise-defined function (also called a piecewise function) is a function which is defined by multiple subfunctions, each subfunction applying to a certain interval of the main function's domain (a subdomain).

Process: (Information). A change or transformation in information, form or state.

Recall: To bring a memory back as an internal experience.

Recognition: The perception that a new input matches a previous one.

RF: Receptive field.

RGC: Retinal ganglion cell.

Scale free network (SFN): A network in which any node can be connected to any other node.

Scintillating: Randomly located neurons that spike in phase with the LFP, pumping it. They are not a constituent of a spike trains.

Small world: A network composed of locally connected nodes. Some nodes (hubs) may be more highly connected. (Strogatz & Watts)

State machine: An abstract machine that can be in one of a finite number of states. The machine is in only one state at a time.

Subgraph: A subset of a graph defined as nodes (vertices) and connections (edges). Used here to describe the network of cortical apertures (areas).

Synchronous: Phase locked.

Topological space: A space defined by the consistency of the neighborhoods of points in that space.

Unique key: An identifier of some entity that is unique within some scope.

VCZ: van Cittert-Zernike theorem. A description of the electric field in one space receiving radiant energy from some other space. Normally the source aperture of radiation emits uncorrelated photons. The VCZ allows calculation of the resulting coherence in the receiving aperture.

VCZ node: An aperture (extended area) that has a coherent phase such that the aperture can be described as coherent (the disordered phase is ignored). Such an aperture can function as a single point node in a graph of synchronously communicating nodes.

Weaver entropy: Entropy of information distributed with some ordering.

References

1 Buzsáki, G., & Schomburg, E. W. (2015). What does gamma coherence tell us about inter-regional neural communication? *Nat. Neurosci., 18*(4), 484-489.

2 Crick, F., & Koch, C. (2003). A framework for consciousness. *Nat. Neurosci., 6*(2), 119-126.

3 El Boustani, S., & Destexhe, A. (2010). Brain dynamics at multiple scales: can one reconcile the apparent low-dimensional chaos of macroscopic variables with the seemingly stochastic behavior of single neurons? *Int. J. Bifurcat. Chaos, 20*(06), 1687-1702.

4 Gray, C. M. (1999). The temporal correlation hypothesis review of visual feature integration: Still alive and well. *Neuron, 24*(1), 31-47.

5 Uhlhaas, P., Pipa, G., Lima, B., Melloni, L., Neuenschwander, S., Nikolic, D., & Singer, W. (2009). Neural synchrony in cortical networks: history, concept and current status. *Front. Integrative Neurosci., 3*(Article 17), 1-19.

6 Wolfe, J. M. & Cave, K. R. (1999). The psychophysical evidence review for a binding problem in human vision. *Neuron, 24*(1), 11-17.

7 Rose, S. (2004). Part I. Introduction: The new brain sciences. In Rees, D., & Rose, S. (Eds.). *The new brain sciences: Perils and prospects*, (pp. 1-14). NY: Cambridge Univ. Press.

8 Churchland, A. K., & Abbott, L. F. (2016). Conceptual and technical advances define a key moment for theoretical neuroscience. *Nat. Neurosci., 19*(3), 348-349.

9 Sejnowski, T. J., Churchland, P. S., & Movshon, J. A. (2014). Putting big data to good use in neuroscience. *Nat. Neurosci., 17*(11), 1440-1441.

10 Yamins, D. L. K., & DiCarlo, J. J. (2016). Using goal-driven deep learning models to understand sensory cortex. *Nat. Neurosci., 19*(3), 356-365.

11 Kuramoto, Y. (1975). Self-entrainment of a population of coupled non-linear oscillators. In *International Symposium on Mathematical Problems in Theoretical Physics* (pp. 420-422). Berlin Heidelberg: Springer.

12 Ehrenstein, W. H., Spillmann, L., & Sarris, V. (2003). Gestalt issues in modern neuroscience. *Axiomathes, 13*(3-4), 433-458.

13 Lappin, J. S., Wason, T. D., & Akutsu, H. (1987). Visual detection of common motion of spatially separate points. *Bull. Psychonomic Soc., 25*(5), 343-343).

14 Lappin, J. & Wason, T. (1991). Chapter 28. The perception of geometrical structure from congruence. In Ellis, S. (Ed.). *Pictorial communications in virtual and real environments*. London: Taylor & Francis. (Previously published in Ellis, S. R., Kaiser, M. K., & Grunwald, A. (Eds.). *Spatial Displays and Spatial Instruments; NASA Conference Publication 10032* (pp. 18-1 - 18-15).

15 Wason, T. (1993). Construction and Evaluation of a Three-Dimensional Display from a Two-Dimensional Projection Surface Based On Theoretical Considerations of Metrification and Affine Space (Doctoral dissertation). North Carolina State University. (UMI No. 9409172)

16 Wason, T., (1998). *U.S. Patent No. 5,751,927.* (Method and apparatus for producing three dimensional displays on a two dimensional surface). Washington,DC: U.S. Patent and Trademark Office.

17 Arenas, A., Dʹiaz-Guilera, A., Kurthsl, J., Moreno, Y., & Zhou, C. (2008). Synchronization in complex networks. *Phys. Rep., 469*(3), 93-153.

18 Kuramoto, Y. (1984). Cooperative dynamics of oscillator community a study based on lattice of rings. *Prog. Theor. Phys. Supp., 79*, 223-240.

19 Weiss, S., Sternklar, S., & Fischer, B. (1987). Double phase-conjugate mirror: analysis, demonstration, and applications. *Opt. Lett., 12*(2), 114-116.

20 Baltes, H., & Peeweeda, H. (1980). Partially coherent sources with phase profile and the Van Cittert-Zernike theorem. *Lettere Al Nuovo Cimento (1971 – 1985), 27*(16), 541-543.

21 Shannon, C. E. (1948). A mathematical theory of communication. *Bell Syst. Tech. J., 27*, 379-423, 623-656 July, October 1948.

22 Shannon, C. E., & Weaver, W. (1963). *The Mathematical Theory of Communication.* (pp. 45-61). University of Illinois Press. (Original work published in 1949)

23 Wittrock, M. C. (1992). Generative learning processes of the brain. *Educ. Psychol., 27*(4), 531-541.

24 Eliasmith, C., Stewart, T. C., Choo, X., Bekolay, T., DeWolf, T., Tang, C., & Rasmussen, D. (2012). A large-scale model of the functioning brain. *Science, 338*(6111), 1202-1205.

25 Merolla, P. A., Arthur, J. V., Alvarez-Icaza, R., Cassidy, A. S., Sawada, J., Akopyan, F., Jackson, B. L., Imam, N., Guo, C., Nakamura, Y., Brezzo, B., Vo, I., Esser, S. K., Appuswamy, R., Taba, B., Amir, A., Flickner, M. D., Risk, W. P., Manohar, R., & Modha, D. S. (2016). A million spiking-neuron integrated circuit with a scalable communication network and interface. *Science, 345*(6197), 668-673.

26 Grossberg, S. (2013). Adaptive resonance theory: How a brain learns to consciously attend, learn, and recognize a changing world. *Neural Networks, 37*, 1–47.

27 Gewaltig, M.-O., & Diesmann, M. (2007) NEST (Neural Simulation Tool), *Scholarpedia, 2*(4), 1430. http://www.nest-initiative.org/Software:About_NEST

28 Kunkel, S., Potjans, T. C., Eppler, J. M., Plesser, H. E., Morrison, A., & Diesmann, M. (2011). Meeting the memory challenges of brain-scale network simulation. *Front. Neuroinformatics, 5*, 35.

29 Kunkel, S., Schmidt, M., Eppler, J. M., Plesser, H. E., Igarashi, J., Masumoto, G., Fukai, T., Ishii, S., Morrison, A., Diesmann, M., & Helias, M. (2013). From laptops to supercomputers: a single highly scalable code base for spiking neuronal network simulations. *BMC Neurosci., 14*(Suppl 1), 163.

30 De Garis, H., Shuo, C., Goertzel, B., & Ruiting, L. (2010). A world survey of artificial brain projects, Part I: Large-scale brain simulations. *Neurocomputing, 74*(1), 3-29.

31 Nature Neuroscience (2011). Focus on computational and systems neuroscience. *Nat. Neurosci., 14*(2), 121.

32 Dehaene, S., Kerszberg, M., & Changeux, J-P (1998). A neuronal model of a global workspace in effortful cognitive tasks. *P. Natl. Acad. Sci., 95*(24), 14529-14534.

33 Hebb, D. O., (1976). Physiological learning theory. *J. Abnorm. Child Psych., 4*(4), 309-314.

34 Hopfield, J. J. & Tank, D. W., (1986). Computing with neural circuits: a model. *Science, 233*(4764), 625-633.

35 Hodges, A. (2012). Beyond Turing's machines. *Science, 336*(6078), 163-164.

36 Wegner, P., & Goldin, D. (2003). Computation beyond Turing machines. *Commun. ACM, 46*(4) 100-102.

37 Goldin, D., Wegner, P., Cooper, S., & Löwe, B. (2005). The Church-Turing thesis: Breaking the myth. *Lect. Notes Comput. Sci., 3526*, 31-64.

38 Hepp, K. (2012). Coherence and decoherence in the brain. *J. Math. Phys., 53*(095222), 5 pages.

39 Baars, B. J., Franklin, S., & Ramsoy, T. Z. (2013). Global workspace dynamics: cortical "binding and propagation" enables conscious contents. *Frontiers in Psychol., 4*, 200.

40 Maass, W. (2011). Liquid state machines: Motivation, theory, and applications. In Cooper, S. B., & Sorbi, A. (Eds.), *Computability in context: Computation and logic in the real world* (pp. 275-296). Singapore: World Scientific.

41 Shagrir, O. (2010). Brains as analog-model computers. *Stud. Hist. Philos. Sci., Part A, 41*(3), 271-279.

42 Penrose, R. (1989). *The emperor's new mind*. Oxford, New York, Melbourne: Oxford University Press.

43 Sporns, O. (2014). Contributions and challenges for network models in cognitive neuroscience. *Nat. Neurosci., 17*(5), 652-660.

44 Basheer, I. A., & Hajmeer, M. (2000). Artificial neural networks: fundamentals, computing, design, and application. *J. Microbiol. Meth., 43*(1), 3-31.

45 Edelman, G. M., & Reeke, G. N. (1982). Selective networks capable of representative transformations, limited generalizations, and associative memory. *P. Natl. Acad. Sci., 79*(6), 2091-2095.

46 Hinton, G. E. (1992). How neural networks learn from experience. *Sci. Am., 267*(3), 145-151.

47 Hinton, G. E. (2011). Machine learning for neuroscience. *Neural Systems & Circuits, 1*(1), 1-2.

48 Hoppensteadt, F. C., & Izhikevich, E. M. (1997). *Weakly connected neural networks*. New York: Springer-Verlag.

49 Song , S., Miller, K. D. & Abbott, L. F. (2000). Competitive Hebbian learning through spike-timing-dependent synaptic plasticity. *Nat. Neuro., 3*(9), 919 - 926.

50 Faugeras, O., Touboul, J. & Cessac, B. (2009). A constructive mean-field analysis of multi-population neural networks with random synaptic weights and stochastic inputs. *Front. Comput. Neurosci., 3*(1), 1-28.

51 Grossberg, S. (2007). Consciousness CLEARS the mind: Brain and consciousness. *Neural Networks, 20*(9), 1040-1053.

52 Felleman, D. J., & Van Essen, D. C. (1991). Distributed hierarchical processing in the primate cerebral cortex. *Cereb. Cortex, 1*(1), 1-47.

53 Hebb, D. O. (1949). *The Organization of Behavior*. New York: Wiley & Sons.

54 Fung, C. C. A., Wong, K. Y. M., & Wu, S. (2009). A moving bump in a continuous manifold: A comprehensive study of the tracking dynamics of continuous attractor neural networks. *Neural Comp., 22*(3), 752-792.

55 Chawanya, T., Aoyagi, T., Nishikawa, I., Okuda, K. & Kuramoto, Y. (1993). A model for feature linking via collective oscillations in the primary visual cortex. *Biol. Cybern., 68*(6), 483-490.

56 Bressloff, P. C. (2005). Spontaneous symmetry breaking in self-organizing neural fields. *Biol. Cybern., 93*(4), 256-274.

57 Tommerdahl, M., Favorov, O. V., & Whitsel, B. L. (2010). Dynamic representations of the somatosensory cortex. *Neuroscience & Biobehavioral Reviews, 34*(2), 160-170.

58 Risi, S., & Stanley, K. O. (2012). An enhanced hypercube-based encoding for evolving the placement, density, and connectivity of neurons. *Artif. Life, 18*(4), 331-363.

59 Modha, D. S., & Singh, R. (2010). Network architecture of the long-distance pathways in the macaque brain. *P. Natl. Acad. Sci., 107* (30), 13485-13490 .

60 Shepherd, GM (ed.) (2004). *The synaptic organization of the brain* (5th ed.). NY: Oxford.

61 Wakana, Setsu, Jiang, Hangyi, Nagae-Poetscher, Lidia M., van Zijl, Peter C. M., & Mori, Susumu (2004). Fiber tract–based atlas of human white matter anatomy. *Radiology, 230*(1), 77-87.

62 Alivisatos, A. P., Chun, M. , Church, G. M., Deisseroth, K., Donoghue, J. P., Greenspan, R. J., McEuen, P. L., Roukes, M. L., Sejnowski, T. J., Weiss, P. S., & Yuste, R. (2013). The brain activity map. *Science, 339*(6125), 1284-1285.

63 Bressloff, P. C. (2009). Stochastic neural field theory and the system-size expansion. *SIAM J. Appl. Math., 70*(5), 1488-1521.

64 Giacomin, G., Luçon, E. & Poquet, C. (2014). Coherence stability and effect of random natural frequencies in populations of coupled oscillators. *J. Dyn. Differ. Equ., 1040*(7294), 333-367.

65 Kellert, S. H. (1993). *In the Wake of Chaos: Unpredictable Order in Dynamical Systems* (p. 62). University of Chicago Press..

66 Wagner, T., Fell, J. & Lehnertz, K. (2010). The detection of transient directional couplings based on phase synchronization. *New J. Phys., 12*, 053031.

67 Bhuiyan, M. A., Pallipuram, V. K., Smith, M. C., Taha, T., & Jalasutram, R. (2010). Acceleration of spiking neural networks in emerging multi-core and GPU architectures. In *2010 IEEE International Symposium on Parallel & Distributed Processing, Workshops and Phd Forum (IPDPSW)* (pp. 1-8). IEEE.

68 Djurfeldt, M., Lundqvist, M., Johansson, C., Rehn, M., Ekeberg, O., & Lansner, A. (2008). Brain-scale simulation of the neocortex on the IBM Blue Gene/L supercomputer. *IBM J. Res. Dev., 52*(1-2), 31-41.

69 Preissl, R., Wong, T. M., Appuswamy, R., Datta, P., Flickner, M., Singh, R., Esser, S. K., McQuinn, E., Risk, W. P., Simon, H. D., & Modha, D. S. (2012). Compass: A scalable simulator for an architecture for cognitive computing. In *Proceedings of the International Conference for High Performance Computing, Networking, Storage, and Analysis* (p. 54). IEEE Computer Society Press.

70 Steck, J. E., Skinner, S. R., Cruz-Cabrera, A. A., Yang, M., & Behrman, E. C. (2007). *Field Computation for Artificial Neural Network Hardware: Nonlinear optical materials*. http://www.scientificcommons.org/42986830.

71 Hodgkin, A. L., Huxley, A. F. & Katz, B. (1952). Measurement of current-voltage relations in the membrane of the giant squid axon of Loligo. *J. Physiol., 116*(4), 424-448.

72 Hodgkin, A. L., & Huxley, A. F. (1952). Currents carried by sodium and potassium ions through the membrane of the giant squid axon of Loligo. *J. Physiol., 116*(4), 449-472.

73 Hodgkin, A. L., & Huxley, A. F. (1952). The components of membrane conductance in the giant squid axon of Loligo. *J. Physiol., 116*(4), 473-496.

74 Hodgkin, A. L., & Huxley, A. F. (1952). The dual effect of membrane potential on sodium conductance in the giant squid axon of Loligo. *J. Physiol., 116*(4), 497-506.

75 Hodgkin, A. L., & Huxley, A. F. (1952). A quantitative description of membrane current and its application to conduction and excitation in nerve. *J. Physiol., 117*(4):500-44.

76 Destexhe, A. (2011). Intracellular and computational evidence for a dominant role of internal network activity in cortical computations. *Curr. Opin. Neurobiol., 21*(5), 717-725.

77 Jehee, J. F. M., Roelfsema, P. R., Deco, G., Murre, J. M. J., & Lamme, V. A. F. (2007). Interactions between higher and lower visual areas improve shape selectivity of higher level neurons—Explaining crowding phenomena. *Brain Res., 1157*(0), 167-176.

78 Lamme, V. A. F., & Roelfsema, P. R. (2000). The distinct modes of vision offered by feedforward and recurrent processing. *Trends Neurosci., 23*(11), 571-579.

79 Contreras, D., Destexhe, A., Sejnowski, T. J., & Steriade, M. (1996). Control of spatiotemporal coherence of a thalamic oscillation by corticothalamic feedback. *Science, 274*(5288), 771–774.

80 Min, B-K. (2010). A thalamic reticular networking model of consciousness. *Theor. Biol. Med. Model., 7*(10), 1-18.

81 Nunez, P. L. (2000). Toward a quantitative description of large-scale neocortical dynamic function and EEG. *Behav. Brain Sci., 23*(3), 371-398.

82 Wendling, F., Ansari-Asl, K., Bartolomei, F., & Senhadji, L. (2009). From EEG signals to brain connectivity: A model-based evaluation of interdependence measures. *J. Neurosci. Meth., 183*(1), 9-18.

83 Muller, L. E., & Destexhe, A. (2012). Propagating waves in thalamus, cortex and the thalamocortical system: experiments and models. *J. Physiol. (Paris), 106*(5), 222-238.

84 Destexhe, A. (2009). Self-sustained asynchronous irregular states and Up/Down states in thalamic, cortical and thalamocortical networks of nonlinear integrate-and-fire neurons. *J. Comput. Neurosci., 27*(3), 493-506,

85 Dehaene, S., Changeux, J.-P., Naccache, L., Sackur, J., & Sergent, C. (2006). Conscious, preconscious, and subliminal processing: a testable taxonomy. *Trends Cogn. Sci., 10*(5), 204-211.

86 Dehaene, S., & Changeux, J.-P. (2011). Experimental and theoretical approaches to conscious processing. *Neuron, 70*(2), 200-227.

87 Henke, K. (2010). A model for memory systems based on processing modes rather than consciousness. *Nat. Rev. Neurosci., 11*(7), 523-532.

88 Drover, J. D., Schiff, N. D., & Victor, J. D. (2010). Dynamics of coupled thalamocortical modules. *J. Comput. Neurosci., 28*(3), 605-616.

89 Destexhe, A., Contreras, D., Sejnowski, T. J., & Steriade, M. (1994). A model of spindle rhythmicity in the isolated thalamic reticular nucleus. *J. Neurophysiol., 72*(2), 803-818.

90 Edelman, G. M., & Gally, J. A. (2001). Degeneracy and complexity in biological systems. *P. Natl. Acad. Sci., 98*(24), 13763-13768.

91 Marder, E., & Taylor, A. L. (2011). Multiple models to capture the variability in biological neurons and networks. *Nat. Neurosci., 14*(2), 133-138.

92 Toro, R., Perron, M., Pike, B., Richer, L., Veillette, S., Pausova, Z., & Paus, T. (2008). Brain size and folding of the human cerebral cortex. *Cereb. Cortex, 18*(10), 2352-2357

93 Henery, C. C., & Mayhew, T. M. (1989). The cerebrum and cerebellum of the fixed human brain: efficient and unbiased estimates of volumes and cortical surface areas. *J. Anat., 167*, 167–180.

94 Pallas, S. L., Roe, A. W. & Sur, M. (1990). Visual projections induced into the auditory pathway of ferrets. I. Novel inputs to primary auditory cortex (AI) from the LP/pulvinar complex and the topography of the MGN-AI projection. *J. Comp. Neurol., 298*(1), 50–68.

95 Sharma J., Angelucci A., Sur M. (2000). Induction of visual orientation modules in auditory cortex. *Nature, 404*(6780), 841-847.

96 Glasser, M. F., Coalson, T. S., Robinson, E. C., Hacker, C. D., Harwell, J., Yacoub, E., Ugurbil, K., Andersson, J., Beckmann, C. F., Jenkinson, M., Smith, S. M., & Van Essen, D. C. (2016). A multi-modal parcellation of human cerebral cortex. *Nature, 536*(7615), 171-178.

97 Power, J. D., Cohen, A. L., Nelson, S. M., Wig, G. S., Barnes, K. A., Church, J. A., Vogel, A. C., Laumann, T. O., Miezin, F. M., Schlaggar, B. L., & Petersen, S. E. (2011). Functional network organization of the human brain. *Neuron, 72*(4), 665-678.

98 Behrens, T. E. J., & Sporns, O. (2012). Human connectomics. *Curr. Opin. Neurobiol., 22*(1), 144-153.

99 Hagmann, P., Cammoun, L., Gigandet, X., Gerhard, S., Grant, P. E., Wedeen, V., Meuli, R., Thiran, J-P, Honey, C. J., & Sporns, O. (2010). MR connectomics: Principles and challenges. *J. Neurosci. Meth., 194*(1), 34-45.

100 Sporns, O., Tononi, G., & Kötter, R. (2005). The Human Connectome: A structural description of the human brain. *PLoS Comput. Biol., 1*(4), e42.

101 Yap, P-T, Wu, G., & Shen, D. (2010). Human Brain Connectomics: Networks, techniques, and applications [Life Sciences]. *IEEE Signal Proc. Mag., 27*(4), 131 - 134.

102 Amunts, K., Malikovic, A., Mohlberg, H., Schormann, T., & Zilles, K. (2000). Brodmann's areas 17 and 18 brought into stereotaxic space—where and how variable? *Neuroimage, 11*(1), 66-84.

103 Stanley, G. B (2013). Reading and writing the neural code. *Nat. Neurosci., 16*(3), 259-263.

104 Anderson, R. C., Pichert, J. W., Goetz, E. T., Schallert, D. L., Stevens, K. V., & Trollip, S. R. (1976). Instantiation of general terms. *J. Verb. Learn. Verb. Be., 15*(6), 667-679.

105 Woodward, P. M., & Lawson, J. D. (1948). The theoretical precision with which an arbitrary radiation-pattern may be obtained from a source of finite size. *J. Institution Electrical Eng. -Part III: Radio and Comm. Eng., 95*(37), 363-370.

106 Fugate, R.Q. (1999). Adaptive Optics for the 21st Century. In *Catching the Perfect Wave: Adaptive Optics and Interferometry for the 21st Century* (Vol. 174) (pp. 55-67).

107 Mandel, L., & Wolf, E. (1967). Terminology in Optics. *J. Opt. Soc. Am., 57,* 854.

108 Mandel, L., & Wolf, E. (1995). *Optical Coherence and Quantum Optics.* Cambridge, UK: Cambridge University Press.

109 Porter, M. B. (2002). Processing goals for the Acoustic Observatory. In *ONR Acoustic Observatory Science Plan Workshop* (pp. 44-55). DTIC.

110 Mountcastle, V. B. (1997). The columnar organization of the neocortex. *Brain, 20*(4) 701–722.

111 Rakic, P (1988). Specification of cerebral cortical areas. *Science, 241*(4862), 170-176.

112 Fox, P. T., Mintun, M. A., Raichle, M. E., Miezin, F. M., Allman, J. M., Van Essen, & D. C. (1986). Mapping human visual cortex with positron emission tomography. *Nature, 323*(6091), 806-809.

113 Cohen, A. L, Fair, D. A, Dosenbach, N. U. F, Miezin, F. M, Dierker, D., Van Essen, D. C, Schlaggar, B. L, & Petersen, S. E. (2008). Defining functional areas in individual human brains using resting functional connectivity MRI. *Neuroimage, 41*(1), 45-57.

114 Kaas, J. H. (1993). Evolution of multiple areas and modules within neocortex. *Perspect. Dev. Neurobiol., 2*(1), 101-107.

115 Wandell, B. A, Brewer, A. A, & Dougherty, R. F (2005). Visual field map clusters in human cortex. *Philos. T. Roy. Soc. B: Biol. Sci., 360*(1456), 693-707.

116 Wandell ,B. A., Dumoulin, S. O., & Brewer, A. A. (2009). Visual Cortex in Humans. In: Squire LR (Ed.) *Encyclopedia of Neuroscience, 10* (pp. 251-257). Oxford: Academic Press.

117 Frostig, R. D., Xiong, Y., Chen-Bee, C. H., Kvašňák, E., & Stehberg, J. (2008). Large-scale organization of rat sensorimotor cortex based on a motif of large activation spreads. *J. Neurosci., 28*(49), 13274-13284.

118 Kajikawa, Y., & Schroeder, C. E. (2011). How local is the local field potential? *Neuron, 72*(5), 847-858.

119 Hubel, D. H. & Wiesel, T. N. (1998). Early exploration of the visual cortex (review). *Neuron, 20*(3), 401–412.

120 Buxhoeveden, D. P., & Casanova, M. F. (2002). The minicolumn hypothesis in neuroscience. *Brain, 125*(5), 935-951.

121 Masino, S. A., Kwon, M. C., Dory, Y., & Frostig, R. D. (1993). Characterization of functional organization within rat barrel cortex using intrinsic signal optical imaging through a thinned skull. *P. Natl. Acad. Sci., 90*(21), 9998-10002.

122 Toga, A. W., Thompson, P. M., Mori, S., Amunts, K., & Zilles, K. (2006). Towards multimodal atlases of the human brain. *Nat. Rev. Neurosci. 7*(12), 952-966.

123 Blinkov, S. M., & Glezer, I. I. (1986). The Human Brain in Figures and Tables: A Quantitative Handbook. NY: Basic Books.

124 Mountcastle V. B. (1957). Modality and topographic properties of single neurons of cat's somatic sensory cortex. *J. Neurophysiol., 20*(4), 408–34.

125 Braendgaard H., Evans, S. M., Howard, C. V., & Gundersen, H. J. (1990). The total number of neurons in the human neocortex unbiasedly estimated using optical disectors. *J. Microsc., 157*(Pt 3):285-304.

126 Horton, J. C. & Adams, D. L. (2005). The cortical column: a structure without a function. *Philos. T. Roy. Soc. B: Biol. Sci., 360*(1456), 837-862.

127 Jones, E. G. (2000). Microcolumns in the cerebral cortex. *P. Natl. Acad. Sci., 97*(10), 5019-5021.

128 Katzel, D., Zemelman, B. V, Buetfering, C., Wolfel, M., & Miesenbock, G. (2011) . The columnar and laminar organization of inhibitory connections to neocortical excitatory cells. *Nat. Neurosci., 14*(1), 100-107.

129 Shepherd, G. M., & Svoboda, K. (2005). Laminar and columnar organization of ascending excitatory projections to layer 2/3 pyramidal neurons in rat barrel cortex. *J. Neurosci., 25*(24):5670-9.

130 Buffalo, E. A., Fries, P., Landman, R., Buschman, T. J., & Desimone, R. (2011). Laminar differences in gamma and alpha coherence in the ventral stream. *P. Natl. Acad. Sci., 108*(27) 11262-11267.

131 Hansen, B. J., & Dragoi, V. (2011). Adaptation-induced synchronization in laminar cortical circuits. *P. Natl. Acad. Sci., 108*(26), 10720-10725.

132 Maier, A., Adams, G. K., Aura, C., & Leopold, D. A. (2010). Distinct superficial and deep laminar domains of activity in the visual cortex during rest and stimulation. *Front. Syst. Neurosci., 4*(31), 1-11.

133 Albright, T. D., Jessell, T. M., Kandel, E. R., & Posner, M. I. (2000). Neural science: Review A century of progress and the mysteries that remain. *Neuron, 25*, S1-S55.

134 McCormick, D.A. (2004). Ch 2. Membrane Properties and Neurotransmitter actions. In Shepherd, G. M. (Ed). *The synaptic organization of the brain* (5th ed) (pp. 39-79). NY: Oxford.

135 Thomson, A. M., & Lamy, C. (2007). Functional maps of neocortical local circuitry. *Front. Neurosci., 1*(1), 19–42.

136 Krieger, P., Kuner, T., & Sakmann, B. (2007). Synaptic connections between layer 5B pyramidal neurons in mouse somatosensory cortex are independent of apical dendrite bundling. *J. Neurosci., 27*(43), 11473-82.

137 Ledergerber, D., & Larkum, M. E. (2010). Properties of layer 6 pyramidal neuron apical dendrites. *J. Neurosci., 30*(39), 13031-13044.

138 Grinvald, A., Lieke, E. E., Frostig, R. D., & Hildesheim, R. (1994). Cortical point-spread function and long-range lateral interactions revealed by real-time optical imaging of macaque monkey primary visual cortex. *J. Neurosci., 14*(5), 2545-2568.

139 Pluta, S., Naka, A., Veit, J., Telian, G., Yao, L., Hakim, R., Taylor, D., & Adesnik, H. (2015). A direct translaminar inhibitory circuit tunes cortical output. *Nat. Neurosci., 18*(11), 1631-1640.

140 Douglas, R., Markram, H., & Martin, K. (2004). Ch 12. Neocortex. In Shepherd, G. M. (Ed.), *The synaptic organization of the brain* (5th ed.) (pp. 499-558). NY: Oxford.

141 Shepherd, G. M. (2011). The microcircuit concept applied to cortical evolution: From three-layer to six-layer cortex. *Front. Neuroanat., 5*(30), 1-15, doi: 10.3389/fnana.2011.00030.

142 Shepherd, G. M. (2004). Ch 1. Introduction to Synaptic Circuits. in Shepherd, G.M. (Ed.), *The synaptic organization of the brain.*, (5th ed.) (pp. 1-38). NY: Oxford.

143 Harris, K. D., & Shepherd, G. M. (2015). The neocortical circuit: Themes and variations. *Nat. Neurosci., 18*(2), 170-181.

144 Deneve, S., & Machens, C. K. (2016). Efficient codes and balanced networks. *Nat. Neurosci., 19*(3), 375-382.

145 Breinig, M. (2013). Coherence. In *Physics 421, Modern Optics*, The University of Tennessee, Department of Physics and Astronomy. http://electron6.phys.utk.edu/optics421/modules/m5/Coherence.htm.

146 McNaught, A. D. & A. Wilkinson, A. (Compiled by) (1997). *IUPAC. Compendium of Chemical Terminology* (2nd ed.) *(the "Gold Book")* (p. 1675.) (XML on-line corrected version: http://goldbook.iupac.org). Blackwell Scientific Publications, Oxford. doi:10.1351/goldbook.

147 Carter, W. H. (1995). Coherence Theory. In Bass, M., (Ed.), *Handbook of Optics* (2nd ed.), Volume 1 (4.1-4.28). NY: McGraw-Hill, NY.

148 Dwyer, J., Lee, H., Martell, A., Stevens, R., Hereld, M., & van Drongelen, W. (2010). Oscillation in a network model of neocortex. *Neurocomputing, 73*(7), 1051-1056.

149 McGinniss, J. (1983). *Fatal Vision*. NY: Signet.

150 Parameshwaran, D., Crone, N. E., & Thiagarajan, T. C. (2012). Coherence potentials encode simple human sensorimotor behavior. *PLoS ONE, 7*(2), e30514.

151 Thiagarajan, T. C., Lebedev, M. A., Nicolelis, M. A. & Plenz, D. (2010). Coherence potentials: Loss-less, all-or-none network events in the cortex. *PLoS Biol., 8*(1), e1000278 (18 pgs).

152 Petermann, T., Thiagarajan, T. C., Lebedev, M. A., Nicolelis, M. A. L., Chialvo, D. R., & Plenz, D. (2009). Spontaneous cortical activity in awake monkeys composed of neuronal avalanches. *P. Natl. Acad. Sci., 106*(37), 15921-15926.

153 Gray, C. M., & McCormick, D. A. (1996). Chattering cells: Superficial pyramidal neurons contributing to the generation of synchronous oscillations in the visual cortex. *Science, 274*(5284), 109-113.

154 Bonifazi, P., Goldin, M., Picardo, M. A., Jorquera, I., Cattani, A., Bianconi, G., Represa, A., Ben-Ari, Y., & Cossart, R. (2009). GABAergic hub neurons orchestrate synchrony in developing hippocampal networks. *Science, 326*(5958), 1419-1424.

155 Wang, X-J (2010). Neurophysiological and Computational Principles of Cortical Rhythms in Cognition. *Physiol. Rev., 90*(3), 1195-1268.

156 Tiesinga, P., Fellous, J. M., & Sejnowski, T. J. (2008). Regulation of spike timing in visual cortical circuits. *Nat. Rev. Neurosci., 9*(2), 97-107.

157 Chialvo, D. R. (2010). Emergent complex neural dynamics. *Nat. Phys., 6*(10), 744-750.

158 Fraiman, D., Balenzuela, P., Foss, J., Chialvo, D. R. (2009). Ising-like dynamics in large-scale functional brain networks. *Phys. Rev. E, 79*(6), 061922 (5).

159 Mirollo, R. E., & Strogatz, S. H. (1990). Synchronization of pulse-coupled biological oscillators. *SIAM J. Appl. Math., 50*(6), 1645-1662.

160 Galán, R. F. (2009). The Phase Oscillator Approximation in Neuroscience: An Analytical Framework to Study Coherent Activity in Neural Networks. In Velazquez, J. L. P., & Wennberg, R. (Eds.), *Coordinated Activity in the Brain* (pp. 65-89). NY: Springer.

161 Nobili, R. (2009). New perspectives in brain information processing. *J. Biol. Phys., 35*(4), 347-360.

162 Jia, X., Smith, M. A., & Kohn, A. (2011). Stimulus selectivity and spatial coherence of gamma components of the local field potential. *J. Neurosci., 31*(25), 9390-9403

163 Estebanez, L., El Boustani, S., Destexhe, A., & Shulz, D. (2012). Correlated input reveals coexisting coding schemes in a sensory cortex. *Nat. Neurosci., 15*(12), 1691-1699.

164 Funke, K., & Wörgötter, F. (1997). On the significance of temporally structured activity in the dorsal lateral geniculate nucleus (LGN). *Prog. Neurobiol., 53*(1), 67-119.

165 Lizier, J. T., Prokopenko, M., & Zomaya, A. Y. (2012). Coherent information structure in complex computation. *Theor. Biosci., 131*(3), 193-203.

166 Li, W., & Packard, N. (1990). The structure of the elementary cellular automata rule space. *Complex Systems, 4*(3), 281-297.

167 Hipp, J. F., Engel, A. K., & Siegel, M. (2011). Oscillatory synchronization in large-scale cortical networks predicts perception. *Neuron, 69*(2), 387-396.

168 Watts, D. J. & Strogatz, S. H. (1998). Collective dynamics of `small-world' networks. *Nature, 393*(6684), 440-442.

169 Galán, R. F. (2008) On how network architecture determines the dominant patterns of spontaneous neural activity. *PLoS ONE, 3*(5): e2148.

170 Abrams, D. M., Pecora, L. M., & Motter, A. E. (2016). Introduction to focus issue: Patterns of network synchronization. *Chaos, 26*(9), 094601.

171 Winful, H. G.. & Rahman, L. (1990). Synchronized chaos and spatiotemporal chaos in arrays of coupled lasers. *Phys. Rev. Lett., 65*(13), 1575-1578.

172 Brandt, S. F., Dellen, B. K., & Wessel, R. (2006). Synchronization from disordered driving forces in arrays of coupled oscillators. *Phys. Rev. Lett., 96*(3), 034104.

173 Winfree, A. T. (1967). Biological rhythms and the behavior of populations of coupled oscillators. *J. Theor. Biol., 16*(1), 15-42. Cited in Strogatz, S. H. (2000). From Kuramoto to Crawford: Exploring the onset of synchronization in populations of coupled oscillators. *Physica D, 143*(1), 1-20.

174 Kuramoto, Y. (1991). Collective synchronization of pulse-coupled oscillators and excitable units. *Physica D, 50*(1), 15-30.

175 Bennett, M. V. L. & Zukin, R. S. (2004). Electrical coupling and neuronal synchronization in the mammalian brain. *Neuron, 41*(4), 495-511.

176 Breakspear, M., Heitmann, S. & Daffertshofer, A. (2010). Generative models of cortical oscillations: Neurobiological implications of the kuramoto model. *Front. Hum. Neurosci., 4*, 190.

177 Buzsáki, G. (2001). Electrical wiring of the oscillating brain, *Neuron, 31*(3), 342-344.

178 Takahashi, N., Kitamura, K., Matsuo, N., Mayford, M., Kano, M., Matsuki, N., & Ikegaya, Y. (2012). Locally synchronized synaptic inputs. *Science, 335*(6066) 353-356.

179 Hata, S., Arai, K., Galán, R. F., & Nakao, H. (2011). Optimal phase response curves for stochastic synchronization of limit-cycle oscillators. *Phys. Rev. E, 84*(016229), 1-10.

180 Stiefel, K. M., Gutkin, B. S., & Sejnowski, T. J. (2009). The effects of cholinergic neuromodulation on neuronal phase-response curves of modeled cortical neurons. *J. Comput. Neurosci., 26*(2), 289-301.

181 Arrott, A., (1957). Criterion for ferromagnetism from observations of magnetic isotherms. *Phys. Rev., 108*(6), 1394-1396.

182 Hong, Y. W., & Scaglione, A. (2005). A scalable synchronization protocol for large scale sensor networks and its applications. *IEEE J. Sel. Area. Comm., 23*(5), 1085-1099.

183 Zeitler, M., Daffertshofer, A., & Gielen, C. C. A. M. (2009). Asymmetry in pulse-coupled oscillators with delay. *Phys. Rev. E, 79*(6) 065203.

184 Ernst, U., Pawelzik, K. , & Geisel, T. (1995). synchronization induced by temporal delays in pulse-coupled oscillators. *Phys. Rev. Lett., 74*(9) 1570-1573.

185 Triplett, B., Klein, D. & Morgansen, K. (2006). Discrete time Kuramoto Models with Delay. In Antsaklis, P. J., & Tabuada, P. (Ed.), *Networked Embedded Sensing and Control: Workshop NESC'05* (pp. 9-23). NY: Springer-Verlag.

186 Onsager, Lars (1944), Crystal statistics. I. A two-dimensional model with an order-disorder transition. *Physical Review, Series II,* **65**(3–4): 117–149.

187 Acebrón, J. A., Bonilla, L. L., Vicente, C. J. P., Ritort, F. & Spigler, R. (2005). The Kuramoto model: A simple paradigm for synchronization phenomena. *Rev. Mod. Phys., 77*(1), 137-185.

188 Kawamura, Y., Nakao, H., Arai, K., Kori, & Kuramoto, Y. (2008). Collective Phase Sensitivity. *Phys. Rev. Lett., 101*(2), 024101.

189 Kuramoto, Y., & Nishikawa, I. (1987). Statistical macrodynamics of large dynamical systems. Case of a phase transition in oscillator communities. *J. Stat. Phys., 49*(3), 569-605.

190 Attia, E. H. (1991). *Large Coherent Arrays for Laser Radars. Phase 1 Report*, Ambler PA: Interspec Inc. DTIC Document ADA245528, http://www.dtic.mil/dtic/tr/fulltext/u2/a245528.pdf.

191 Gastrein, P., Campanac, É., Gasselin, C., Cudmore, R. H., Bialowas, A., Carlier, E., Fronzaroli-Molinieres, L., Ankri, N., & Debanne, D. (2011). The role of hyperpolarization-activated cationic current in spike-time precision and intrinsic resonance in cortical neurons in vitro. *J. Physiol., 589*(15), 3753-3773.

192 Hansel, D., Mato, G., & Meunier, C. (1995). Synchrony in excitatory neural networks. *Neural Comput., 7*(2), 307-337.

193 Hutcheon, B., & Yarom, Y. (2000). Resonance, oscillation and the intrinsic frequency preferences of neurons. *Trends Neurosci, 23*(5), 216-222.

194 Tkacik, G., Schneidman, E., Berry, Michael J. II, Bialek, W. (2006). Ising models for networks of real neurons. *arXiv preprint q-bio/0611072v1*

195 Strogatz, S. H. (2000). From Kuramoto to Crawford: Exploring the onset of synchronization in populations of coupled oscillators. *Physica D, 143*(1), 1-20.

196 Kuramoto, Y., & Battogtokh, D. (2002). Coexistence of coherence and incoherence in nonlocally coupled phase oscillators. *Nonlinear Phenomena in Complex Systems*, 5(4), 380-385.

197 Okuda, K., & Kuramoto, Y. (1991). Mutual entrainment between populations of coupled oscillators. *Prog. Theor. Phys., 86*(6), 1159-1176.

198 King, P. D., Zylberberg, J., & Deweese, M. R. (2013). Inhibitory interneurons decorrelate excitatory cells to drive sparse code formation in a spiking model of V1. *J. Neurosci., 33*(13):5475-5485.

199 Yoshimura, Y., & Callaway, E. M. (2005). Fine-scale specificity of cortical networks depends on inhibitory cell type and connectivity. *Nat. Neurosci., 8*(11), 1552-1559.

200 Lovett-Barron, M., Turi, G. F., Kaifosh, P., Lee, P. H., Bolze, F., Sun, X.-H., Nicoud, J.-F., Zemelman, B. V., Sternson, S. M., & Losonczy, A. (2012). Regulation of neuronal input transformations by tunable dendritic inhibition. *Nat. Neurosci., 15*(3), 423-430.

201 Couey, J. J., Witoelar, A., Zhang, S.-J., Zheng, K., Ye, J., Dunn, B., Czajkowski, R., Moser, M.-B., Moser, E. I., Roudi, Y., & Witter, M. P. (2013). Recurrent inhibitory circuitry as a mechanism for grid formation. *Nat. Neurosci., 16*(3), 318-324.

202 Gutzler, T., Hillman, T. R., Alexandrov, S. A., & Sampson, D. D. (2010). Coherent aperture-synthesis, wide-field, high-resolution holographic microscopy of biological tissue. *Opt. Lett., 35*(8), 1136-1138.

203 Monnier, J. D. (2003). Optical interferometry in astronomy. *Rep. Prog. Phys., 66*(5), 789.

204 Rabb, D., Jameson, D., Stokes, A., & Stafford, J. (2010). Distributed aperture synthesis. *Opt. Express, 18*(10), 10334-10342.

205 Steinberg, B. D. (1976). Principles of Aperture and Array System Design, including random and adaptive arrays (pp. 71-94). NY: John Wiley & Sons.

206 Schneidman, E., Berry, M. J., Segev, R., & Bialek, W. (2006). Weak pairwise correlations imply strongly correlated network states in a neural population. *Nature, 440*(7087), 1007--1012.

207 Truccolo, W., Hochberg, L. R., & Donoghue, J. P. (2010). Collective dynamics in human and monkey sensorimotor cortex: predicting single neuron spikes. *Nat. Neurosci. 13*(1), 105-111.

208 Yu, S., Huang, D., Singer, W., & Nikolic, D. (2008). A small world of neuronal synchrony. *Cereb. Cortex, 12*(2), 891-901.

209 Peters, A., Palay, S. L., & Webster, HdeF. (1991). *The fine structure of the nervous system: Neurons and their supporting cells* (3rd ed.)(pp. 101-211). USA: Oxford University Press.

210 Eugenin, E. A., Basilio, D., Sáez, J. C., Orellana, J. A., Raine, C. S., Bukauskas, F., Bennett, M. V. L., & Berman, J. W. (2012). The role of gap junction channels during physiologic and pathologic conditions of the human central nervous system. *J. Neuroimmune. Pharmacol., 7*(3), 499-518.

211 Hill, E. J., Jiménez-González, C., Tarczyluk, M., Nagel, D. A., Coleman, M. D., & Parri, H. R. (2012). NT2 derived neuronal and astrocytic network signalling. *PLoS ONE, 7*(5), e36098.

212 Sun, W., McConnell, E., Pare, J.-F., Xu, Q., Chen, M., Peng, W., Lovatt, D., Han, X., Smith, Y., & Nedergaard, M. (2013). Glutamate-dependent neuroglial calcium signaling differs between young and adult brain. *Science, 339*(6116), 197-200.

213 Destexhe, A., & Sejnowski, T. J. (2002). The initiation of bursts in thalamic neurons and the cortical control of thalamic sensitivity. *Philos. T. Roy. Soc. B: Biol. Sci., 357*(1428), 1649-1657.

214 Di Garbo, A. (2009). Dynamics of a minimal neural model consisting of an astrocyte, a neuron, and an interneuron. *J. Biol. Phys., 35*(4), 361-382.

215 Skardal, P. S., & Restrepo, J. G. (2012). Hierarchical synchrony of phase oscillators in modular networks. *Phys. Rev. E, 85*(1), 016208 (8 pages).

216 Agnati, L. F., Guidolin, D., Guescini, M., Genedani, S., & Fuxe, K. (2010). Understanding wiring and volume transmission. *Brain Res. Rev., 64*(1), 137-159.

217 Földy, C., Lee, S.-H., Morgan, R. J. & Soltesz, I. (2010). Regulation of fast-spiking cell synapses by the chloride channel ClC-2. *Nat. Neurosci., 13*(9), 1047–1049.

218 Goda, Y., & Sabatini, B.L. (2011). Synaptic function and regulation: Editorial overview. *Curr. Opin. Neurobiol., 21*(2), 1-3.

219 Gundelfinger, E. D., & tom Dieck, S. (2000). Molecular organization of excitatory chemical synapses in the mammalian brain. *Naturwissenschaften, 87*(12), 513-523.

220 Hofer, S. B., Ko, H., Pichler, B., Vogelstein, J., Ros, H., Zeng, H., Lein, E., Lesica, N. A., & Mrsic-Flogel, T. D. (2011). Differential connectivity and response dynamics of excitatory and inhibitory neurons in visual cortex. *Nat. Neurosci., 14*(8) 1045-1052.

221 Klemann, C. J. H. M., & Roubos, E. W. (2011). The gray area between synapse structure and function—Gray's synapse types I and II revisited. *Synapse, 65*(11), 1222-1230.

222 Peters, A., Palay, S. L., & Webster, HdeF. (1991). *The fine structure of the nervous system: Neurons and their supporting cells* (3rd ed.)(pp. 138-211). USA: Oxford University Press.

223 Brunel, N. (2016). Is cortical connectivity optimized for storing information? *Nat. Neurosci., 19*(5), 749-755.

224 Coggan, J. S., Bartol, T. M., Esquenazi, E., Stiles, J. R., Lamont, S., Martone, M. E., Berg, D. K., Ellisman, M. H., Sejnowski, T. J. (2005). Evidence for ectopic neurotransmission at a neuronal synapse. *Science, 309*(5733), 446-451.

225 Bean, B. P. (2007). The action potential in mammalian central neurons. *Nat. Rev. Neurosci., 8*(6), 451-465.

226 Froemke, R. C., & Dan, Y. (2001). Spike-timing-dependent synaptic modification induced by natural spike trains. *Nature, 416*(6879), 433-438.

227 Holt, G. R., & Koch, C. (1999). Electrical interactions via the extracellular potential near cell bodies. *J. Comput. Neurosci., 6*(2) 169-184.

228 McCaig, C. D., Song, B., & Rajnicek, A. M. (2009). Electrical dimensions in cell science. *J. Cell Sci., 122*(23), 4267-4276.

229 Han, X., Chen, M., Wang, F., Windrem, M., Wang, S., Shanz, S., Xu, Q., Oberheimm N. A., Bekar, L., Betstadt, S., Silva, A. J., Takano, T., Goldman, S. A., & Nedergaard, M. (2013). Forebrain engraftment by human glial progenitor cells enhances synaptic plasticity and learning in adult mice. *Cell stem cell, 12*(3), 342-353.

230 Walker, F. D., & Hild, W. J. (1969). Neuroglia electrically coupled to neurons. *Science, 165*(3893), 602-603.

231 Peters, A., Palay, S. L., & Webster, HdeF. (1991). *The fine structure of the nervous system: Neurons and their supporting cells* (3rd ed.)(pp. 203-210). USA: Oxford University Press.

232 Curti, S., Hoge, G., Nagy, J. I., & Pereda, A. E. (2012). Electrical transmission between mammalian neurons is supported by a small fraction of gap junction channels. *J. Membrane. Biol., 245*(5-6), 283-290.

233 Pereda, A. E., Curti, S., Hoge, G., Cachope, R., Flores, C. E., & Rash, J. E. (2013). Gap junction-mediated electrical transmission: Regulatory mechanisms and plasticity. *Biochimica et Biophysica Acta (BBA) – Biomembranes, 1828*(1), 134-146.

234 Nagy, J. I., Ionescu, A.-V., Lynn, B. D., & Rash, J. E. (2003). Coupling of astrocyte connexins Cx26, Cx30, Cx43 to oligodendrocyte Cx29, Cx32, Cx47: Implications from normal and connexin32 knockout mice. *Glia, 44*(3), 205-218.

235 Hameroff, S. (2010). The "conscious pilot"—dendritic synchrony moves through the brain to mediate consciousness. *J. Biol. Phys., 36*(1), 71-93.

236 Houades, V., Koulakoff, A., Ezan, P., Seif, I., & Giaume, C. (2008). Gap junction-mediated astrocytic networks in the mouse barrel cortex. *J. Neurosci., 28*(20), 5207-5217.

237 Lee, H. S., Ghetti, A., Pinto-Duarte, A., Wang, X., Dziewczapolski, G., Galimi, F., Huitron-Resendiz, S., Piña-Crespo, J. C., Roberts, A. J., Verma, I. M., Sejnowski, T. J., & Heinemann, S. F. (2014). Astrocytes contribute to gamma oscillations and recognition memory. *P. Natl. Acad. Sci., 111*(32), E3343-E3352.

238 Oberheim, N. A., Goldman, S. A., & Nedergaard, M. (2012). Heterogeneity of astrocytic form and function. *Methods Mol. Biol., 814*, 23–45.

239 Halassa, Michael M., Fellin, T., Takano, H., Dong, J.-H., & Haydon, P. G. (2007). Synaptic islands defined by the territory of a single astrocyte. *J. Neurosci., 27*(24), 6473-6477.

240 Newman, E. A. (2005). Calcium increases in retinal glial cells evoked by light-induced neuronal activity. *J. Neurosci., 25*(23), 5502-5510.

241 Steinhäuser, C., Seifert, G., & Bedner, P. (2012). Astrocyte dysfunction in temporal lobe epilepsy: K+ channels and gap junction coupling. *Glia, 60*(8), 1192-1202.

242 Østby, I., Øyehaug, L., Einevoll, G. T., Nagelhus, E. A., Plahte, E., Zeuthen, T., Lloyd, C. M., Ottersen, O. P., & Omholt, S. W. (2009). Astrocytic mechanisms explaining neural-activity-induced shrinkage of extraneuronal space. *PLoS Comput. Biol., 5*(1) e1000272.

243 Syková, E. (2005). Glia and volume transmission during physiological and pathological states. *J. Neural Transm., 112*(1), 137-147.

244 Bloomfield, S. A., & Volgyi, B. (2009). The diverse functional roles and regulation of neuronal gap junctions in the retina. *Nat. Rev. Neurosci., 10*(7), 495-506.

245 Vervaeke, K., Lőrincz, A., Nusser, Z., & Silver, R. A. (2012). Gap junctions compensate for sublinear dendritic integration in an inhibitory network. *Science, 335*(6076), 1624-1628.

246 Spruston, N. (2001). Axonal gap junctions send ripples through the hippocampus. *Neuron, 31*(5), 669-671.

247 Deans, M. R., Gibson, J. R., Sellitto, C., Connors, B. W., & Paul, D. L. (2001). Synchronous activity of inhibitory networks in neocortex requires electrical synapses containing connexin36. *Neuron, 31*(3), 477-485.

248 Haas, J. S., Zavala, B., & Landisman, C. E. (2011). Activity-dependent long-term depression of electrical synapses. *Science, 334*(6054), 389-393.

249 Traub, R. D., Kopell, N., Bibbig, A., Buhl, E. H., LeBeau, F. E. N., & Whittington, M. A. (2001). Gap junctions between interneuron dendrites can enhance synchrony of gamma oscillations in distributed networks. *J. Neurosci., 21*(23), 9478-9486.

250 Veruki, M. Lin, & Hartveit, E. (2002). AII (rod) amacrine cells form a network of electrically coupled interneurons in the mammalian retina. *Neuron, 33*(6), 935-946.

251 Baumann, N., & Pham-Dinh, D. (2001). Biology of oligodendrocyte and myelin in the mammalian central nervous system. *Physiol. Rev., 81*(2), 871-927.

252 Katz, B., & Schmitt, O. (1940). Electric interaction between two adjacent nerve fibers. *J. Physiol., 97*, 471-488.

253 Syková, E., & Nicholson, C. (2008). Diffusion in brain extracellular space. *Physiol. Rev., 88*(4), 1277-1340.

254 Anastassiou, C. A., Montgomery, S. M., Barahona, M., Buzsáki, G., & Koch, C. (2010). The effect of spatially inhomogeneous extracellular electric fields on neurons. *J. Neurosci., 30*(5), 1925-1936.

255 Barros, C. S., Franco, S. J., & Müller, U. (2011). Extracellular matrix: Functions in the nervous system. *Cold Spring Harbor Perspectives in Biology, 3*(1), a005108. Cold Spring Harbor Laboratory Press.

256 Zimmermann, D. R., & Dours-Zimmermann, M. T. (2008). Extracellular matrix of the central nervous system: from neglect to challenge. *Histochem. Cell. Biol., 130*(4), 635–653.

257 Simard, M., & Nedergaard, M. (2004). The neurobiology of glia in the context of water and ion homeostasis. *Neuroscience, 129*(4), 877-96.

258 Weyand, S., & Iwata, S. (2010). Old gate gets a new look. *Science, 329*(5988), 151-152.

259 Anastassiou, C. A., Perin, R., Markram, H., & Koch, C. (2011). Ephaptic coupling of cortical neurons. *Nat. Neurosci., 14*(2), 217-223.

260 Blot, A., & Barbour, B. (2014). Ultra-rapid axon-axon ephaptic inhibition of cerebellar Purkinje cells by the pinceau. *Nat. Neurosci., 17*(2), 289-295.

261 Brückner, G., Szeöke, S., Pavlica, S., Grosche, J., & Kacza, J. (2006). Axon initial segment ensheathed by extracellular matrix in perineuronal nets. *Neuroscience, 138*(2), 365-375.

262 Ogawa, Y., & Rasband, M. N. (2008). The functional organization and assembly of the axon initial segment. *Curr. Opin. Neurobiol., 18*(3), 307-313.

263 Colbert, C. M., & Johnston, D. (1996). Axonal action-potential initiation and Na+ channel densities in the soma and axon initial segment of subicular pyramidal neurons. *J. Neurosci., 16*(21), 6676-6686.

264 Yu, Y., Shu, Y., & McCormick, D. A. (2008). Cortical action potential backpropagation explains spike threshold variability and rapid-onset kinetics. *J. Neurosci., 28*(29): 7260–7272.

265 Dityatev, A., Schachner, M., & Sonderegger, P. (2010). The dual role of the extracellular matrix in synaptic plasticity and homeostasis. *Nat. Rev. Neurosci., 11*(11), 735-746.

266 Frischknecht, R., Heine, M., Perrais, D., Seidenbecher, C. I., Choquet, D., & Gundelfinger, E. D. (2009). Brain extracellular matrix affects AMPA receptor lateral mobility and short-term synaptic plasticity. *Nat. Neurosci., 12*(7), 897-904.

267 Wang, D., & Fawcett, J. (2012). The perineuronal net and the control of CNS plasticity. *Cell Tissue Res., 349*(1), 147-160.

268 Holthoff, K., & Witte, O. W. (1996) Intrinsic optical signals in rat neocortical slices measured with near-infrared dark-field microscopy reveal changes in extracellular space. *J. Neurosci., 16*(8), 2740–2749.

269 Theodosis, D. T., Piet, R., Poulain, D. A., & Oliet, S. H. R. (2004). Neuronal, glial and synaptic remodeling in the adult hypothalamus: functional consequences and role of cell surface and extracellular matrix adhesion molecules. *Neurochem. Int., 45*(4), 491-501.

270 Turley, E. A., Hossain, M. Z., Sorokan, T., Jordan, L. M., & Nagy, J. I. (1994). Astrocyte and microglial motility in vitro is functionally dependent on the hyaluronan receptor RHAMM. *Glia, 12*(1), 68-80.

271 Bedard, C., Rodrigues, S., Roy, N., Contreras, D., & Destexhe, A. (2010). Evidence for frequency-dependent extracellular impedance from the transfer function between extracellular and intracellular potentials. *J. Comp. Neurosci., 29*, 389-403, 2010.

272 Greschner, M., Shlens, J., Bakolitsa, C., Field, G. D., Gauthier, J. L., Jepson, L. H., Sher, A., Litke, A. M., & Chichilnisky, E. J. (2011). Correlated firing among major ganglion cell types in primate retina. *J. Physiol., 589*(1), 75-86.

273 Newman, M. E., & Watts, D. J. (1999). Scaling and percolation in the small-world network model. *Phys. Rev. E, 60*(6), 7332.

274 Motter, A. E., & Yang, Y. (2017). The unfolding and control of network cascades. *Physics Today, 70*(1), 32-39.

275 Gray, C. M., & Singer, W. (1989). Stimulus-specific neuronal oscillations in orientation columns of cat visual cortex. *P. Natl. Acad. Sci., 86*(5), 1698-1702.

276 Casanova, M. F., Buxhoeveden, D., & Gomez, J. (2003). Disruption in the inhibitory architecture of the cell minicolumn: implications for autisim. *The Neuroscientist, 9*(6), 496-507.

277 Helmstaedter, M., Sakmann, B., & Feldmeyer, D. (2009). Neuronal correlates of local, lateral, and translaminar inhibition with reference to cortical columns. *Cerebral Cortex, 19*(4), 926-937.

278 Somogyi, Peter, Freund, Tamas F., Hodgson, Anthony J., Somogyi, Jozsef, Beroukas, Dimitra, & Chubb, Ian W. (1985). Identified axo-axonic cells are immunoreactive for GABA in the hippocampus visual cortex of the cat. *Brain Res., 332*(1), 143-149.

279 Douglas, R., Markram, H, & Martin, K. (2004). Neocortex. in Shepherd, G. M. (Ed.) (2004). *Synaptic organization of the brain.* (5th ed.) (p. 521). NY: Oxford University Press.

280 Hubel, D. H., & Wiesel, T. N. (1974). Sequence regularity and geometry of orientation columns in the monkey striate cortex. *J. Comp. Neurol., 158*(3), 267-293.

281 DeFelipe, J. (1999). Chandelier cells and epilepsy. *Brain Res., 122*(10), 1807-1822.

282 Peters, A., Proskauer, C. C., & C. E. Ribak (1982). Chandelier cells in rat visual cortex. *J. Comp. Neurol., 206*(4), 397-416.

283 Marin-Padilla, M. (1987). The chandelier cell of the human visual cortex: A Golgi study. *J. Comp. Neurol., 256*(1), 61-70.

284 Jiang, X., Shen, S., Cadwell, C. R., Berens, P., Sinz, F., Ecker, A. S., Patel, S., & Tolias, A. S. (2015). Principles of connectivity among morphologically defined cell types in adult neocortex. *Science, 350*(6264), aac9462.

285 Taniguchi, H., Lu, J., & Huang, Z. J. (2013). The spatial and temporal origin of chandelier cells in mouse neocortex. *Science, 339*(6115), 70-74.

286 DeFelipe, J., López-Cruz, P. L., Benavides-Piccione, R., Bielza, C., Larrañaga, P., Anderson, S., ... & Ascoli, G. A. (2013). New insights into the classification and nomenclature of cortical GABAergic interneurons. *Nat. Rev. Neurosci., 14*(3), 202-216.

287 Markram, H., Toledo-Rodriguez, M., Wang, Y., Gupta, A., Silberberg, G., & Wu, C. (2004). Interneurons of the neocortical inhibitory system. *Nat. Rev. Neurosci., 5*(10), 793-807.

288 Stepanyants, A., Martinez, L. M., Ferecskó, A. S., & Kisvárday, Z. F. (2009). The fractions of short- and long-range connections in the visual cortex. *P. Natl. Acad. Sci., 106*(9) 3555-3560.

289 Wang, S., Chandrasekaran, L., Fernandez, F. R., White, J. A., & Canavier, C. C. (2012). Short conduction delays cause inhibition rather than excitation to favor synchrony in hybrid neuronal networks of the entorhinal cortex. *PLoS Comput. Biol., 8*(1), e1002306 1-19.

290 Leuba, G., Kraftsik, R., & Saini, K. (1998). Quantitative distribution of parvalbumin, calretinin, and calbindin D-28k immunoreactive neurons in the visual cortex of normal and Alzheimer cases. *Exp. Neurol., 152*(2), 278-291.

291 Debanne, D., Campanac, E., Bialowas, A., Carlier, E., & Alcaraz, G., (2011). Axon physiology. *Physiol. Rev., 91*(2), 555-602.

292 Peters, A., Palay, S. L., & Webster, HdeF. (1991). *The fine structure of the nervous system: Neurons and their supporting cells* (3rd ed.)(pp. 101-115). USA: Oxford University Press.

293 Boiko, T., Van Wart, A., Caldwell, J. H., Levinson, S. R., Trimmer, J. S., & Matthews, G. (2003). Functional specialization of the axon initial segment by isoform-specific sodium channel targeting. *J. Neurosci., 23*(6), 2306-2313.

294 Brackenbury, W. J., Calhoun, J. D., Chen, C., Miyazaki, H., Nukina, N., Oyama, F., Ranscht, B., & Isoma L. L. (2010). Functional reciprocity between Na+ channel Nav1.6 and ß1 subunits in the coordinated regulation of excitability and neurite outgrowth. *Proc. Natl. Acad. Sci., 107*(5), 2283–2288.

295 Dulla, C. G., & Huguenard, J. R. (2009). Who let the spikes out? *Nat. Neurosci., 12*(8), 959-960.

296 Hedstrom, K. L., & Rasband, M. N. (2006). Intrinsic and extrinsic determinants of ion channel localization in neurons. *J. Neurochem., 98*(5):1345-52.

297 Kuba, H., Oichi, Y., & Ohmori, H,. (2010). Presynaptic activity regulates Na(+) channel distribution at the axon initial segment. *Nature, 465*(7301), 1075-8.

298 Lai, H. C., & Jan, L. Y. (2006). The distribution and targeting of neuronal voltage-gated ion channels. *Nat. Rev. Neurosci., 7*(7), 548-562.

299 Van Wart, A., Trimmer, J. S., & Matthews, G. (2007). Polarized distribution of ion channels within microdomains of the axon initial segment. *J. Comp. Neurol., 500*(2), 339-352.

300 Colbert, C. M., & Pan, E. (2002). Ion channel properties underlying axonal action potential initiation in pyramidal neurons. *Nat. Neurosci., 5*(6), 533-538.

301 Hu, W., Tian, C., Li, T., Yang, M., Hou, H., & Shu, Y. (2009). Distinct contributions of Nav1.6 and Nav1.2 in action potential initiation and backpropagation. *Nat. Neurosci., 12*(8), 996-1002.

302 Kole, M. H. P., Ilschner, S. U., Kampa, B. M., Williams, S. R., Ruben, P. C., & Stuart, G. J. (2008). Action potential generation requires a high sodium channel density in the axon initial segment. *Nat. Neurosci., 11*(2), 178-186.

303 Palay, S. L., Sotelo, C., Peters, A. ,& Orkand, P. M. (1968). The axon hillock and the initial segment. *J. Cell Biol., 38*(1), 193-201.

304 Luscher, H. R, & Larkum, M. E. (1998). Modeling action potential initiation and back-propagation in dendrites of cultured rat motoneurons. *J. Neurophysiol., 80*(2), 715-729.

305 Somogyi, P., Nunzi, M. G., Gorio, A., & Smith, A. D. (1983). A new type of specific interneuron in the monkey hippocampus forming synapses exclusively with the axon initial segments of pyramidal cells. *Brain Research* **259**(1), 137-142.

306 Dugladze, Tamar, Schmitz, Dietmar, Whittington, Miles A., Vida, Imre, & Gloveli, Tengis (2012). Segregation of Axonal and Somatic Activity During Fast Network Oscillations. *Science, 336*(6087), 1458-1461.

307 Pouille, F., & Scanziani, M. (2001). Enforcement of temporal fidelity in pyramidal cells by somatic feed-forward inhibition. *Science, 293*(5532), 1159-1163.

308 Swadlow, H. A. (2003). Fast-spike interneurons and feedforward inhibition in awake sensory neocortex. *Cereb. Cortex, 13*(1), 2-32.

309 Clark, B. D., Goldberg, E. M., & Rudy, B. (2009). Electrogenic tuning of the axon initial segment. *The Neuroscientist, 15*(6), 651-668.

310 Goldberg, E. M., Clark , B. D., Zagha, E., Nahmani, M., Erisir, A., & Rudy, B. (2008). K+ Channels at the axon initial segment dampen near-threshold excitability of neocortical fast-spiking GABAergic interneurons. *Neuron, 58*(3), 387-400.

311 Kole, M. H. P., Letzkus, J. J., & Stuart, G. J. (2007). Axon initial segment Kv1 channels control axonal action potential waveform and synaptic efficacy. *Neuron, 55*(4) 633-647.

312 Shah, M. M., Migliore, M., Valencia, I., Cooper, E. C., & Brown, D. A. (2008). Functional significance of axonal Kv7 channels in hippocampal pyramidal neurons. *P. Natl. Acad. Sci., 105*(22), 7869-7874.

313 Price, C. J., Scott, R., Rusakov, D. A., & Capogna, M. (2008). GABAB receptor modulation of feedforward inhibition through hippocampal neurogliaform cells. *J. Neurosci., 28*(27), 6974-6982. Cited in Stern, P. (2008). Fine tuning of spike timing. *Science, 321*(5887), 318.

314 Hirsch, J. A., & Martinez, L. M. (2006). Laminar processing in the visual cortical column. *Curr. Opin. Neurobiol., 16*(4), 377-384.

315 Dávid, C., Schleicher, A., Zuschratter, W., & Staiger, J. F. (2007). The innervation of parvalbumin-containing interneurons by VIP-immunopositive interneurons in the primary somatosensory cortex of the adult rat. *Eur. J. Neurosci., 25*(8), 2329-2340.

316 Staiger, J. F., Zilles, K. and Freund, T. F. (1996). Distribution of GABAergic elements postsynaptic to ventroposteromedial thalamic projections in layer IV of rat barrel cortex. *Eur. J. Neurosci., 8*(11), 2273–2285.

317 Esser, S. K., Hill, S., & Tononi, G. (2009). Breakdown of effective connectivity during slow wave sleep: investigating the mechanism underlying a cortical gate using large-scale modeling. *J. Neurophysiol., 102*(4), 2096-2111.

318 Ahonen, T., Rahtu, E., Ojansivu, V., & Heikkila, J. (2008). Recognition of blurred faces using local phase quantization. *19th International Conference on Pattern Recognition* (pp. 1-4). IEEE.

319 Frien, A., & Eckhorn, R. (2000). Functional coupling shows stronger stimulus dependency for fast oscillations than for low-frequency components in striate cortex of awake monkey. *Eur. J. Neurosci., 12*(4), 1466-1478.

320 Douglas, R. J., & Martin, K. A. C. (1990). Control of neuronal output by inhibition at the axon initial segment. *Neural Comput., 2*(3), 283-292.

321 Szabadics, J., Varga, C., Molnar, G., Olah, S., Barzo, P., & Tamas, G. (2006). Excitatory effect of GABAergic axo-axonic cells in cortical microcircuits. *Science, 311*(5758), 233-235.

322 Reato, D., Rahman, A., Bikson, M., & Parra, L. C. (2010). Low-intensity electrical stimulation affects network dynamics by modulating population rate and spike timing. *J. Neurosci., 30*(45) 5067-15079.

323 Kettenmann, H., & Ranson, B. R. (1988). Electrical coupling between astrocytes and between oligodendrocytes studied in mammalian cell cultures. *Glia, 1*(1), 64-73.

324 Bullmore, E., & Sporns, O. (2009). Complex brain networks: Graph theoretical analysis of structural and functional systems. *Nat. Rev. Neurosci., 10*(3), 186-198.

325 Tagliazucchi, E., Balenzuela, P., Fraiman, D., & Chialvo, D. R. (2011). Point process analysis of large-scale brain fMRI dynamics. *arXiv preprint arXiv:1107.4572.*

326 Yu, S., Yang, H., Nakahara, H., Santos, G. S., Nikolić, D., & Plenz, D. (2011). Higher-order interactions characterized in cortical activity. *J. Neurosci., 31*(48), 17514-17526.

327 Bartfeld, E., & Grinvald, A. (1992). Relationships between orientation-preference pinwheels, cytochrome oxidase blobs, and ocular-dominance columns in primate striate cortex. *P. Natl. Acad. Sci., 89*(24), 11905-11909.

328 Hoppensteadt, F. C., & Izhikevich, E. M. (1998), Thalamo-cortical interactions modeled by weakly connected oscillators: Could the brain use FM radio principles? *Biosystems, 48*(1–3). 85-94.

329 Berens, P., Keliris, G. A., Ecker, A. S., Logothetis, N. K., & Tolias, A. S. (2008). Comparing the feature selectivity of the gamma-band of the local field potential and the underlying spiking activity in primate visual cortex. *Frontiers in Systems Neuroscience, 2*(2).

330 beim Graben, P., & Rodrigues, S. (2012). Coupling continuous neural networks to the electromagnetic field in nervous tissue. *arXiv preprint arXiv: 1006.5841v5.*

331 Berens, P., Logothetis, N. K., & Tolias, A. S. (2016). Local field potentials, BOLD and spiking activity – relationships and physiological mechanisms. *Nature* 2010 (2016), 58. (original work published in *Nat. Precedings*, 2010. Retrieved from http://precedings.nature.com/documents/5216/version/1/files/npre20105216-1.pdf.

332 Buzsáki, G., Anastassiou, C. A., & Koch, C. (2012). The origin of extracellular fields and currents -- EEG, ECoG, LFP and spikes. *Nat. Rev. Neurosci., 13*(6), 407-420.

333 Waldert, S., Lemon, R. N., & Kraskov, A. (2013). Influence of spiking activity on cortical local field potentials. *J. Physiol., 591*(21), 5291-5303.

334 Destexhe, A., & Bedard, C. (2012). Do neurons generate monopolar current sources?. *J. Neurophysiol., 108*(4), 953-955.

335 Riera, J. J., Ogawa, T., Goto, T., Sumiyoshi, A., Nonaka, H., Evans, A., ... & Kawashima, R. (2012). Pitfalls in the dipolar model for the neocortical EEG sources. *J. Neurophysiol., 108*(4), 956-975.

336 Zanos, S., Zanos, T. P., Marmarelis, V. Z., Ojemann, G. A., & Fetz, E. E. (2012). Relationships between spike-free local field potentials and spike timing in human temporal cortex. *J. Neurophysiol., 107*(7), 1808-1821.

337 Khodagholy, D., Gelinas, J. N., Thesen, T., Doyle, W., Devinsky, O., Malliaras, G. G., & Buzsáki, G. (2015). NeuroGrid: recording action potentials from the surface of the brain. *Nat. Neurosci., 18*(2), 310-315.

338 Agarwal, G., Stevenson, I. H., Berényi, A., Mizuseki, K., Buzsáki, G., & Sommer, F. T. (2014). Spatially distributed local fields in the hippocampus encode rat position. *Science, 344*(6184), 626-630.

339 Mendoza-Halliday, D., Torres, S., & Martinez-Trujillo, J. C. (2014). Sharp emergence of feature-selective sustained activity along the dorsal visual pathway. *Nat. Neurosci., 17*(9), 1255-1262.

340 Denker, M., Zehl, L., Brochier, T., Riehle, A., & Grün, S. (2012). Comparing the spatio-temporal organization of joint spiking and local field potential oscillations in motor cortex. *BMC Neuroscience, 13*(1), 1.

341 Reimann, M. W., Anastassiou, C. A., Perin, R., Hill, S. L., Markram, H., & Koch, C.(2013). A biophysically detailed model of neocortical local field potentials predicts the critical role of active membrane currents. *Neuron, 79*(2), 375-390.

342 Bair, W., & Koch, C. (1996). Temporal precision of spike trains in extrastriate cortex of the behaving Macaque monkey. *Neural Compuation, 8*(6), 1185–1202.

343 Mazzoni, A., Kayser, C., Murayama, Y., Martinez, J., Quiroga, R. Q., Logothetis, N. K., & Panzeri, S. (2011). Local field potential phase and spike timing convey information about different visual features in primary visual cortex. *BMC Neuroscience, 12*(1), 1.

344 Masquelier, T., Hugues, E., Deco, G., & Thorpe, S. J. (2009). Oscillations, phase-of-firing coding, and spike timing-dependent plasticity: an efficient learning scheme. *J Neurosci., 29*(43), 13484-93.

345 Grün, S., Diesmann, M., & Aertsen, A. (2010). Unitary Event Analysis. In Grün, S. & Rotter, S. (Eds.), *Analysis of Parallel Spike Trains, Springer Series in Computational Neuroscience 10*(7), (pp. 191-220). NY: Springer.

346 Nawrot, M. P. (2010). Analysis and Interpretation of Interval and Count Variability in Neural Spike Trains. In Grün, S., & Rotter, S. (Eds.), *Analysis of Parallel Spike Trains. Springer Series in Computational Neuroscience, 3*(7) (pp. 37-58). NY: Springer.

347 Fries, P., Nikolić, D., & Singer, W. (2007). The gamma cycle. *Trends Neurosci., 30*(7), 309-316.

348 Fröhlich, F., & McCormick, D. A. (2010). Endogenous electric fields may guide neocortical network activity. *Neuron, 67*(1), 129-143.

349 Chandler, D. J., Gao, W.-J., & Waterhouse, B. D. (2014). Heterogeneous organization of the locus coeruleus projections to prefrontal and motor cortices. *P. Natl. Acad. Sci., 111*(18), 6816-6821.

350 Haider, B., Hausser, M., & Carandini, M. (2013). Inhibition dominates sensory responses in the awake cortex. *Nature, 493*(7430), 97–100.

351 Boudreau, J. C. (1965). Neural volleying: upper frequency limits detectable in the auditory system. *Nature, 208*(5016), 1237-1238.

352 Engel, A. K., König, P., Gray, C. M., & Singer, W. (1990). Stimulus-dependent neuronal oscillations in cat visual cortex: inter-columnar interaction as determined by cross-correlation analysis. *Eur. J. Neurosci., 2*(7), 588-606.

353 Timofeev, I., Bazhenov, M., Seigneur, J., & Sejnowski, T. J. (2012). Neuronal Synchronization and Thalamocortical Rhythms during Sleep, Wake, and Epilepsy. In: Noebels, J. L., Avoli, M., Rogawski, M. A., Olsen, R. W., Delgado-Escueta, A. V. (Eds.), *Jasper's Basic Mechanisms of the Epilepsies* (4th ed.) (pp. 157-175). Bethesda, MD: National Center for Biotechnology Information.

354 Peyrache, A., Dehghani, N., Eskandar, E. N., Madsen, J. R., Anderson, W. S., Donoghue, J. A., ... & Destexhe, A. (2012). Spatiotemporal dynamics of neocortical excitation and inhibition during human sleep. *P. Natl. Acad. Sci., 109*(5), 1731-1736.

355 Huupponen, E., Kulkas, A., Saastamoinen, A., Tenhunen, M., & Himanen, S. (2011). Identification of deep sleep and awake with computational EEG measures. *J. Med. Syst.*, 35(6), 1413-1420.

356 Paz, J. T., Bryant, A. S., Peng, K., Fenno, L., Yizhar, O., Frankel, W. N., Deisseroth, K., & Huguenard, J. R. (2011). A new mode of corticothalamic transmission revealed in the Gria4-/- model of absence epilepsy. *Nat. Neurosci.*, *14*(9), 1167-1173.

357 Wang, Q., Webber, R. M. & Stanley, G. B. (2010). Thalamic synchrony and the adaptive gating of information flow to cortex. *Nat. Neurosci.*, *13*(12) 1534-1541.

358 Destexhe, A., Contreras, D., & Steriade, M. (1999). Cortically-induced coherence of a thalamic-generated oscillation. *Neuroscience, 92*(2), 427-443.

359 Nir, Y., Mukamel, R., Dinstein, I., Privman, E., Harel, M., Fisch, L., Gelbard-Sagiv, H., Kipervasser, S., Andelman, F., Neufeld, M. Y., Kramer, U., Arieli, A., Fried, I., & Malach, R. (2008). Interhemispheric correlations of slow spontaneous neuronal fluctuations revealed in human sensory cortex. *Nat. Neurosci.*, *11*(9), 1100-1108.

360 Ermentrout, G. B. & Kleinfeld, D. (2001). Traveling electrical waves in cortex. *Neuron, 29*(1), 33-44.

361 Cheng-yu T. L., Poo, M.-m., & Dan, Y. (2009). Burst spiking of a single cortical neuron modifies global brain state. *Science, 324*(5927), 643-646.

362 Katzner, S., Nauhaus, I., Benucci, A., Bonin, V., Ringach, D. Z. L., & Carandini, M. (2009). Local origin of field potentials in visual cortex. *Neuron, 61*(1), 35–41.

363 Okun, M., Naim, A., & Lampl, I. (2010). The subthreshold relation between cortical local field potential and neuronal firing unveiled by intracellular recordings in awake rats. *J. Neurosci.*, *30*(12), 4440-4448.

364 Volgushev, M., Chauvette, S., & Timofeev, I. (2011). Long-range correlation of the membrane potential in neocortical neurons during slow oscillation. *Prog. Brain Res.*, *193*, 181-199.

365 Varela, F., Lachaux, J.-P., Rodriguez, E., & Martinerie, J. (2001). The brainweb: Phase synchronization and large-scale integration. *Nat. Rev. Neurosci.*, *2*(4), 229-239.

366 Plenz, D., & Thiagarajan, T. C. (2007). The organizing principles of neuronal avalanches: Cell assemblies in the cortex? *Trends Neurosci., 30*(3), 101-110.

367 Lovett-Barron, M., & Losonczy, A. (2013). Circuits supporting the grid. *Nat. Neurosci., 16*(3), 255-257.

368 Schmidt-Hieber, C., & Hausser, M. (2013). Cellular mechanisms of spatial navigation in the medial entorhinal cortex. *Nat. Neurosci., 16*(3), 325-331.

369 Johansson, C., Rehn, M., & Lansner, A. (2006). Attractor neural networks with patchy connectivity. *Neurocomputing, 69*(7), 627-633.

370 Lundqvist, M., Rehn, M., Djurfeldt, M., & Lansner, A. (2006). Attractor dynamics in a modular network model of neocortex. *Network: Computation in Neural Systems,* 17(3), 253-276.

371 Wills, T. J., Lever, C., Cacucci, F., Burgess, N., & O'Keefe, J. (2005). Attractor dynamics in the hippocampal representation of the local environment. *Science, 308*(5723) 873-876.

372 Fraser, A. M., & Swinney, H. L. (1986). Independent coordinates for strange attractors from mutual information. *Phys. Rev. A, 33*(2), 1134-1140.

373 Grossberg, S. & Pearson, L. R. (2008). Laminar cortical dynamics of cognitive and motor working memory, sequence learning and performance: Toward a unified theory of how the cerebral cortex works. *Psychol. Rev.,* 115(3), 677-732.

374 Rosenblum, M. G., Pikovsky, A. S., & Kurths, J. (1996). Phase synchronization of chaotic oscillators. *Phys. Rev. Lett.,* 76(11), 1804-1807.

375 Masamizu, Y., Tanaka, Y. R., Tanaka, Y. H., Hira, R., Ohkubo, F., Kitamura, K., Isomura, Y., Okada, T., Matsuzaki, M. (2014). Two distinct layer-specific dynamics of cortical ensembles during learning of a motor task. *Nat. Neurosci.,* 17(7), 987-994.

376 Mizuseki, K., Diba, K., Pastalkova, E., & Buzsáki, G. (2011). Hippocampal CA1 pyramidal cells form functionally distinct sublayers. *Nat. Neurosci.,* 14(9), 1174-1181.

377 Bollimunta, A., Mo, J., Schroeder, C. E., & Ding, M. (2011). Neuronal mechanisms and attentional modulation of corticothalamic alpha oscillations. *J. Neurosci., 31*(13), 4935-4943.

378 Burns, S. P., Xing, D., & Shapley, R. M. (2011). Is gamma-band activity in the local field potential of V1 cortex a "clock" or filtered noise? *J. Neurosci., 31*(26), 9658-9664.

379 Behrman, E. C., Niemel, J., Steck, J. E., & Skinner, S. R. (1996). A quantum dot neural network. *Proceedings of the 4th Workshop on Physics of Computation* (pp. 22-24). Boston, MA: New England Complex Systems Institute.

380 Weiler, N., Wood, L., Yu, J., Solla, S. A., & Shepherd, G. M. (2008). Top-down laminar organization of the excitatory network in motor cortex. *Nat. Neurosci., 11*(3):360-6.

381 Wang, Y., Jin, J., Kremkow, J., Lashgari, R., Komban, S. J., & Alonso, J. M. (2015). Columnar organization of spatial phase in visual cortex. *Nat. Neurosci., 18*(1), 97-103.

382 Tomassy, G. S., Berger, D. R., Chen, H.-H., Kasthuri, N., Hayworth, K. J., Vercelli, A., Seung, H. S., Lichtman, J.W., & Arlotta, P. (2014). Distinct profiles of myelin distribution along single axons of pyramidal neurons in the neocortex. *Science, 344*(6181), 319-324.

383 Constantinople, C. M., & Bruno, R. M. (2013). Deep cortical layers are activated directly by thalamus. *Science, 340*(6140), 1591-1594.

384 Buzsáki, G., & Draguhn, A. (2004). Neuronal oscillations in cortical networks. *Science, 304*(5679), 1926-1929.

385 Marcus, G., Marblestone, A., & Dean, T. (2014). The atoms of neural computation. *Science, 346*(6209), 551-552.

386 Hull, C. L. (1930). Simple trial and error learning: A study in psychological theory. *Psychol. Rev. 37*(3), 241-256.

387 Hull, C. L. (1930). Knowledge and purpose as habit mechanisms. *Psychol. Rev., 37*(6), 511-525.

388 Debanne, D., Bialowas, A., & Rama, S. (2013). What are the mechanisms for analogue and digital signalling in the brain? *Nat. Rev. Neurosci., 14*(1), 63-69.

389 Nesse, W. H., Maler, L., & Longtin, A. (2010). Biophysical information representation in temporally correlated spike trains. *P. Natl. Acad. Sci., 107*(51), 21973-21978.

390 Butts, D. A., Weng, C., Jin, J., Yeh, C.-I, Lesica, N. A., Alonso, J.-M., & Stanley, G. B. (2007). Temporal precision in the neural code and the timescales of natural vision. *Nature, 449*(7158), 92-95.

391 Gollisch, T., & Meister, M. (2008). Rapid neural coding in the retina with relative spike latencies. *Science, 319*(5866) 1108-1111.

392 Gawne, T. J., McClurkin, J. W., Richmond, B. J., & Optican, L. M. (1991). Lateral geniculate neurons in behaving primates. III. Response predictions of a channel model with multiple spatial-to-temporal filters. *J. Neurophysiol., 66*(3), 809-823.

393 McClurkin, J. W., Gawne, T. J., Richmond, B. J., Optican, L. M., & Robinson, D. L. (1991). Lateral geniculate neurons in behaving primates. I. Responses to two-dimensional stimuli. *J. Neurophysiol., 66*(3), 777-793.

394 McClurkin, J. W., Gawne, T. J., Optican, L. M., & Richmond, B. J. (1991). Lateral geniculate neurons in behaving primates. II. Encoding of visual information in the temporal shape of the response. *J. Neurophysiol., 66*(3), 794-808.

395 McClurkin, J. W., Optican, L. M., Richmond, B. J., & Gawne, T. J. (1991). Concurrent processing and complexity of temporally encoded neuronal messages in visual perception. *Science, 253*(5020), 675-677.

396 Wang, X., Vaingankar, V., Sanchez, C. S., Sommer, F. T., & Hirsch, J. A. (2011). Thalamic interneurons and relay cells use complementary synaptic mechanisms for visual processing. *Nat. Neurosci., 14*(2), 224-231.

397 Wolfart, J., Debay, D., Le Masson, G., Destexhe, A., & Bal, T. (2005). Synaptic background activity controls spike transfer from thalamus to cortex. *Nat. Neurosci., 8*(12), 1760-1767.

398 Warland, D. K., Reinagel, P., & Meister, M. (1997). Decoding visual information from a population of retinal ganglion cells. *J. Neurophysiol., 78*(5), 2336-2350.

399 Chichilnisky, E. J. (2001). A simple white noise analysis of neuronal light responses. *Network: Computation in Neural Systems, 12*(2) 199-213.

400 Ala-Laurila, P., Greschner, M., Chichilnisky, E. J., & Rieke, F. (2011). Cone photoreceptor contributions to noise and correlations in the retinal output. *Nat. Neurosci., 14*(10), 1309-1316.

401 Shlens, J., Field, G. D., Gauthier, J. L., Greschner, M., Sher, A., Litke, A. M., & Chichilnisky, E. J. (2009). The structure of large-scale synchronized firing in primate retina. *J. Neurosci., 29*(15), 5022-5031.

402 Wong, R. O. L. (1999). Retinal waves and visual system development. *Ann. Rev. Neurosci., 22*(1), 29-47.

403 Padmanabhan, K., & Urban, N. N. (2010). Intrinsic biophysical diversity decorrelates neuronal firing while increasing information content. *Nat. Neurosci., 13*(10) 1276-1282.

404 Harris, K. D., Henze, D. A., Hirase, H., Leinekugel, X., Dragoi, G., Czurko, A., & Buzsáki, G. (2002). Spike train dynamics predicts theta-related phase precession in hippocampal pyramidal cells. *Nature, 417*(6890), 738-741.

405 Bonnevie, T., Dunn, B., Fyhn, M., Hafting, T., Derdikman, D., Kubie, J. L, Roudi, Y., Moser, E. I., & Moser, M.-B. (2013). Grid cells require excitatory drive from the hippocampus. *Nat. Neurosci., 16*(3), 309-317.

406 Heys, J. G., MacLeod, K. M., Moss, C. F. M. & Hasselmo, M. E. (2013). Bat and rat neurons differ in theta-frequency resonance despite similar coding of space. *Science, 340*(6130), 363-367.

407 Sreenivasan, S., & Fiete, I. (2011). Grid cells generate an analog error-correcting code for singularly precise neural computation. *Nat. Neurosci., 14*(10), 1330-1337.

408 Kalluri, R., Xue, J., & Eatock, R. A. (2010). Ion channels set spike timing regularity of mammalian vestibular afferent neurons. *J. Neurophysiol., 104*(4), 2034-2051.

409 Nemenman, I., Bialek, W. (2004). Entropy and information in neural spike trains: Progress on the sampling problem. *Phys. Rev. E, 69*(5), 056111 (6 pgs).

410 Strong, S. P., Koberle, R., de Ruyter van Steveninck, R. R., & Bialek, W. (1998). Entropy and information in neural spike trains. *Physical Review Lett., 80*(1), 197-200.

411 Borst, A., Theunissen, F. E. (1999). Information theory and neural coding. *Nat. Neurosci., 2*(11), 947-57.

412 Chomiak, T., Peters, S., & Hu, B. (2008). Functional architecture and spike timing properties of corticofugal projections from rat ventral temporal cortex. *J. Neurophysiol., 100*(1), 327-335.

413 Alecu, L., Frezza-Buet, H., & Alexandre, F. (2011). Can self-organization emerge through dynamic neural fields computation? *Connect. Sci., 23*(1), 1-31.

414 Amari, S.-i. (1977). Dynamics of pattern formation in lateral-inhibition type neural fields. *Biol. Cybern. 27*(2), 77-87.

415 Coombes, S. (2005). Waves, bumps, and patterns in neural field theories. *Biol. Cybern., 93*(2), 91-108.

416 Coombes, S., Venkov, N. A., Shiau, L., Bojak, I., Liley, D. T. J., & Laing, C. R. (2007). Modeling electrocortical activity through improved local approximations of integral neural field equations. *Phys. Rev. E, 76*(5), 051901-051908.

417 Freeman, W. J., & Vitiello, G. (2008). Vortices in brain waves. *Int. J. Mod. Phys. B, 24*(17), 3269-3295.

418 Freeman, W. J. (2009). Vortices in brain activity: Their mechanism and significance for perception. *Neural Networks, 22*(5-6), 491-501.

419 Freestone, D. R., Aram, P., Dewar, M., Scerri, K., Grayden, D. B., & Kadirkamanathan, V. (2011). A data-driven framework for neural field modeling. *NeuroImage, 56*(3), 1043-1058.

420 Goodman, J. W. (2005). Two-Dimensional Sampling Theory. In *Introduction to Fourier Optics* (3rd ed.). (pp. 22-30). Greenwood Village, CO: Robert & Co..

421 Linaro, D., Storace, M. & Mattia, M. (2011). Inferring network dynamics and neuron properties from population recordings. *Front. Comput. Neurosci., 5*(43), 1-17.

422 Wu, Si, Amari, Shun-ichi, & Nakahara, Hiroyuki (2002). Population coding and decoding in a neural field: A computational study. *Neural Comput., 14*(5), 999-1026.

423 Wu, W., Tiesinga, P. H, Tucker, T. R, Mitroff, S. R, & Fitzpatrick, D. (2011). Dynamics of population response to changes of motion. *J. Neurosci., 31*(36), 12767-12777.

424 Yger, P., El Boustani, S., Destexhe, A., & Frégnac, Y. (2011). Topologically invariant macroscopic statistics in balanced networks of conductance-based integrate-and-fire neurons. *J. Comput. Neurosci. 31*(2), 229-245.

425 Graf, A. B. A., Kohn, A., Jazayeri, M., & Movshon, J. A. (2011). Decoding the activity of neuronal populations in macaque primary visual cortex. *Nat. Neurosci., 14*(2), 239-245.

426 Finn, E. S., Shen, X., Scheinost, D., Rosenberg, M. D., Huang, J., Chun, M. M., ... & Constable, R. T. (2015). Functional connectome fingerprinting: identifying individuals using patterns of brain connectivity. *Nat. Neurosci.,* 18(11), 1664-1671.

427 de Gennes, P. G. (1988). Transport of Polymers. In Guyon, E., Nadal, J.-P., Pomeau, Y. (Eds.), *Proceedings of the NATO Conference on Disorder and Mixing, NATO ASI Series 152* (pp. 203-214). Netherlands: Springer.

428 Sahimi, M. (1993). Fractal and superdiffusive transport and hydrodynamic dispersion in heterogeneous porous media. *Transport Porous Med., 13*(1), 3-40.

429 Casasent, D., & Psaltis, D. (1976). Position, rotation, and scale invariant optical correlation. *Appl. Opt., 15*(7), 1795-1799.

430 DiCarlo, J. J., & Cox, D. D.(2007). Untangling invariant object recognition. *Trends Cogn. Sci., 11*(8) 333-341.

431 Lappin, J. S., & Wason, T. D. (1991). Chapter 28. The perception of geometrical congruence. In Ellis, S. R., Kaiser, M. K., & Grunwald, A. J. (Eds.), *Pictorial communications in virtual and real environments* (pp. 425-448). Bristol, PA: Taylor & Francis.

432 Cutting, J. E. (1987). Rigidity in cinema seen from the front row, side aisle. *J. Exp. Psychol. Human, 13*(3), 323-334.

433 Weaver, W. (1948). Science and Complexity. *Am. Sci., 36*(536), 1-11.

434 Tkačik, G., Marre, O., Amodei, D., Schneidman, E., Bialek, W. & Berry, M. J. II (2014). Searching for collective behavior in a large network of sensory neurons. *PLoS Comput. Biol. 10*(1). e1003408.

435 Pfeiffer, F., Weitkamp, T., Bunk, O., & David, C. (2006). Phase retrieval and differential phase-contrast imaging with low-brilliance x-ray sources. *Nature Phys., 2*(4), 258-261.

436 Kamondi, A., Acsady, L., Wang, X. J., & Buzsáki, G. (1998). Theta oscillations in somata and dendrites of hippocampal pyramidal cells in vivo: activity-dependent phase-precession of action potentials. *Hippocampus, 8*(3), 244–261.

437 Burwick, T. (2008). Temporal coding: Assembly formation through constructive interference. *Neural Comput., 20*(7), 1796-1820.

438 Borisyuk, R. M., & Hoppensteadt, F. C. (1998). Memorizing and recalling spatial–temporal patterns in an oscillator model of the hippocampus. *Biosystems, 48*(1-3) 3-10.

439 Quiroga, R. Q. (2012). Concept cells: the building blocks of declarative memory functions. *Nat. Rev. Neurosci., 13*(8), 587-597.

440 Moreno, Y. V´azquez-Prada, M., & F. Pacheco, A. 2 (2004). Fitness for synchronization of network motifs. *Physica A, 343*(C), 279-287.

441 Tsodyks, M., Uziel, A., & Markram, H. (2000). Synchrony generation in recurrent networks with frequency-dependent synapses. *J Neurosci, 20*(1), 825-835.

442 Kuramoto, Y., & Battogtokh, D. (2002). Coexistence of coherence and incoherence in nonlocally coupled phase oscillators. *Nonlinear Phenomena in Complex Systems, 5*(4), 380-385.

443 Okuda, K., & Kuramoto, Y. (1991). Mutual entrainment between populations of coupled oscillators. *Prog. Theor. Phys., 86*(6), 1159-1176.

444 Niyogi, R. K., & English, L. Q. (2009). Learning-rate-dependent clustering and self-development in a network of coupled phase oscillators. *Phys. Rev. E, 80*(6), 066213.

445 Sompolinsky, H., Golomb, D., & Kleinfeld, D. (1990). Global processing of visual stimuli in a neural network of coupled oscillators. *P. Natl. Acad. Sci., 87*(18), 7200-7204.

446 Baran, I., Stewart, M. P., Kampes, B. M., Perski, Z., & Lilly, P. (2003). A modification to the Goldstein radar interferogram filter. *IEEE T. Geosci. Remote, 41*(9), 2114-2118.

447 Moelans, N., Blanpain, B., & Wollants, P. (2008). An introduction to phase-field modeling of microstructure evolution. *Calphad, 32*(2), 268-294.

448 Steinbach, I., Pezzolla, F., Nestler, B., Seeßelberg, M., Prieler, R., Schmitz, G.J., & Rezende, J. L. L. (1996). A phase field concept for multiphase systems. *Physica D, 94*(3), 135-147.

449 Grochow, J. A., & Kellis, M. (2007). Network motif discovery using subgraph enumeration and symmetry-breaking. In *Annual International Conference on Research in Computational Molecular Biology.* (pp. 92-106). Berlin Heidelberg: Springer.

450 Milo, R., Shen-Orr, S., Itzkovitz, S., Kashtan, N., Chklovskii, D., & Alon, U. (2002). Network motifs: simple building blocks of complex networks. *Science, 298*(5594):824–827.

451 Sporns, O., Kötter, R. (2004) Motifs in Brain Networks. *PLoS Biol, 2*(11): e369.

452 Crair, M. C., Ruthazer, E. S., Gillespie, D. C., & Stryker, M. P. (1997). Ocular dominance peaks at pinwheel center singularities of the orientation map in cat visual cortex. *J Neurophysiol., 77*(6), 3381-3385.

453 Hubel, D. H., & Wiesel, T. N. (1963). Shape and arrangement of columns in cat's striate cortex. *J. Physiol., 165*(3), 559-568.

454 Nauhaus, I., Nielsen, K. J., Disney, A. A., & Callaway, E. M. (2012). Orthogonal micro-organization of orientation and spatial frequency in primate primary visual cortex. *Nat. Neurosci., 15*(12), 1683-1690.

455 Reichl, L., Löwel, S., & Wolf, F. (2009). Pinwheel stabilization by ocular dominance segregation. *Phys. Rev. Lett., 102*(20), 208102.

456 Metherell, A. F. (1968). The relative importance of phase and amplitude in acoustical holography. *Descriptive Note : Research Communication no. 63*, Douglas Aircraft Co. Inc., Huntington Beach Calif., Advanced Research Labs.

457 Montemezzani, G. (2002). Holography and optical phase conjugation. Lecture held at ETH Zürich.

458 Yarrow, S., Challis, E., & Seriès, P. (2012). Fisher and Shannon information in finite neural populations. *Neural Comput., 24*(7), 1740-1780.

459 Royer, S., Zemelman, B. V., Losonczy, A., Kim, J., Chance, F., Magee, J. C., & Buzsáki, G. (2012). Control of timing, rate and bursts of hippocampal place cells by dendritic and somatic inhibition. *Nat. Neurosci., 15*(5), 769-775.

460 Schroeder, C. E., & Lakatos, P. (2009). Low-frequency neuronal oscillations as instruments of sensory selection. *Trends Neurosci., 32*(1), 9-18.

461 Harris, K. D., Csicsvari, J., Hirase, H., Dragoi, G., & Buzsáki, G. (2003). Organization of cell assemblies in the hippocampus. *Nature, 424*(6948), 552-556.

462 Cohen, M. R., & Kohn, A. (2011). Measuring and interpreting neuronal correlations. *Nat. Neurosci., 14*(7) 811-819.

463 Tanaka, H.-A., Hasegawa, A., Mizuno, H., & Endo, T. (2002). Synchronizability of distributed clock oscillators. *IEEE T. Circuits-I, 49*(9), 1271-1278.

464 Chen, L.-Q. (2002). Phase-field models for microstructure evolution. *Ann. Rev. Mater. Res., 32*(1), 113-140.

465 Sabatini, S. P., Solari, F., & Secchi, L. (2004), A continuum-field model of visual cortex stimulus-driven behaviour: emergent oscillations and coherence fields. *Neurocomputing 57*, 411-433.

466 Fix, G. J. (1982). *Phase field methods for free boundary problems. Paper 32,* Department of Mathematical Sciences, CMU. http://repository.cmu.edu/math/32.

467 Ko, H., Hofer, S. B., Pichler, B., Buchanan, K. A., Sjostrom, P. J., & Mrsic-Flogel, T. D. (2011). Functional specificity of local synaptic connections in neocortical networks. *Nature, 473*(7345), 87–91.

468 Koffka, K. (1935). Force Fields: Gestalt concept of mechanisms which produce visual organization. In *Principles of Gestalt Psychology.* NY: Harcourt, Brace. Cited in Kaufman, L., (1974). *Sight and Mind* (pp. 464-473). NY: Oxford.

469 Cunningham, J. P., & Yu, B. M. (2014). Dimensionality reduction for large-scale neural recordings. *Nat. Neurosci., 17*(11), 1500-1509.

470 Brown, B. R., Lohmann, A. W. (1969). Computer-generated binary holograms. *IBM J. Res. Dev., 13*(2), 160.

471 Mesgarani, N., Cheung, C., Johnson, K., & Chang, E. F. (2014). Phonetic feature encoding in human superior temporal gyrus. *Science, 343*(6174), 1006-1010.

472 Horikawa, T., Tamaki, M., Miyawaki, Y., & Kamitani, Y. (2013). Neural decoding of visual imagery during sleep. *Science, 340*(6132), 639-642.

473 Macke, J., Berens, P., & Bethge, M. (2011). Statistical analysis of multi-cell recordings: Linking population coding models to experimental data. *Front. Comput. Neurosci., 5*(35), 1-2.

474 Baruchi. I., Grossman, D., Volman, V., Shein, M., Hunter, J., Towle, V. L., & Ben-Jacob, E. (2006). Functional holography analysis: Simplifying the complexity of dynamical networks. *Chaos, 16*(1), 015112.

475 Pollen, D. A., & Ronner, S. F. (1981). Phase relationships between adjacent simple cells in the visual cortex. *Science, 212*(4501), 1409-1411.

476 Lesburguères, E., Gobbo, O. L., Alaux-Cantin, S., Hambucken, A., Trifilieff, P., & Bontempi, B. (2011). Early tagging of cortical networks is required for the formation of enduring associative memory. *Science, 331*(6019), 924-928.

477 Tsanov, M., & Manahan-Vaughan, D. (2008). Synaptic plasticity from visual cortex to hippocampus: systems integration in spatial information processing. *Neuroscientist, 14*(6) 584-597.

478 Kuhl, B. A., Rissman, J., & Wagner, A. D. (2012). Multi-voxel patterns of visual category representation during episodic encoding are predictive of subsequent memory. *Neuropsychologia, 50*(4), 458-469.

479 Baldauf, D., & Desimone, R. (2014). Neural mechanisms of object-based attention. *Science, 344*(6182), 424-427.

480 Cudmore, R. H., Fronzaroli-Molinieres, L., Giraud, P., & Debanne, D. (2010). Spike-time precision and network synchrony are controlled by the homeostatic regulation of the D-type potassium current. *J. Neurosci., 30*(38), 12885-12895.

481 Alkire, M. T., Hudetz, A. G., & Tononi, G. (2008). Consciousness and anesthesia. *Science, 322*(5903) 876-880.

482 Reinhold, K., Lien, A. D., & Scanziani, M. (2015). Distinct recurrent versus afferent dynamics in cortical visual processing. *Nat. Neurosci., 18*(12), 1789–1797.

483 Sherman, S. M. (2006). Thalamus. *Scholarpedia, 1*(9):1583. doi:10.4249/scholarpedia.1583

484 Ciszak, M., Montina, A., & Arecchi, F. (2009). Spike synchronization of chaotic oscillators as a phase transition. *Cogn. Process., 39*(10), 33-39.

485 Usrey, W. M., & Reid, R. C. (1999). Synchronous activity in the visual system. *Annu. Rev. Physiol., 61*(1), 435-456.

486 Tiesinga, P. H. E., & Sejnowski, T. J. (2010). Mechanisms for phase shifting in cortical networks and their role in communication through coherence. *Frontiers in Human Neuroscience, 4* , 196.

487 Câteau, H., Kitano, K., & Fukai, T. (2008). Interplay between a phase response curve and spike-timing-dependent plasticity leading to wireless clustering, *Phys. Rev. E, 77*(5), 051909.

488 Nadasdy, Z. (2010). Binding by asynchrony: The neuronal phase code. *Frontiers in Neuroscience, 4*(51) 1-11.

489 Burgess, N., Barry, C. & O'Keefe, J. (2007), An oscillatory interference model of grid cell firing. *Hippocampus, 17*(9), 801–812.

490 Cooper, G. R. J., & Cowan, D. R. (2006). Enhancing potential field data using filters based on the local phase. *Comput. Geosci., 32*(10), 1585-1591.

491 Singer, W. (2009). Distributed processing and temporal codes in neuronal networks. *Cognitive Neurodynamics, 3*(3), 189-196.

492 Apollonov, V. V., Derzhavin, S. I., Kislov, V. I., Kazakov, A. A., Koval', Y. P., Kuz'minov, V. V., Mashkovskii, D. A., & Prokhorov, A. M. (1998). Spatial phase locking of linear arrays of 4 and 12 wide-aperture semiconductor laser diodes in an external cavity. *Quantum Electron., 28*(3), 257.

493 Pribram, K. (1991). Brain and Perception: Holonomy and Structure in Figural Processing. New Jersey: Lawrence Eribaum Associates.

494 Owechko, Y. ; Dunning, G. ; Nordin, G. ; Soffer, B. H. (1992). *Stimulated Photorefractive Optical Neural Networks. Final rept. AD-A258 825*, 140 pgs. DTIC.

495 Burke, W. J. & Sheng, P. (1977). Crosstalk noise from multiple thick-phase holograms. *J. Appl. Phys., 48*(2), 681 – 685.

496 Gonçalves, C., Pizolato, J. C., Cirino, G. A., & Neto, L. G. (2007). White light computer-generated phase hologram. In *Frontiers in Optics, Diffractive Micro- and Nanostructures for Sensing and Information Processing IV* (p. FThV5). Optical Society of America.

497 Yariv, A., & Kwong, S. K. (1986). Associative memories based on message-bearing optical modes in phase-conjugate resonators. *Opt. Lett., 11*(3), 186-188.

498 Steinberg, B. (1982). Properties of phase synchronizing sources for a radio camera. *IEEE T. Antenn. Propag., 30*(6), 1086-1092.

499 Izhikevich, E. M., Gally, J. A., & Edelman, G. M. (2004). Spike-timing dynamics of neuronal groups. *Cereb. Cortex, 14*(8), 933-944.

500 Hynes, R. O. (2009), The extracellular matrix: Not just pretty fibrils. *Science, 326*(5957), 1216-1219.

501 Min, R., & Nevian, T. (2012). Astrocyte signaling controls spike timing-dependent depression at neocortical synapses. *Nat. Neurosci., 15*(5), 746-753.

502 Fell, J., & Axmacher, N. (2011). The role of phase synchronization in memory processes. *Nat. Rev. Neurosci., 12*(2) 105-118.

503 Wolf, M. M., Verstraete, F., Hastings, M. B., & Cirac, J. I. (2008). Area laws in quantum systems: mutual information and correlations. *Phys. Rev. Lett., 100*(7), 070502.

504 Brunel, N., & Nadal, J.-P. (1998). Mutual information, Fisher information, and population coding. *Neural Comput., 10*(7), 1731-1757.

505 Romera, E., & Dehesa, J. S. (2004). The Fisher–Shannon information plane, an electron correlation tool. *J. Chem. Phys., 120*(19), 8906-8912.

506 Mitchell, M. (1996). Computation in cellular automata: A selected review. *Nonstandard Computation*, 95-140.

507 Callaway, E. M. (2005). Structure and function of parallel pathways in the primate early visual system. *J. Physiol., 566*(1), 13-19.

508 Bullier, J. (2001). Integrated model of visual processing. *Brain Res. Rev., 36*(2–3), 96-107.

509 Eglen, S. J., & Le Novère, N. (2012). Cellular spacing: Analysis and modelling of retinal mosaics. In *Computational Systems Neurobiology* (pp. 365-385). Netherlands: Springer.

510 Smith, S. L., & Hausser, M. (2010). Parallel processing of visual space by neighboring neurons in mouse visual cortex. *Nat. Neurosci., 13*(9), 1144-1149.

511 Schwartz, E. L. (1977). Spatial mapping in the primate sensory projection: Analytic structure and relevance to perception. *Biol. Cybern., 25*(4), 181-194.

512 Tootell, R. B. H., Mendola, J. D., Hadjikhani, N. K., Ledden, P. J., Liu, A. K., Reppas, J. B., Sereno, M. I., & Dale, A. M. (2007). Functional analysis of V3A and related areas in human visual cortex. *J. Neurosci., 17*(18), 7060-7078.

513 Kuffler, S. W. (1953). Discharge patterns and functional organization of the mammalian retina. *J. Neurophysiol., 1*(16) 37-68.

514 Rodieck, R. W. (1965). Quantitative analysis of cat retinal ganglion cell response to visual stimuli. *Vision Res., 5*(12) 583-601.

515 Wilson, H. R., & Giese, S. C. (1977). Threshold visibility of frequency gradient patterns. *Vision Res., 17*(10), 1177-1190.

516 Marr, D. (1982). *Vision* (pp. 54-68). San Francisco: W. H. Freeman & Co.

517 Wilson, H. R. & Bergen, J. R. (1979). A four mechanism model for threshold spatial vision. *Vision Res., 19*(1), 19-32.

518 Dacey, D., Packer, O. S.., Diller, L., Brainard, D., Peterson, B., & Lee, B. (2000). Center surround receptive field structure of cone bipolar cells in primate retina. *Vision Res., 40*(14) 1801- 1811.

519 Marr, D. & Hildreth, E. (1980). Theory of edge detection. *P. Roy. Soc. Lond. B Bio., 207*(1167), 187-217.

520 Weliky, M., Bosking, W. H., & Fitzpatrick, D. (1996). A systematic map of direction preference in primary visual cortex. *Nature, 379*(6567), 725-728.

521 Georgeson, M. A., May, K. A., Freeman, T. C. A., & Hesse, G. S. (2007). From filters to features: Scale–space analysis of edge and blur coding in human vision. *J. Vision, 7*(13), 7-7.

522 Racine, R. (1996). The telescope point spread function. *Publ. Astron. Soc. Pacific, 108*(726), 699-705.

523 Hung, C. P., Ramsden, B. M., & Roe, A. W. (2007). A functional circuitry for edge-induced brightness perception. *Nat. Neurosci., 10*(9), 1185-1190.

524 Kaschube, M., Schnabel, M., Löwel, S., Coppola, D. M., White, L. E., & Wolf, F. (2010). Universality in the evolution of orientation columns in the visual cortex. *Science, 330*(6007), 1113-1116.

525 Henriksson, L., Hyvärinen, A., & Vanni, S. (2009). Representation of cross-frequency spatial phase relationships in human visual cortex. *J. Neurosci., 29*(45), 14342-14351.

526 Tanigawa, H., Lu, H. D., Roe, A. W. (2010). Functional organization for color and orientation in macaque V4. *Nat. Neurosci., 13*(12), 1542-1548.

527 Schmid, A. M. (2008). The processing of feature discontinuities for different cue types in primary visual cortex. *Brain Res., 1238*, 59-74.

528 Onat, S., Nortmann, N., Rekauzke, S., König, P., & Jancke, D. (2011). Independent encoding of grating motion across stationary feature maps in primary visual cortex visualized with voltage-sensitive dye imaging. *NeuroImage, 55*(4), 1763-1770.

529 Liu, B.-h., Li, P., Sun, Y. J., Li, Y.-t., Zhang, L. I., & Tao, H. W. (2010). Intervening inhibition underlies simple-cell receptive field structure in visual cortex. *Nat. Neurosci., 13*(1), 89-96.

530 Monier, C., Chavane, F., Baudot, P., Graham, L.J., & Frégnac, Y. (2003). Orientation and direction selectivity of synaptic inputs in visual cortical neurons-A diversity of combinations produces spike tuning. *Neuron, 37*(4), 663-680.

531 Saygin, Z. M., Osher, D. E, Koldewyn, K., Reynolds, G., Gabrieli, J. D. E., & Saxe, R. R. (2012). Anatomical connectivity patterns predict face selectivity in the fusiform gyrus. *Nat. Neurosci., 15*(2), 321-327.

532 Yu, Y.-C., He, S., Chen, S., Fu, Y., Brown, K. N., Yao, X.-H., Ma, J., Gao, K. P., Sosinsky, G. E., Huang, K., & Shi, S.-H. (2012). Preferential electrical coupling regulates neocortical lineage-dependent microcircuit assembly. *Nature, 486*,113–117.

533 Paik, S.-B., & Ringach, D. L. (2011). Retinal origin of orientation maps in visual cortex. *Nat. Neurosci., 14*(7), 919-925.

534 Huberman, A. D., Feller, M. B., & Chapman, B. (2008). Mechanisms underlying development of visual maps and receptive fields. *Annu. Rev. Neurosci., 31*(1), 479-509.

535 Richards, B. A., Voss, O. P., & Akerman, C. J. (2010). GABAergic circuits control stimulus-instructed receptive field development in the optic tectum. *Nat. Neurosci., 13*(9), 1098-1106.

536 Stepanyants, A., Hirsch, J. A., Martinez, L. M., Kisvárday, Z. F., Ferecskó, A. S., & Chklovskii, D. B. (2008). Local potential connectivity in cat primary visual cortex. *Cereb. Cortex, 18*(1), 13-28.

537 Eroglu, C., & Barres, B. A. (2010). Regulation of synaptic connectivity by glia. *Nature, 468*(7321), 223-231.

538 Okada, D., Ozawa, F., & Inokuchi, K. (2009). Input-specific spine entry of soma-derived Vesl-1S protein conforms to synaptic tagging. *Science, 324*(904), 904-909.

539 McManus, J. N. J., Li, W., & Gilbert, C. D. (2011). Adaptive shape processing in primary visual cortex. *P. Natl. Acad. Sci., 108*(24), 9739-9746.

540 Chen, X., Han, F., Poo, M.-m., & Dan, Y. (2007). Excitatory and suppressive receptive field subunits in awake monkey primary visual cortex (V1). *P. Natl. Acad. Sci., 104*(48), 19120-19125.

541 Marr, D. (1982). *Vision* (pp. 61-68). San Francisco: W. H. Freeman & Co.

542 Newman, E. A. (2004). Glial modulation of synaptic transmission in the retina. *Glia, 47*(3), 268-74.

543 Wässle, H., Puller, C., Müller, F., & Haverkamp, S. (2009). Cone contacts, mosaics, and territories of bipolar cells in the mouse retina. *J. Neurosci., 29*(1), 106-117.

544 Vaney, D. I., Sivyer, B., & Taylor, W. R. (2012). Direction selectivity in the retina: symmetry and asymmetry in structure and function. *Nat. Rev. Neurosci., 13*(3), 194-208.

545 Soo, F. S., Schwartz, G. W., Sadeghi, K., & Berry, M. J. (2011). Fine spatial information represented in a population of retinal ganglion cells. *J. Neurosci., 31*(6) 2145-2155.

546 Murphy, G. J., & Rieke, F. (2008). Signals and noise in an inhibitory interneuron diverge to control activity in nearby retinal ganglion cells. *Nat. Neurosci., 11*(3), 318 - 326.

547 Zhang, A.-J., Jacoby, R., & Wu, S. M.(2011). Light- and dopamine-regulated receptive field plasticity in primate horizontal cells. *J. Comp. Neurol., 519*(11), 2125-2134.

548 Rossi, E. A, & Roorda, A. (2010). The relationship between visual resolution and cone spacing in the human fovea. *Nat. Neurosci., 13*(2), 156-157.

549 Buracas, G. T., Zador, A. M., DeWeese, M. R., & Albright, T. D. (1998). Efficient discrimination of temporal patterns by motion-sensitive neurons in primate visual cortex. *Neuron. 20*(5):959-69.

550 Li, L.-y., Li, Y.-t., Zhou, M., Tao, H., W., & Zhang, L. I. (2013). Intracortical multiplication of thalamocortical signals in mouse auditory cortex. *Nat. Neurosci., 16*(9), 1179-1181.

551 Temereanca, S., & Simons, D. J. (2004). Functional topography of corticothalamic feedback enhances thalamic spatial response tuning in the somatosensory whisker/barrel system. *Neuron, 41*(4), 639-651.

552 Afraz, A., Boyden, E. S., & DiCarlo, J. J. (2015). Optogenetic and pharmacological suppression of spatial clusters of face neurons reveal their causal role in face gender discrimination. *P. Natl. Acad. Sci., 112*(21), 6730-6735.

553 Ivakin, E. V., Koptek, V. G., Lazaruk, A. M., Petrovich, I. P., & Rubanov, A. S. (1979). Phase conjugation of light fields as a result of nonlinear interaction in saturable media. *JETP Lett., 30*(10), 613-616.

554 Psaltis, D., Brady, D., Gu, X.G., & Lin, S. (1990). Holography in artificial neural networks. *Nature, 343*(6256), 325-330.

555 Wood, G. L. (2000). Phase conjugation. In Waynant, R. W., & Ediger, M. N. (Eds.), *Electro-Optics Handbook* (pp. 14.1-14.40). New York: McGraw-Hill.

556 Woods, D., & Naughton, T. J. (2009). Optical computing. *Appl. Math. Comput., 215*(4), 1417-1430.

557 Yariv, A. (1978). Phase conjugate optics and real-time holography. *IEEE J. Quantum Electron., 14*(9), 650-660.

558 Fong, B. H., Colburn, J. S., Ottusch, J. J., Visher, J. L., & Sievenpiper, D. F. (2010). Scalar and tensor holographic artificial impedance surfaces. *IEEE T. Antenn. Propag., 58*(10), 3212-3221.

559 Watson, A. B., Yang, G. Y., Solomon, J. A., & Villasenor, J. (1997). Visibility of wavelet quantization noise. *IEEE T. Image Process., 8*(6), 1164-1175.

560 Fries, P. (2005). A mechanism for cognitive dynamics: Neuronal communication through neuronal coherence. *Trends Cogn. Sci., 9*(10), 474-480.

561 McCollough, C. (2000). Do McCollough effects provide evidence for global pattern processing? *Percept. Psychophys., 62*(2), 350-362.

562 Darmopil, S., Hjerling-Leffler, J., Ernfors, P., Winblad, B., Diamond, M. C., Eriksson, P. S., Bogdanovic, N. M., Abdul H., & Zhu, S. W. (2002). Environmental enrichment and the brain. *Prog. Brain Res., 138*, 109-133.

563 Nelson, S. B., & Turrigiano, G. G. (2008). Strength through diversity. *Neuron, 60*(3), 477-482.

564 Rust, N. C., & Stocker, A. A. (2010). Ambiguity and invariance: Two fundamental challenges for visual processing. *Curr. Opin. Neurobiol., 20*(3), 382-388.

565 Tuytelaars, T. & Mikolajczyk, K. (2008). Local invariant feature detectors: A survey. *Found. Trends Comput. Graph. Vis. 3*(3), 177-280.

566 Turk-Browne, N. B. (2013). Functional interactions as big data in the human brain. *Science, 342*(6158) 580-584.

567 Moldakarimov, S. B., McClelland, J. L., & Ermentrout, G. B. (2006). A homeostatic rule for inhibitory synapses promotes temporal sharpening and cortical reorganization. *P. Natl. Acad. Sci., 103*(44), 16526-16531.

568 Gong, P., & van Leeuwen, C. (2009). Distributed dynamical computation in neural circuits with propagating coherent activity patterns. *PLoS Comput. Biol. 5*(12), http://dx.doi.org/10.1371%2Fjournal.pcbi.1000611 e1000611.

569 Hughes, S. W., & Crunelli, V. (2007). Just a phase they're going through: The complex interaction of intrinsic high-threshold bursting and gap junctions in the generation of thalamic α and θ rhythms. *Int. J. Psychophys., 64*(1), 3-17.

570 El Boustani, S., Yger, P., Frégnac, Y., & Destexhe, A. (2012). Stable learning in stochastic network states. *J. Neurosci., 32*(1), 194-214, 2012.

571 Turrigiano, G. G., & Nelson, S. B. (2004). Homeostatic plasticity in the developing nervous system. *Nat. Rev. Neurosci., 5*(2), 97-107.

572 Bender, K. J., & Trussell, L. O. (2009). Axon initial segment Ca2+ channels influence action potential generation and timing. *Neuron, 61*(2), 259-271.

573 Bailey, C. H., Giustetto, M., Huang, Y.-Y., Hawkins, R. D., & Kandel, E. R. (2000). Is heterosynaptic modulation essential for stabilizing hebbian plasiticity and memory. *Nat. Rev. Neurosci., 1*(1), 11-20.

574 Castillo, P. E., Chiu, C. Q., & Carroll, R. C. (2011). Long-term plasticity at inhibitory synapses. *Curr. Opin. Neurobiol., 21*(2), 328-338.

575 McCann, C. M., Tapia, J. C., Kim, H., Coggan, J. S., & Lichtman, J. W. (2008). Rapid and modifiable neurotransmitter receptor dynamics at a neuronal synapse in vivo. *Nat. Neurosci., 11*(7), 807-815.

576 Toth, A. B., Terauchi, A., Zhang, L. Y., Johnson-Venkatesh, E. M., Larsen, D. J., Sutton, M. A., & Umemori, H. (2013). Synapse maturation by activity-dependent ectodomain shedding of SIRP[alpha]. *Nat. Neurosci., 16*(10), 1417-1425.

577 Dunwiddie, T., & Lynch, G. (1987). Long-term potentiation and depression of synaptic responses in the rat hippocampus: Localization and frequency dependency. *J. Physiol., 276*(1), 353-367.

578 Woodin, M. A., Ganguly, K., & Poo, M.-m. (2003). Coincident pre- and postsynaptic activity modifies GABAergic synapses by postsynaptic changes in Cl– transporter activity. *Neuron, 39*(5), 807-820.

579 Johnston, D., Christie, B. R., Frick, A., Gray, R., Hoffman, D. A., Schexnayder, L. K., Watanabe, S., & Yuan, L.-L. (2003). Active dendrites, potassium channels and synaptic plasticity. *Philos. T. Roy. Soc. B, 358*(1432)667-674.

580 Wang, D. O., Kim, S. M., Zhao, Y., Hwang, H., Miura, S. K., Sossin, W. S. & Martin, K. C. (2009). Synapse- and stimulus-specific local translation during long-term neuronal plasticity. *Science, 324*(5934), 1536-1540.

581 Lisman, J., & Spruston, N. (2005). Postsynaptic depolarization requirements for LTP and LTD: A critique of spike timing-dependent plasticity. *Nat. Neurosci., 8*(7), 839-841.

582 Pi, H. J., Otmakhov, N., Lemelin, D., De Koninck, P., & Lisman, J. (2010). Autonomous CaMKII can promote either long-term potentiation or long-term depression, depending on the state of T305/T306 phosphorylation. *J. Neurosci., 30*(26), 8704-8709.

583 Wang, H.-X., Gerkin, R. C., Nauen, D. W., & Bi, G. Q. (2005). Coactivation and timing-dependent integration of synaptic potentiation and depression. *Nat. Neurosci., 8*(2), 187-193.

584 Ito, M. (1989). Long-term depression. *Annu. Rev. Neurosci., 12*(1), 85-102.

585 Artola, A., & Singer, W. (1987). Long-term potentiation and NMDA receptors in rat visual cortex. *Nature, 330*(6149), 649-652.

586 Sjöström, P. J., Rancz, E. A., Roth, A., & Häusser, M. (2008). Dendritic excitability and synaptic plasticity. *Physiol. Rev., 88*(2), 769-840.

587 Bittner, K. C., Grienberger, C., Vaidya, S. P., Milstein, A. D., Macklin, J. J., Suh, J., Tonegawa, S., & Magee, J. C. (2015). Conjunctive input processing drives feature selectivity in hippocampal CA1 neurons. *Nat. Neurosci., 18*(8), 1133-1142.

588 Ullian, E. M., Sapperstein, S. K., Christopherson, K. S., & Barres, B. A. (2001). Control of synapse number by glia. *Science, 291*(5504), 657-661.

589 Allen, N. J. & Barres, B. A. (2009). Neuroscience: Glia—more than just brain glue. *Nature, 457*(7230), 675-677.

590 Jehee, J. F. M., & Murre, J. M. J. (2008). The scalable mammalian brain: Emergent distributions of glia and neurons. *Biol. Cybern., 98*(5), 439-445.

591 Herculano-Houzel, S. (2009). The human brain in numbers: A linearly scaled-up primate brain. *Front. Hum. Neurosci., 3*(31), 1-11.

592 Diamond, M. C., Scheibel, A. B., Murphy Jr., G. M., & Harvey, T. (1985). On the brain of a scientist: Albert Einstein. *Exp. Neurol., 88*(1), 198-204.

593 Nicholson, C. (2010). Neuron–glia interactions: Parting the waves. *Nat. Rev. Neurosci., 11*(10), 666.

594 Perea, G., Navarrete, M., & Araque, A. (2009). Tripartite synapses: Astrocytes process and control synaptic information. *Trends Neurosci., 32*(8), 421-421.

595 Panatier, A., Vallée, J., Haber, M., Murai, K. K., Lacaille, J.-C., & Robitaille, R. (2011). Astrocytes are endogenous regulators of basal transmission at central synapses. *Cell, 146*(5), 785-798.

596 Grosche, A., & Reichenbach, A. (2013). Developmental refining of neuroglial signaling? *Science, 339*(6116), 152-153.

597 Grossfeld, R. M., Hargittai, P. T. & Lieberman, E. M. (1995). Glutamate-mediated neuron-glia signaling in invertebrates and vertebrates. In Vernadakis A. & Roots, B. I. (Eds.), *Neuron-Glia Interrelation in Phylogeny* (pp. 129-159). Totowa, NJ: Humana Press.

598 Santelloa, M. & Volterra, A. (2009). Protein trafficking, targeting, and interaction at the glutamate synapse: Synaptic modulation by astrocytes via Ca2+-dependent glutamate release. *Neuroscience, 158*(1), 253-259.

599 Newman, E., & Reichenbach, A. (1996). The Müller cell: A functional element of the retina. *Trends Neurosci., 19*(8), 307-312.

600 Price, C. J., Scott, R., Rusakov, D. A., & Capogna, M. (2008). GABAB receptor modulation of feedforward inhibition through hippocampal neurogliaform cells. *J. Neurosci., 28*(27), 6974-6982.

601 Schummers, J., Yu, H., & Sur, M. (2008). Tuned responses of astrocytes and their influence on hemodynamic signals in the visual cortex. *Science, 320*(5883), 1638-1643.

602 Orkand R. K., & Opava S. C. (1994). Glial function in homeostasis of the neuronal microenvironment. *News Physiol. Sci., 9*(6), 265-267.

603 Pozo, K., & Goda, Y. (2010). Unraveling mechanisms of homeostatic synaptic plasticity. *Neuron, 66*(3), 337-351.

604 Olypher, A. V., & Prinz, A. A. (2010). Geometry and dynamics of activity-dependent homeostatic regulation in neurons. *J Comput Neurosci., 28*(3), 361-74.

605 Friedman, A. K., Walsh, J. J., Juarez, B., Ku, S. M., Chaudhury, D., Wang, J., Li, X., Dietz, D. M., Pan, N., Vialou, V. F., Neve, R. L., Yue, Z., & Han, M.-H. (2014). Enhancing depression mechanisms in midbrain dopamine neurons achieves homeostatic resilience. *Science, 344*(6181), 313-319.

606 Service, R. F. (2014). The brain chip. *Science, 345*(6197), 614-616.

607 Versace, M., & Chandler, B. (2010). The brain of a new machine. *IEEE Spectrum, 47*(12), 30-37.

608 Hameroff, S., Nip, A., Porter, M., & Tuszynski, J. (2002). Conduction pathways in microtubules, biological quantum computation, and consciousness. *Biosystems, 64*(1-3) 149-168.

609 Litt, A., Eliasmith, C., Kroon, F. W., Weinstein, S., & Thagard, P. (2006). Is the brain a quantum computer? *Cognitive Sci., 30*(3), 593-603.

610 Insel, T. R., Landis, S. C., & Collins, F. S. (2013). The NIH BRAIN initiative. *Science, 340*(6133), 687-688.

611 Munakata, Y., & McClelland, J. L. (2003). Connectionist models of development. *Developmental Sci., 6*(4), 413-429.

612 Katz, L. C., Weliky, M., & Crowley, J. C. (2000). Visual cortex: New Perspectives. In Gazzaniga, M. S. (Ed.), *The New Cognitive Neurosciences* (2nd ed.)(pp. 199-212). Cambridge, Massachusetts: A Bradford Book, The MIT Press.

613 Markram, H. (2012). The Human Brain Project. *Sci. Am., 306*(6), 50–55.

614 McClelland, J. L., Botvinick, M. M., Noelle, D. C., Plaut, D. C., Rogers, T. T., Seidenberg, M. S., & Smith, L. B. (2010). Letting structure emerge: Connectionist and dynamical systems approaches to cognition. *Trends Neurosci., 14*(8), 348-356.

615 Maass, W. (2007). Liquid computing. Computation and logic in the real world. *Lect. Notes Comput. Sc., 4497*, Springer, 507-516.

616 Grzyb, B. J., Jaume, I., Chinellato, E., Wojcik, G. M., & Kaminski, W. A. (2009). Which model to use for the liquid state machine? *2009 IJCNN*, 1018-1024. IEEE.

617 He, Y., & Evans, A. (2010). Graph theoretical modeling of brain connectivity. *Curr. Opin. Neurol., 23*(4), 1-10.

618 Schultz, S. R., & Panzeri, S. (2001). Temporal correlations and neural spike train entropy. *Phys. Rev. Lett., 86*(25), 5823-5826.

619 Buzsáki, G., & Wang, X.-J. (2012). Mechanisms of gamma oscillations. *Annu. Rev. Neurosci., 35*, 203-225.

620 Bal, T., Debay, D., & Destexhe, A. (2000). Cortical feedback controls the frequency and synchrony of oscillations in the visual thalamus. *J. Neurosci., 20*(19) 7478-7488.

621 Jones, E. G. (2002). Thalamic circuitry and thalamocortical synchrony. *Philos. T. Roy. Soc. B, 357*(1428), 1659-1673.

622 Desimone, R. (2009). Neural synchrony and selective attention. *2009 IJCN*, 683-684. IEEE.

623 Gregoriou, G. G., Gotts, S. J., Zhou, H., & Desimone, R. (2009). High-frequency, long-range coupling between prefrontal and visual cortex during attention. *Science, 324*(5931), 1207-1210.

624 Yamamoto, J., Suh, J., Takeuchi, D., & Tonegawa, S. (2014). Successful execution of working memory linked to synchronized high-frequency gamma oscillations. *Cell, 157*(4), 845-857.

625 Liebe, S., Hoerzer, G. M., Logothetis, N. K., & Rainer, G. (2012). Theta coupling between V4 and prefrontal cortex predicts visual short-term memory performance. *Nat. Neurosci., 15*(3), 456-462.

626 Koch, C., & Crick, F. (1994). Some Further Ideas Regarding the Neuronal Basis of Awareness. In Koch, C. & Davis, J. L. (Eds.), *Large-scale neuronal theories of the brain* (pp. 93-110). Cambridge, MA: MIT Press.

627 Crick, F. & Koch, C. (1990). Some reflections on visual awareness. In *Cold Spring Harbor symposia on quantitative biology (Vol. 55)*(pp. 953-962). Cold Spring Harbor Press.

628 Welberg, L. (2010). Learning and memory: It's all in the timing. *Nat. Rev. Neurosci., 11*(5), 296-296.

629 Colgin, L. L., & Moser, E. I. (2009). Hippocampal theta rhythms follow the beat of their own drum. *Nat. Neurosci., 12*(12), 1483-1484.

630 Goutagny, R., Jackson, J., & Williams, S. (2009). Self-generated theta oscillations in the hippocampus. *Nat. Neurosci., 12*(12), 1491-1493.

631 Hartwich, K., Pollak, T., & Klausberger, T. (2009). Distinct firing patterns of identified basket and dendrite-targeting interneurons in the prefrontal cortex during hippocampal theta and local spindle oscillations. *J. Neurosci., 29*(30), 9563-9574.

632 Alonso, J. M. M. & Swadlow, H. A. (2015). Thalamus controls recurrent cortical dynamics. *Nat. Neurosci., 18*(12), 1703–1704.

633 Fischer, I., Raul, V. , Buldu, J. M., Peil, M., Mirasso, C. R., Torrent, M. C., & Garcia-Ojalvo, J. (2006). Zero-lag long-range synchronization via dynamical relaying. *Phys. Rev. Lett., 97*(12), 123902.1-123902.4.

634 Bal, T., & McCormick, D. A. (1993). Mechanisms of oscillatory activity in guinea-pig nucleus reticularis thalami in vitro: A mammalian pacemaker. *J. Physiol., 468*(1), 669-691.

635 Long, M. A., Landisman, C. E., & Connors, B. W. (2004). Small clusters of electrically coupled neurons generate synchronous rhythms in the thalamic reticular nucleus. *J. Neurosci., 24*(2), 341-349.

636 Fridman, P. (2010). The self-cohering tied-array. *Astron. Astrophys., 510*(0004-6361).

637 Frank,T. D., Daffertshofer,A., Peper,C. E., Beek, P. J., & Haken H. (2000). Towards a comprehensive theory of brain activity: Coupled oscillator systems under external forces. *Physica D, 144*(1-2), 62-86.

638 Hong, Y. W., & Scaglione, A. (2004). Distributed change detection in large scale sensor networks through the synchronization of pulse-coupled oscillators. *2004 Int. Conf. Acoust. Spee., Proceedings, 3 (ICASSP'04)* (pp. 869-872). IEEE.

639 Ehm, W., Bach, M., & Kornmeier, J. (2011). Ambiguous figures and binding: EEG frequency modulations during multistable perception. *Psychophysiology, 48*(4), 547-558.

640 Barrett, A. B., Murphy, M., Bruno, M.-A., Noirhomme, Q, Boly, M., Laureys, S., & Seth, A. K. (2012). Granger causality analysis of steady-state electroencephalographic signals during propofol-induced anaesthesia. *PLoS ONE* **7**(1): e29072.

641 Boveroux, P., Vanhaudenhuyse, A., Bruno, M. A., Noirhomme, Q., Lauwick, S., Luxen, A., Degueldre, C., Plenevaux, A., Schnakers, C., Phillips, C. & Brichant, J. F. (2010). Breakdown of within-and between-network resting state functional magnetic resonance imaging connectivity during propofol-induced loss of consciousness. *Anesthesiology, 113*(5), 1038-1053.

642 Zhang, N., Rane, P., Huang, W., Liang, Z., Kennedy, D., Frazier, J. A., & King, J. (2010). Mapping resting-state brain networks in conscious animals. *J. Neurosci. Meth., 189*(2), 186-96.

643 Melloni, L., Molina, C., Pena, M., Torres, D., Singer, W., Rodriguez, E. (2007). Synchronization of neural activity across cortical areas correlates with conscious perception. *J. Neurosci. 27*(11):2858-65.

644 Fox, M. D., Snyder, A. Z., Vincent, J. L., Corbetta, M., Van Essen, D. C., & Raichle, M. E. (2005). The human brain is intrinsically organized into dynamic, anticorrelated functional networks. *P. Natl. Acad. Sci., 102*(27), 9673-9678.

645 Womelsdorf, T., Valiante, T. A., Sahin, N. T., Miller, K. J, & Tiesinga, P. (2014). Dynamic circuit motifs underlying rhythmic gain control, gating and integration. *Nat. Neurosci., 17*(8), 1031-1039.

646 Englert, A., Kinzel, W., , Aviad, Y., Butkovski, M., Reidler, I., Zigzag, M., Kanter, I., & Rosenbluh, M. (2010). Zero lag synchronization of chaotic systems with time delayed couplings. *Phys. Rev. Lett., 104*(11), 114102.

647 Gollo, L. L., Mirasso. C., Sporns, O., & Breakspear, M. (2014). Mechanisms of zero-lag synchronization in cortical motifs. *PLoS Comput. Biol., 10*(4), e1003548.

648 Cronin-Golomb, M., Fischer, B., Kwong, S.-K., White, J. O., & Yariv, A. (1985). Nondegenerate optical oscillation in a resonator formed by two phase-conjugate mirrors. *Opt. Lett., 10*(7), 353-355.

649 Lam, J. F., & Brown, W. P. (1980). Optical resonators with phase-conjugate mirrors. *Opt. Lett., 5*(2), 61-63.

650 Liu, S. R. (1993). Spatiotemporal behavior and nonlinear dynamics in a phase conjugate resonator. *Technical Report, NASA-CR-4567, NAS 1.26:4567*, ID: 19940019056.

651 Yeh, P. (1985). Theory of phase-conjugate oscillators. *J. Opt. Soc. Am. A, 2*(5), 727-730.

652 Hoppensteadt, F. C., & Izhikevich, E. M. (1996). Synaptic organizations and dynamical properties of weakly connected neural oscillators. *Biol. Cybern., 75*(2) 117-127.

653 Fries, P. (2009). Neuronal Gamma-Band Synchronization as a Fundamental Process in Cortical Computation. *Annu. Rev. Neurosci., 31*(1) 209-224.

654 Link, S., Luković, I., & Mogin, P. (2010). *Performance evaluation of natural and surrogate key database architectures* (pp. 1-15). School of Engineering and Computer Science, Victoria University of Wellington.

655 Blaze, M. (1993, December). A cryptographic file system for UNIX. In *Proceedings of the 1st ACM conference on Computer and communications security* (pp. 9-16). ACM.

656 Knuth, D. (1973). *The Art of Computer Programming, volume 3, Sorting and Searching* (pp. 506–542). Reading, MA: Addison-Wesley.

657 Ayzenshtat, I., Meirovithz, E., Edelman, H., Werner-Reiss, U., Bienenstock, E., Abeles, M., & Slovin, H. (2010). Precise spatiotemporal patterns among visual cortical areas and their relation to visual stimulus Processing. *J. Neurosci., 30*(33), 1232-11245.

658 Zadeh, L. A. (1988). Fuzzy logic. *Computer, 21*(4), 83-93.

659 Advisory Committee to the Air Force Systems Command (NAS-NRC) Washington DC (1968). van Cittert-Zernike theorem in short form. In *Woods Hole Summer Study* (pp. 5-6). http://oai.dtic.mil/oai/oai?verb=getRecord&metadataPrefix=html&identifier=AD0680806.

660 Carter, W. H. & Wolf, E. (1977). Coherence and radiometry with quasihomogeneous planar sources. *J. Opt. Soc. Am., 67*(6), 785-796.

661 Roychoudhuri, C. & Lefebvre, K. R. (1995). Van Cittert-Zernike theorem for introductory optics course using the concept of fringe visibility. In *Proc. SPIE, 2525* (pp. 148-160). SPIE.

662 Mesulam, M. M. (1998). From sensation to cognition. *Brain, 121*(6), 1013-1052.

663 Huang, Y., Qiu, Y., & Huang, Z. (2008). Universal description of the optical coherence. In *Proceedings of SPIE, the International Society for Optical Engineering* (Vol. 6837) (pp. 683713-1-683713-6). International Society for Optical Engineering; 1999.

664 Buzsáki, G., & Diba, K. (2010). Ch **7**. Oscillation-supported Information Processing and Transfer at the Hippocampus–Entorhinal–Neocortical Interface. Dynamic Coordination in the Brain. In von der Malsburg, C., Phillips, W. A., & Singer, W. (Eds.), *Dynamic Coordination in the Brain: From Neurons to Mind, Strüngmann Forum Report* (Vol. 5) (pp. 101-113). Cambridge, MA: The MIT Press.

665 MacLean, J. N., Watson, B. O., Aaron, G. B., & Yuste, R. (2005). Internal dynamics determine the cortical response to thalamic stimulation. *Neuron, 48*(5), 811-823.

666 Renart, A., de la Rocha, J., Bartho, P., Hollender, L., Parga, N., Reyes, A., & Harris, K. D. (2010). The asynchronous state in cortical circuits. *Science, 327*(5965), 587-590.

667 Sperandio, I., Chouinard, P. A., & Goodale, M. A. (2012). Retinotopic activity in V1 reflects the perceived and not the retinal size of an afterimage. *Nat. Neurosci., 15*(4), 540-542.

668 Yli-Krekola, A. (2007). *A bio-inspired computational model of covert attention and learning* (M. S. Thesis). Helsinkji University of Technology.

669 Abeles, M., Hayon, G., & Lehmann, D. (2004). Modeling compositionality by dynamic binding of synfire chains. *J. Comput. Neurosci., 17*(2), 179-201.

670 Barbarossa, S., & Celano, F. (2005). Self-organizing sensor networks designed as a population of mutually coupled oscillators. In *2005 IEEE 6th Workshop on Signal Processing Advances in Wireless Communications* (pp. 475-479). IEEE.

671 Wright, J. J., Robinson, P. A., Rennie, C. J., Gordon, E., Bourke, P. D., Chapman, C. L., Hawthorn, N., Lees, G. J., & Alexander, D. (2001). Toward an integrated continuum model of cerebral dynamics: The cerebral rhythms, synchronous oscillation and cortical stability. *BioSystems, 63*(1-3), 71-88.

672 Wong, Y. T., Fabiszak, M. M., Novikov, Y., Daw, N. D., & Pesaran, B. (2016). Coherent neuronal ensembles are rapidly recruited when making a look-reach decision. *Nat. Neurosci., 19*(2), 327-334.

673 Cappe, C., Morel, A., Barone, P., & Rouiller, E. M. (2009). The thalamocortical projection systems in primate: An anatomical support for multisensory and sensorimotor interplay. *Cereb. Cortex, 19*(9), 2025-2037.

674 Akam, T., Oren, I., Mantoan, L., Ferenczi, E., & Kullmann, D. M. (2012). Oscillatory dynamics in the hippocampus support dentate gyrus-CA3 coupling. *Nat. Neurosci., 15*(5), 763-768.

675 Salinas, E., & Sejnowski, T. J. (2001). Correlated neuronal activity and the flow of neural information. *Nat. Rev. Neurosci., 2*(8), 539-550.

676 Watrous, A. J, Tandon, N., Conner, C. R., Pieters, T., & Ekstrom, A. D. (2013). Frequency-specific network connectivity increases underlie accurate spatiotemporal memory retrieval. *Nat. Neurosci., 16*(3), 349-356.

677 Sehatpour, P., Molholm, S., Schwartz, T. H., Mahoney, J. R., Mehta, A. D., Javitt, D. C., Stanton, P. K., & Foxe, J. J. (2008). A human intracranial study of long-range oscillatory coherence across a frontal–occipital–hippocampal brain network during visual object processing. *P. Natl. Acad. Sci., 105*(11), 4399-4404.

678 Ramsey, L. E. (2010). *Phase Locking in high gamma during a speech task source.* Faculty of Medicine Theses, Neuroscience and Cognition, Utrecht University.

679 Sporns, O. (2002). Graph Theory Methods For The Analysis Of Neural Connectivity Patterns. In Kötter, R. (ed.) *Neurosciences Databases: A practical guide* (pp. 179-187). Norwell, MA: Kluwer Academic Publishers.

680 Bondy, J. A., & Murty, U. S. R. (1976). Basic Graph Theory. in *Graph theory with applications* (Vol. 290) (pp. 1-31). London: Macmillan.

681 Watts, D. J. & Strogatz, S. H. (1998). Collective dynamics of `small-world' networks. *Nature, 393*(6684), 440-442.

682 Hwang, K., Hallquist, M. N., Luna, B. (2013). The development of hub architecture in the human functional brain network. *Cereb. Cortex, 23*(10), 2380-2393.

683 van den Heuvel, M. P., Kahn, R. S., Goñi, J., & Sporns, O. (2012). High-cost, high-capacity backbone for global brain communication. *P. Natl. Acad. Sci., 109*(28), 11372-11377.

684 van den Heuvel M. P., Sporns, O. (2011). Rich-club organization of the human connectome. *J. Neurosci., 31*(44), 15775-86.

685 Zamora-Lopez, G., Zhou, C., & Kurths, J. (2011). Exploring brain function from anatomical connectivity. *Frontiers in Neuroscience, 5*(83), 1-11.

686 Zuo, X.-N., Ehmke, R., Mennes, M., Imperati, D., Castellanos, F. X., Sporns, O., & Milham, M. P. (2012). Network centrality in the human functional connectome. *Cereb. Cortex, 22*(8), 1862-1875.

687 Kravitz, D. J., Saleem, K. S., Baker, C. I., & Mishkin, M. (2011). A new neural framework for visuospatial processing. *Nat. Rev. Neurosci., 12*(4), 217-230.

688 Barthélemy, M. (2011). Spatial networks. *Phys. Rep., 499*(1), 1-101.

689 Sneppen, K., & Rosvall, M. (2006). A communication perspective on network topologies (Physics of non-equilibrium systems: Self-organized structures and dynamics far from equilibrium). *Prog. Theor. Phys. Supp.,* (165), 103-118.

690 Scannell, J. W., Burns, G. A. P. C., Hilgetag, C. C., O'Neil, M. A., & Young, M. P. (1999). The connectional organization of the cortico-thalamic system of the cat. *Cereb. Cortex, 9*(3), 277-299.

691 Young, M. P. (1992). Objective analysis of the topological organization of the primate cortical visual system. *Nature, 358*(6382), 152-5.

692 Steinke, G. K., & Galán, R. F. (2011). Brain rhythms reveal a hierarchical network organization. *PLoS Comput. Biol., 7*(10): e1002207.

693 Deco, G., Jirsa, V. K., & McIntosh, A. R. (2011). Emerging concepts for the dynamical organization of resting-state activity in the brain. *Nat. Rev. Neurosci., 12*(1), 43-56.

694 Axmacher, N., Schmitz, D. P., Wagner, T., Elger, C. E. & Fell, J. (2008). Interactions between medial temporal lobe, prefrontal cortex, and inferior temporal regions during visual working memory: A combined intracranial EEG and functional magnetic resonance imaging study. *J. Neurosci., 28*(29), 7304-7312.

695 Tommerdahl, M., Tannan, V., Holden, J. K., & Baranek, G. T. (2008). Absence of stimulus-driven synchronization effects on sensory perception in autism: Evidence for local underconnectivity? *Behav. Brain Funct., 4*(1), 1-9.

696 Briggs, F., & Usrey, W. M. (2008). Emerging views of corticothalamic function. *Curr. Opin. Neurobiol., 18*(4), Sensory systems, 403-407.

697 Ferrarelli, F., & Tononi, G. (2011). The thalamic reticular nucleus and schizophrenia. *Schizophrenia Bull., 37*(2), 306-315.

698 Ohara, P. T., & Lieberman, A. R. (1985). The thalamic reticular nucleus of the adult rat: Experimental anatomical studies. *J. Neurocytol., 14*(3), 365-411.

699 Sherman, S. M., & Guillery, R. W. (2002). The role of the thalamus in the flow of information to the cortex. *Philos. T. Roy. Soc. B, 357*(1428), 1695-1708.

700 Vicente, R., Gollo, L. L., Mirasso, C. R., Fischer, I., & Pipa, G. (2008). Dynamical relaying can yield zero time lag neuronal synchrony despite long conduction delays. *Proc. Natl. Acad. Sci., 105*(44): 17157–17162.

701 Jin, J., Wang, Y., Swadlow, H. A., & Alonso, J. M. (2011). Population receptive fields of ON and OFF thalamic inputs to an orientation column in visual cortex. *Nat. Neurosci., 14*(2) 232-238.

702 Crick, F. (1984). Function of the thalamic reticular complex: The searchlight hypothesis. *Proc. Natl. Acad. Sci., 81*(14), 4586– 4590.

703 McClurkin, J. W., Optican, L. M., & Richmond, B. J. (1994). Cortical feedback increases visual information transmitted by monkey parvocellular lateral geniculate nucleus neurons. *Visual Neurosci., 11*(03), 601-617.

704 Wang, S., Bickford, M. E., van Horn, S. C., Erisir, A., Godwin, D. W., & Sherman, S. M. (2001). Synaptic targets of thalamic reticular nucleus terminals in the visual thalamus of the cat. *J. Comp. Neurol., 440*(4), 321–341.

705 Pinault, D., & Deschênes, M. (1998). Anatomical evidence for a mechanism of lateral inhibition in the rat thalamus. *European J. Neurosci., 10*(11), 3462–3469.

706 Landisman, C. E., Long, M. A., Beierlein, M., Deans, M. R., & Connors, B. W. (2002). Electrical synapses in the thalamic reticular nucleus. *J. Neurosci., 22*(3), 1002-1009.

707 Minlebaev, M., Colonnese, M., Tsintsadze, T., Sirota, A., & Khazipov, R. (2011). Early gamma oscillations synchronize developing thalamus and cortex. *Science, 334*(6053), 226-229.

708 Budd, J. M. L., Kovács, K., Ferecskó, A. S., Buzás, P., Eysel, U. T., & Kisvárday, Z. F. (2010). Neocortical axon arbors trade-off material and conduction delay conservation. *PLoS Comput. Biol., 6*(3), e1000711.

709 Salami, M., Itami, C., Tsumoto, T., & Kimura, F. (2003). Change of conduction velocity by regional myelination yields constant latency irrespective of distance between thalamus and cortex. *P. Natl. Acad. Sci., 100*(10), 6174-6179.

710 Lien, A. D., & Scanziani, M. (2013). Tuned thalamic excitation is amplified by visual cortical circuits. *Nat. Neurosci., 16*(9), 1315-1323.

711 Li, Y.-t., Ibrahim, L. A., Liu, B.-h., Zhang, L. I., & Tao, H. W. (2013). Linear transformation of thalamocortical input by intracortical excitation. *Nat. Neurosci., 16*(9), 1324-1330.

712 Sherman, S. M., & Guillery, R. W. (2011). Distinct functions for direct and transthalamic corticocortical connections. *J. Neurophysiol., 106*(3), 1068-1077.

713 Treisman, A. M., & Gelade, G. (1980). A feature-integration theory of attention. *Cognitive Psychol., 12*(1), 97-136.

714 Gollo, L. L., Mirasso, C., & Villa, A. E. P. (2010). Dynamic control for synchronization of separated cortical areas through thalamic relay. *NeuroImage, 52*(3) 947-955.

715 Zhang, J., Ackman, J. B., Xu, H.-P., & Crair, M. C. (2012). Visual map development depends on the temporal pattern of binocular activity in mice. *Nat. Neurosci., 15*(2), 298-307.

716 Krupa, D. J., Ghazanfar, A. A., & Nicolelis, M. A. L. (1999). Immediate thalamic sensory plasticity depends on corticothalamic feedback. *P. Natl. Acad. Sci., 96*(14), 8200-8205.

717 Sporns, O. (2011). The human connectome: A complex network. *Ann. NY Acad. Sci., 1224*(1), 109-125.

718 Mohajerani, M. H., Chan, A. W., Mohsenvand, M., LeDue, J., Liu, R., McVea, D. A., Boyd, J. D., Wang, Y. T., Reimers, M., & Murphy, T. H. (2013). Spontaneous cortical activity alternates between motifs defined by regional axonal projections. *Nat. Neurosci., 16*(10), 1426-1435.

719 Cappe, C., Morel, A., & Rouiller, E. M. (2007). Thalamocortical and the dual pattern of corticothalamic projections of the posterior parietal cortex in macaque monkeys. *Neuroscience, 146*(3), 1371-1387.

720 Guillery, R. W., & Harting, J. K. (2003). Structure and connections of the thalamic reticular nucleus: Advancing views over half a century. *J. Comp. Neurol., 463*(4), 360-371.

721 Saalmann, Y. B., Pinsk, M. A., Wang, L., Li, X., & Kastner, S. (2012).The pulvinar regulates information transmission between cortical areas based on attention demands. *Science, 337*(6095), 753-756.

722 Zhou, B. B., & Roy, R. (2007). Isochronal synchrony and bidirectional communication with delay-coupled nonlinear oscillators. *Phys. Rev. E, 75*(2)l 026205 (5 pages).

723 Santhanam, M. S., & Arora, S. (2007). Zero delay synchronization of chaos in coupled map lattices. *Phys. Rev. E, 76*(2), 026202.

724 Vicente, R., Mirasso, C. R., & Fischer, I. (2007). Simultaneous bidirectional message transmission in a chaos-based communication scheme. *Opt. Lett., 32*(4), 403-405.

725 Chawla, D., Friston, K. J., & Lumer, E. D. (2001). Zero-lag synchronous dynamics in triplets of interconnected cortical areas. *Neural Networks, 14*(6–7), 727-735.

726 Sadeghi, S. & Valizadeh, A. (2014). Synchronization of delayed coupled neurons in presence of inhomogeneity. J. *Comput. Neurosci., 36*(1), 55-66.

727 Poulet, J. F. A., Fernandez, L. M. J., Crochet, S., & Petersen, C. C. H. (2012). Thalamic control of cortical states. *Nat. Neurosci., 15*(3), 370-372.

728 Behrens, T. E. J., & Sporns, O. (2011). Human connectomics. *Curr. Opin. Neurobiol., 22*, 1–10.

729 Wright, J. J. (2011). Attractor dynamics and thermodynamic analogies in the cerebral cortex: synchronous oscillation, the background EEG, and the regulation of attention. B. Math. Biol., 73(2), 436-457.

730 Pluim, J. P., Maintz, J. A., & Viergever, M. A. (2003). Mutual-information-based registration of medical images: A survey. *IEEE T. Med. Imaging, 22*(8), 986-1004.

731 Barlow, H. B., Kaushal, T. P., & Mitchison, G. J. (1989). Finding minimum entropy codes. *Neural Comput., 1*(3), 412-423.

732 Schneidman, E., Still, S., Berry, M. J., & Bialek, W. (2003). Network information and connected correlations. *Phys. Rev. Lett., 91*(23), 238701.

733 Weaver, W. (1963). Introductory note on the general setting of the analytical communication studies. In Shannon, C. E., & Weaver, W. (1963). *The Mathematical Theory of Communication* (pp. 1-28). University of Illinois Press. (Original work published 1949)

734 Sluga, D., & Lotric, U. (2013). Generalized information-theoretic measures for feature selection. In *11ᵗʰ International Conference on Adaptive and Natural Computing Algorithms, ICANNGA 2013* (Vol. 7824) (pp. 189-197). Berlin Heidelberg: Springer.

735 Tononi, G., Edelman, G. M., & Sporns, O. (1998). Complexity and coherency: Integrating information in the brain. *Trends Cogn. Sci., 2*(12), 474-484.

736 Prideaux, J. (2000). A comparison between Karl Pribram's "Holographic Brain Theory" and more conventional models of neuronal computation. Virginia Commonwealth University. http://www.acsa2000.net/bcngroup/jponkp

737 Paninski, L. (2003). Estimation of entropy and mutual information. *Neural Comput., 15*(6), 1191-1253.

738 Viola, P., & Wells III, W. M. (1997). Alignment by maximization of mutual information. *Int.J. Comput. Vision, 24*(2), 137-154.

739 Schreiber, T. (2000). Measuring information transfer. *Phys. Rev. Lett., 85*(2), 461-464.

740 Papana, A., Kugiumtzis, D., & Larsson, P. G. (2011). Reducing the bias of causality measures. *Phys. Rev. E, 83*(3), 036207.

741 Georgiev, G., & Georgiev, I. (2002). The least action and the metric of an organized system. *Open Systems and Information Dynamics, 9*(4), 371-380.

742 Granger, C. W. J. (1988). Some recent development in a concept of causality. *J. Econometrics, 39*(1–2), 199-211.

743 Hutchinson, G. E. (1948). Circular causal systems in ecology. *Ann. NY Acad. Sci., 50*, 221-246.

744 Kosko, B. (1988). Bidirectional associative memories. *IEEE T. Sys. Man Cyb., 18*(1), 49-60.

745 Wang, M., Gibbons, J., & Wu, N. (2011). Incremental updates for efficient bidirectional transformations. *ACM SIGPLAN Notices, 46*(9), 392-403).

746 Finkel, L. H., Yen, S.-C., & Menschik, E. (1998). Synchronization: The Computational Currency of Cognition. In *ICANN 98* (pp. 23-40). London: Springer.

747 Weiss, S., Sternklar, S., & Fischer, B. (1987). Double phase-conjugate mirror: analysis, demonstration, and applications. *Opt. Lett., 12*(2), 114-116.

748 Segev, M., Engin, D., Yariv, A., & Valley, G. C. (1993). Temporal evolution of photorefractive double phase-conjugate mirrors. *Opt. Lett., 18*(21), 1828-1830.

749 Sternklar, S., Weiss, S., & Fischer, B. (1987). Optical information processing with the double phase conjugate mirror. *Opt. Eng., 26*(5), 265423-265423.

750 Långsjö, J. W., Alkire, M. T., Kaskinoro, K., Hayama, H., Maksimow, A., Kaisti, K. K., Aalto, S., Aantaa, R., Jääskeläinen, S. K., Revonsuo, A., & Scheinin, H. (2012). Returning from oblivion: Imaging the neural core of consciousness. *J. Neurosci., 32*(14), 4935-4943.

751 Beni, G., & Wang, J. (1993). Swarm Intelligence in Cellular Robotic Systems. In Dario, P., Sandini, G., Aebisher, P., (Eds.), *Robots and Biological Systems: Towards a New Bionics?* (pp. 703-712). Berlin Heidelberg: Springer.

752 Beni, G. (2004, July). Order by disordered action in swarms. In *Proceedings of the 2004 International Workshop on Swarm Robotics* (pp. 153-171). Berlin Heidelberg: Springer.

753 Couzin, I. D., Krause, J., James, R., Ruxton, G. D., & Franks, N. R. (2002). Collective memory and spatial sorting in animal groups. *J. Theor. Biol., 218*(1), 1-11.

754 Moussaid, M., Garnier, S., Theraulaz, G., & Helbing, D. (2009). Collective information processing and pattern formation in swarms, flocks, and crowds. *Topics in Cognitive Science, 1*(3), 469-497.

755 Pikovsky, A., Rosenblum, M., & Kurths, J. (2003). *Synchronization: A Universal Concept in Nonlinear Sciences* (Cambridge Nonlinear Science Series, Vol. 12) (pp. 102-104). Cambridge University Press.

756 Greer, D. S. (2008). The computational manifold approach to consciousness and symbolic processing in the cerebral cortex. In *Cognitive Informatics, 2008. ICCI 2008. 7th IEEE International Conference on* (pp. 94-103). IEEE.

757 Tang, Y. Y., Rothbart, M. K., & Posner, M. I. (2012). Neural correlates of establishing, maintaining, and switching brain states. *Trends Cogn. Sci., 16*(6), 330-337.

758 Zikopoulos, B., & Barbas, H. (2006). Prefrontal projections to the thalamic reticular nucleus form a unique circuit for attentional mechanisms. *J. Neurosci., 26*(28), 7348-7361.

759 Wurtz, R. H., McAlonan, K., Cavanaugh, J., & Berman, R. A. (2011). Thalamic pathways for active vision. *Trends Cogn. Sci., 15*(4), 177-184.

760 Dhamala , M., Assisi ,C. G., Jirsa, V. K. Steinberg, F. L. & Kelso, J. A. S. (2007). Multisensory integration for timing engages different brain networks. *NeuroImage, 34*(2), 764-773.

761 Martinez-Conde, S., Macknik, S. L., & Hubel, D. H. (2004). The role of fixational eye movements in visual perception. *Nat. Rev. Neurosci., 5*(3), 229-240.

762 Lee, A. C. H., & Rudebeck, S. R. (2010). Investigating the interaction between spatial perception and working memory in the human medial temporal lobe. *J. Cognitive Neurosci., 22*(12), 2823-2835.

763 Murray, E. A., Bussey, T. J., & Saksida, L. M. (2007), Visual perception and memory: A new view of medial temporal lobe function in primates and rodents. *Annu. Rev. Neurosci., 30*(1),99-122.

764 Sevush, S. (2006). Single-neuron theory of consciousness. *J. Theor. Biol., 238*(3), 704-725.

765 Quiroga, R. Q., Kreiman, G., Koch, C., & Fried, I. (2012). Sparse but not 'grandmother-cell' coding in the medial temporal lobe. *Trends Cogn. Sci., 12*(3), 87-91.

766 Gross, C. G. (2002). Genealogy of the "Grandmother Cell." *The Neuroscientist, 8*(5), 512-518.

767 Bassett, D. S., & Gazzaniga, M. S. (2011). Understanding complexity in the human brain. *Trends Cogn. Sci., 15*(5), 200-209.

768 Gaffan, D. (2005). Widespread cortical networks underlie memory and attention. *Science, 309*(5744), 2172-2173.

769 Power, J. D., Cohen, A. L., Nelson, S. M., Wig, G. S., Barnes, K. A., Church, J. A., Vogel, A. C., Laumann, T. O., Miezin, F. M., Schlaggar, B. L., & Petersen, S. E. (2011). Functional network organization of the human brain. *Neuron, 72*(4), 665-678.

770 Squire, L. R., & Zola-Morgan, S. (1991). The medial temporal lobe memory system. *Science, 253*(5026), 1380 - 1386.

771 Galuba, W., & Girdzijauskas, S. (2009). Distributed Hash Table. In Liu, L., & Özsu, M. T. (Eds.), *Encyclopedia of Database Systems* (pp. 903-904). NY: Springer.

772 Kaufman, M. T., Churchland, M. M., Ryu, S. I., & Shenoy, K. V. (2014). Cortical activity in the null space: Permitting preparation without movement. *Nat. Neurosci., 17*(3), 440-448.

773 Guo, Z. V., Li, N., Huber, D., Ophir, E., Gutnisky, D., Ting, J. T., Feng, G., Svoboda, K. (2014). Flow of cortical activity underlying a tactile decision in mice. *Neuron, 81*(1), 179-94.

774 Li, N., Chen, T. W., Guo, Z. V., Gerfen, C. R., & Svoboda, K. (2015). A motor cortex circuit for motor planning and movement. *Nature, 519*(7541), 51-56.

775 Siegel, M., Buschman, T. J., & Miller, E. K. (2015). Cortical information flow during flexible sensorimotor decisions. *Science, 348*(6241), 1352-1355.

776 Orban, G. A., Van Essen, D., & Vanduffel, W. (2004). Comparative mapping of higher visual areas in monkeys and humans. *Trends Cogn. Sci.*, *8*(7), 315-324.

777 Nassi, J. J. & Callaway, E. M. (2009). Parallel processing strategies of the primate visual system. *Nat. Rev. Neurosci.,* *10*(5), 360-372.

778 Chen, C.-M., Lakatos, P., Shah, A. S. Mehta, A. D. Givre, S. J. Javitt, D. C. Schroeder, C. E. (2007). Functional Anatomy and Interaction of Fast and Slow Visual Pathways in Macaque Monkeys. *Cereb. Cortex, 17*(7) 1561-1569.

779 Van Essen, D. C., Anderson, C. H., & Felleman, D. J. (1992). Information processing in the primate visual system: an integrated systems perspective. *Science, 255*(5043), 419-423.

780 Modha, D. S., & Singh, R. (2010). Network architecture of the long-distance pathways in the macaque brain. *P. Natl. Acad. Sci., 107*(30) 13485-13490.

781 Kravitz, D. J., Saleem, K. S., Baker, C. I., Ungerleider, L. G., & Mishkin, M. (2013). The ventral visual pathway: An expanded neural framework for the processing of object quality. *Trends Cogn. Sci., 17*(1), 26–49.

782 Orban, G. A. (2008). Higher order visual processing in Macaque extrastriate cortex. *Physiol. Rev., 88*(1), 59-89.

783 Hubel, D. H. (1982). Evolution of ideas on the primary visual cortex, 1955–1978: A biased historical account. Bioscience Reports, 2(7), 435-469.

784 Kourtzi, Z., & Kanwisher, N. (2001). Representation of perceived object shape by the human lateral occipital complex. *Science, 293*(5534), 1506-1509.

785 Gauthier, J. L, Field, G. D, Sher, A., Greschner, M., Shlens, J., Litke, A. M., & Chichilnisky, E. J. (2009). Receptive fields in primate retina are coordinated to sample visual space more uniformly. *PLoS Biol., 7*(4), e1000063.

786 Li, K. Y., Tiruveedhula, P., & Roorda, A. (2010). Intersubject variability of foveal cone photoreceptor density in relation to eye length. *Invest. Ophthalmol. Vis. Sci., 51*(12), 6858-6867.

787 Liu, Y. S., Stevens, C. F., & Sharpee, T. O. (2009). Predictable irregularities in retinal receptive fields. *P. Natl. Acad. Sci., 106*(38), 16499-16504.

788 Curcio, C. A., Sloan, K. R., Kalina, R. E., & Hendrickson, A. E. (1990). Human photoreceptor topography. *J. Comp. Neurol., 292*(4), 497-523.

789 Steinberg, R. H., Reid, M., & Lacy, P. L. (1973). The distribution of rods and cones in the retina of the cat (Felis domesticus). *J. Comp. Neurol., 148*(2), 229-248.

790 Mowafy, L., Lappin, J. S., Anderson, B. l., & Mauk, D. l. (1992). Temporal factors in the discrimination of coherent motion. *Percept.ion & Psychophys., 52*(5), 508-518.

791 Julesz, B. (1971). *Foundations of Cyclopean perception.* Chicago: University of Chicago Press.

792 Nielsen, K. R. K., & Poggio, T. (1984). Vertical image registration in stereopsis. *Vision Res., 24*(10), 1133-1140.

793 Nauhaus,I., Benucci, A., Carandini, M. & Ringach, D. L. (2008). Neuronal selectivity and local map structure in visual cortex. *Neuron, 57*(5), 673–679.

794 Jarosiewicz, B., Schummers, J., Malik, W. Q., Brown, E. N., & Sur, M. (2012). Functional biases in visual cortex neurons with identified projections to higher cortical targets. *Curr Biol., 22*(4), 269-77.

795 Pelphrey, K. A., Morris, J. P., Michelich, C. R., Allison, T., & McCarthy, G. (2005). Functional anatomy of biological motion perception in posterior temporal cortex: An fMRI study of eye, mouth and hand movements. *Cereb. Cortex, 15*(12), 1866-1876.

796 Weisel, T. N., & Hubel, D. H. (1965). Comparison of the effects of unilateral and bilateral eye closure on cortical unit responses in kittens. *J. Neurophysiol., 28*(6), 1029-1040.

797 Kanizsa, G. (1976). Subjective contours. *Sci. Am., 234*(4), 48-52.

798 Grammer, K. & Thornhill, R. (1994). Human (Homo sapiens) facial attractiveness and sexual selection: The role of symmetry and averageness. *J. Comp. Psychol., 108*(3), 233-242.

799 Rhodes, G., Proffitt, F., Grady, J. M., & Sumich, A. (1998). Facial symmetry and the perception of beauty. *Psychon. B. Rev., 5*(4), 659-669.

800 Chaudhuri, R., & Fiete, I. (2016). Computational principles of memory. *Nat. Neurosci., 19*(3), 394-403.

801 Cutsuridis, V., Cobb, S., & Graham, B. (2008). Encoding and retrieval in a CA1 microcircuit model of the hippocampus. In Kurková, V., Neruda, R., Koutník, J., (Eds.), *18th International Conference on Artificial Neural Networks - ICANN 2008. Proceedings, Part II* (pp. 238-247). Berlin Heidelberg: Springer.

802 Neves, G., Cooke, S. F., & Bliss, T. V. P. (2008). Synaptic plasticity, memory and the hippocampus: A neural network approach to causality. *Nat. Rev. Neurosci., 9*(1), 65-75.

803 Buzsáki, G., & Chrobak, J. J, (2005). Synaptic plasticity and self-organization in the hippocampus. *Nat. Neurosci., 8*(11), 1418-1420.

804 Serre, T., Wolf, L., & Poggio, T. (2005). Object recognition with features inspired by visual cortex. In *2005 IEEE Computer Society Conference on Computer Vision and Pattern Recognition (CVPR'05)* (Vol. 2) (pp. 994-1000). IEEE.

805 Lowe, D., Lee, S.-W., Bülthoff, H., & Poggio, T. (2000). Towards a computational model for object recognition in IT cortex. In *Proceedings of the First IEEE International Workshop on Biologically Motivated Computer Vision* (pp. 20-31). Berlin Heidelberg: Springer.

806 Neven, H., & Aertsen, A. (1992). Rate coherence and event coherence in the visual cortex: A neuronal model of object recognition. *Biol. Cybern., 67*(4), 309-322.

807 Yariv, A., & Pepper, D. M. (1977). Amplified reflection, phase conjugation, and oscillation in degenerate four-wave mixing. *Opt. Lett., 1*(1), 16-18.

808 Barrouillet, P., Bernardin, S., & Camos, V. (2004). Time constraints and resource sharing in adults' working memory spans. *J. Exp. Psychol. Gen., 133*(1), 83-100.

809 Walker, M. P. (2005). A refined model of sleep and the time course of memory formation. *Behav. Brain. Sci., 28*(01), 51-64.

810 Zhang, Z., You, Z., & Chu, D. (2014). Fundamentals of phase-only liquid crystal on silicon (LCOS) devices. *Light. Sci. Appl., 3*,(10), e213.

811 Chen, X., Gabitto, M., Peng, Y., Ryba, N. J. P., & Zuker, Charles S. (2011). A gustotopic map of taste qualities in the mammalian brain. *Science, 333*(6047), 1262-1266.

812 Maurer, C., Schwaighofer, A., Jesacher, A., Bernet, S., & Ritsch-Marte, M. (2008). Suppression of undesired diffraction orders of binary phase holograms. *Appl. Optics, 47*(22), 3994-3998.

813 Popescu, A. T., Popa, D., & Pare, D. (2009). Coherent gamma oscillations couple the amygdala and striatum during learning. *Nat. Neurosci., 12*(6), 801-807.

814 Stevens, M. C. (2009). The developmental cognitive neuroscience of functional connectivity. *Brain Cognition, 70*(1), 1–12.

815 Tomasi, D., & Volkow, N. D. (2011). Association between functional connectivity hubs and brain networks. *Cereb. Cortex, 21*(9), 2003-2013.

816 Singer, W., & Gray, C. M. (1995). Visual feature integration and the temporal correlation hypothesis. *Annu. Rev. Neurosci., 18*(1), 555-586.

817 Keller, C. J., Bickel, S., Entz, L., Ulbert, I., Milham, M. P., Kelly, C., & Mehta, A. D. (2011). Intrinsic functional architecture predicts electrically evoked responses in the human brain. *P. Natl. Acad. Sci., 108*(25), 10308-10313.

818 Eichenbaum, H., Yonelinas, A.P., & Ranganath, C. (2007). The medial temporal lobe and recognition memory. *Annu. Rev. Neurosci., 30*(1), 123-152.

819 Tsivilis, D., Vann, S. D., Denby, C., Roberts, N., Mayes, A. R., Montaldi, D., & Aggleton, J. P. (2008). A disproportionate role for the fornix and mammillary bodies in recall versus recognition memory. *Nat. Neurosci., 11*(7), 834-842.

820 Mitchell, A. S., & Gaffan, D. (2008). The magnocellular mediodorsal thalamus is necessary for memory acquisition, but not retrieval. *J. Neurosci., 28*(1), 258-263.

821 Nadel, L., & Moscovitch, M. (2001). The hippocampal complex and long-term memory revisited. *Trends Cogn. Sci., 5*(6), 228-230.

822 Wang, S.-H., Teixeira, C. M., Wheeler, A. L., & Frankland, P. W. (2009). The precision of remote context memories does not require the hippocampus. *Nat. Neurosci., 12*(3), 253-255.

823 Aggleton, J. P., & Brown, M. W. (1999). Episodic memory, amnesia, and the hippocampal--anterior thalamic axis. *Behav. Brain Sci., 22*(03), 425-444.

824 Jezek, K., Henriksen, E. J., Treves, A., Moser, E. I., & Moser, M.-B. (2011). Theta-paced flickering between place-cell maps in the hippocampus. *Nature, 478*(7368), 246-249.

825 Salazar, R. F., Dotson, N. M., Bressler, S. L., & Gray, C. M. (2012). Content-specific fronto-parietal synchronization during visual working memory. *Science, 338*(6110), 1097-1100.

826 Liu, D., Gu, X., Zhu, J., Zhang, X., Han, Z., Yan, W., Cheng, Q., Hao, J., Fan, H., Hou, R., Chen, Z., Chen, Y., & Li, C. T. (2014). Medial prefrontal activity during delay period contributes to learning of a working memory task. *Science, 346*(6208), 458-463.

827 Yassa, M. A., & Stark, C. E. L. (2008). Multiple signals of recognition memory in the medial temporal lobe. *Hippocampus, 18*(9), 945-954.

828 Carr, M. F., Jadhav, S. P., & Frank, L. M (2011). Hippocampal replay in the awake state: A potential substrate for memory consolidation and retrieval. *Nat. Neurosci., 14*(2), 147-153.

829 Payne, J., Chambers, A. M., & Kensinger, E. A. (2012). Sleep promotes lasting changes in selective memory for emotional scenes. *Frontiers in Integrative Neuroscience, 6*, 108-119.

830 Boyce, R., Glasgow, S. D., Williams, S., & Adamantidis, A. (2016). Causal evidence for the role of REM sleep theta rhythm in contextual memory consolidation. *Science, 352*(6287), 812-816.

831 Maingret, N., Girardeau, G., Todorova, R., Goutierre, M., & Zugaro, M. (2016). Hippocampo-cortical coupling mediates memory consolidation during sleep. *Nat. Neurosci., 19*(7), 959-964.

832 Ji, D., & Wilson, M. A. (2007). Coordinated memory replay in the visual cortex and hippocampus during sleep. *Nat. Neurosci., 10*(1), 100-107.

833 Izquierdo, I., Bevilaqua, L. R. M., Rossato, J. I., Bonini, J. S., Medina, J. H., & Cammarota, M. (2006). Different molecular cascades in different sites of the brain control memory consolidation. *Trends Neurosci., 29*(9), 496-505.

834 Mather, M. (2007). Emotional arousal and memory binding: an object-based framework. *Perspec. Psychol. Sci., 2*(1), 33-52.

835 Hu, P., Stylos-Allan, M., & Walker, M. P. (2006). Sleep facilitates consolidation of emotional declarative memory. *Psychol. Sci., 17*(10), 891-898.

836 Mitchell, A. S. & Gaffan, D. (2008). The magnocellular mediodorsal thalamus is necessary for memory acquisition, but not retrieval. *J. Neurosci., 28*(1), 258-263.

837 Behrens, C. J., van den Boom, L. P., de Hoz, L., Friedman, A., & Heinemann, U. (2005). Induction of sharp wave-ripple complexes in vitro and reorganization of hippocampal networks. *Nat. Neurosci., 8*(11), 1560-1567.

838 Candès, E., Romberg, J., & Tao, T. (2006). Robust uncertainty principles: Exact signal reconstruction from highly incomplete frequency information. *IEEE Trans. Inf. Theory, 52*(2), 489–509.

839 Afraz, A., Yamins, D. L. K., & DiCarlo, J. J. (2014). Neural Mechanisms Underlying Visual Object Recognition. In *Cold Spring Harbor Symposia on Quantitative Biology* (Vol. 79, pp. 99- 107). Cold Spring Harbor Laboratory Press.

840 Lopez-Aranda, M. F., Lopez-Tellez, J. F., Navarro-Lobato, I., Masmudi-Martin, M., Gutierrez, A., & Khan, Z. U. (2009). Role of Layer 6 of V2 Visual Cortex in Object-Recognition Memory. *Science, 325*(5936), 87-89.

841 Cichy, R. M., Pantazis, D., & Oliva, A. (2014). Resolving human object recognition in space and time. *Nat. Neurosci., 17*(3), 455-462.

842 Strogatz, S. H., Mirollo, R. E., & Matthews, P. C. (1992). Coupled nonlinear oscillators below the synchronization threshold: Relaxation by generalized Landau damping. *Phys. Rev. Lett., 68*(18), 2730-2733.

843 Shrii, M. M., Senthilkumar, D. V., & Kurths, J. (2012). Delay-induced synchrony in complex networks with conjugate coupling. *Phys. Rev. E, 85*(5), 057203-1-057203-5.